CAMBRIDGE STUDIES IN MODERN OPTICS: 3

General Editors

P. L. KNIGHT
Optics Section, Imperial College of Science and Technology

W. J. FIRTH
Department of Physics, Strathclyde University

S. D. SMITH, FRS
Department of Physics, Heriot-Watt University

FABRY–PEROT INTERFEROMETERS

FABRY–PEROT INTERFEROMETERS

G. HERNANDEZ

Geophysics Program
University of Washington,
Seattle, Washington

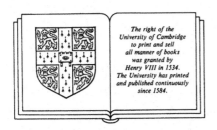

The right of the
University of Cambridge
to print and sell
all manner of books
was granted by
Henry VIII in 1534.
The University has printed
and published continuously
since 1584.

CAMBRIDGE UNIVERSITY PRESS

Cambridge

New York   New Rochelle

Melbourne   Sydney

Published by the Press Syndicate of the University of Cambridge
The Pitt Building, Trumpington Street, Cambridge CB2 1RP
32 East 57th Street, New York, NY 10022, USA
10 Stamford Road, Oakleigh, Melbourne 3166, Austalia

First published 1986
First paperback edition 1988 (with corrections)

British Library cataloguing in publication data
Hernandez, G.
Fabry–Perot interferometers.—(Cambridge studies in modern optics; 3)
1. Interferometer
I. Title
535'.4   QC411

Library of Congress cataloguing in publication data
Hernandez, G.
Fabry–Perot interferometers.
Bibliography
Includes indexes.
1. Interferometer. I. Title.
QC411.H47 1986 535".4 85—24307

ISBN 0 521 32238 3 hard covers
ISBN 0 521 36812 X paperback

Transferred to digital printing 2002

....wholly out of impalpable materials, air, and the prismatic interferences of light, ingeniously focused by mirrors upon empty space. But you *do* it, that's the queerness!

William James

Contents

Chapter 3

Chapter 4

Chapter 5

Chapter 6

Chapter 7

Chapter 8

Bibliography

Glossary

Indexes

Preface

Although the Fabry-Perot interferometer is one of the more useful spectroscopic instruments available, there exists no locus where a contemporary presentation on this instrument is obtainable. The present text, which began as a short review and then grew beyond its original scope, attempts to present a broad overview of the Fabry-Perot as a spectroscopic tool.

The mathematical aspects of the Fabry-Perot are emphasized in the text, as would be expected for a spectroscopic tool which is used for quantitative measurements. The criteria for the choice of reference material for the text are rather arbitrary, in that the original and/or (in the author's opinion) relevant contemporary references are included. Thus any sins of omission or commission are to be blamed squarely on this writer.

I would like to thank Dr G. Vanasse for having provided the original impetus to write, to Professor G. G. Shepherd, Dr S. Silverman, Dr D. Rees, Dr D. A. Jennings, Dr K. M. Evenson and my wife Donna for reading the manuscript and providing valuable comments and suggestions. The help of Mr J. L. Smith in carrying a very large share of the author's experimental investigations, while the manuscript inched along, is most appreciated. The support and help of the Library staff, in particular Ms Jean Bankhead, Ms Lorna Kent, Ms Debra Losey, Ms Victoria Schneller and Ms Jane Watterson is gratefully acknowledged. Then there are those colleagues who in many ways, beyond enumeration, have made it possible to write the present text. They are Dr R. G. Roble, Dr M. Gadsden, Dr J. N. Howard, Dr T. S. Cress, Dr P. Giacomo, Professor P. B. Hays, Dr R. J. Sica, Professor M. A. Biondi, Professor G. J. Romick, Dr C. R. Burnett, Mr A. E. Sapp, Mr D. Hoge, and Ms B. Sloan.

G. H.
Boulder, Colorado

Introduction

The Fabry-Perot interferometer is a very powerful, yet versatile, spectroscopic tool. The basic property distinguishing the Fabry-Perot from other spectroscopic devices is simply that, for a given resolving power, the Fabry-Perot is the most luminous instrument available. This is correct when the comparison is made with other devices which give a spectrum directly as part of their operation. For Fourier (or multiplex) spectroscopy, the Fabry-Perot is competitive with the more commonly used Michelson interferometer.

The above, necessarily brief, introduction indicates the reason for the continued use of this device in both high-resolution studies and low-resolution investigations of extremely faint radiation sources.

In the following text an attempt is made to describe the Fabry-Perot and its properties, with an emphasis on the mathematical aspects, after a historical account is presented (Chapter 1) on the device and disciplines where it has been useful. The mathematical framework used to describe the ideal Fabry-Perot is found in the next chapter (Chapter 2), where the definitions of the basic parameters used elsewhere in the text are to be found. The next two sections (Chapters 3 and 4) are the natural continuation of the quantitative treatment of a Fabry-Perot for the optimum determination of Doppler shifts and widths in the presence of the (unavoidable) signal noise and the use of multiple Fabry-Perot devices in series to enhance some of the properties of a single device, respectively.

Chapter 5 presents the quantitative arguments for the existence of the highest luminosity-resolving power product for the Fabry-Perot, followed by the Connes spherical variation of the Fabry-Perot, which is a compensated device having a luminosity-resolving power product which increases with increasing resolving power. The behavior of the (practical) finite-size Fabry-Perot and the transient temporal behavior of the device are related, since they share the common attribute of a limited number of interfering light beams; thus, the two cases are discussed using the same mathematical framework. This treatment naturally leads to the description of the Fabry-Perot in terms of transfer functions, rather than in the classical terms usually employed.

The effects of emitting and absorbing species in the etalon cavity are illustrated in Chapter 6, ending with a discussion of the prime contemporary use of the Fabry-Perot, namely the laser. This device is discussed in terms of electromagnetic waves, cavity modes and diffraction effects, since the classical treatment is not a suitable medium of presentation.

The next section (Chapter 7) illustrates the preparation, construction, operation and alignment of practical Fabry-Perot devices. A discussion on mirror substrates, the mirrors themselves, their mountings, materials for construction, detectors, alignment methods, etc. is followed by an example on how to select an appropriate Fabry-Perot, thus bringing together all the developments of the earlier chapters into a practical application. The last section, Chapter 8, illustrates some of the seldom used configurations and arrangements of the Fabry-Perot which, incidentally, serve as proof of some of the previously defined characteristics of this device. The use of the Fabry-Perot in Fourier, or multiplex, spectroscopy is explored in some detail in order to show the relationship between the Fabry-Perot and the two-beam Michelson interferometer.

Although an author would prefer the text to be read in the order it was written, the reading order of the text depends, as would be expected, on the interests of the reader. For those readers who would like to find whether or not the Fabry-Perot would be useful for their needs, familiarization with the notation of Chapter 2, followed by Chapters 5 and 1 is the simpler approach. In all cases, familiarization with Chapter 2 and its notation is suggested for more efficient reading of the other sections. For those who need not be convinced on the usefulness of the device, but want to know how others have built and operated Fabry-Perot spectrometers, Chapter 7 with its example is nearly self-sufficient, except for the notation. For those in the field who would like to browse, the Contents list will be the best road map.

Chapter 1

Historical perspective

The present chapter attempts to give a reasonable, but by no means exhaustive, historical development of the multiple beam interferometer known as the Fabry-Perot.

The Fabry-Perot interferometer consists of two parallel flat semi-transparent mirrors separated by a fixed distance. This arrangement is called an etalon (Fabry and Perot, 1897) or interference gauge (Rayleigh, 1906). A monochromatic light wave incident upon an etalon at an arbitrary angle to the normal of the mirror surfaces will undergo multiple reflections within the mirrors. The intensity distribution of the etalon-reflected and etalon-transmitted interfering beams is found to be, because of the circular symmetry of the device, a set of bright concentric rings, or fringes, on a dark background for the transmission case, and a complementary set of dark fringes on a light background for the reflection case. The angular diameter of these fringes is dependent on the spacing between the etalon mirrors and the inverse wavelength (i.e., wavenumber) of the radiation. Thus, the basic function of a Fabry-Perot device is to transform wavelength into an angular displacement; however, in this process, the etalon adds something of its own to the resultant fringes, and this will be discussed in a later section.

The multiple-beam interferometer is a natural outgrowth of the observations of Newton (1730) and Fizeau (1862) of equal-spacing interference phenomena and of Haidinger's (1849) observations of equal-inclination interference phenomena in thin

FIGURE 1.1. Charles Fabry (1867 - 1945). The Bettmann Archive, Inc.

films. The fringes arising from equal spacing and equal inclination are usually called Fizeau and Haidinger fringes, respectively. The investigations of Lummer (1884) on the properties of light upon reflection from plane-parallel glass plates describes the reflection multiple-beam-interferometer, while Bouloch's (1893) report describes the instrument, the theory and experimental use of the transmission multiple-beam-interferometer. A description of Bouloch's device for metrological use was given by Fabry and Perot (1896, 1897). Fabry and his coworkers, in particular Perot and Buisson, fully explored the behavior, characteristics and use of this device to the extent that it is called the Fabry-Perot interferometer. The original contributions of Lummer and Bouloch have not been appreciated fully, and the suggestion by Duffieux (1969) that this device be called the Bouloch-Fabry interferometer has not acquired wide usage.

About this time, a variant of the Fabry-Perot, based on the Herschel reflection prism, was introduced by Lummer and Gehrcke (1903) and bears the name of these authors. Although this Lummer-Gehrcke plate was a strong contender with the Fabry-Perot (Lummer and Gehrcke, 1904; Perot and Fabry, 1904c; Gehrcke, 1905; Gehrcke and von Baeyer, 1906; Nagaoka and Takamine, 1912; Bogros, 1930), technical difficulties in its construction, as well as its inflexibility, soon placed this device

Although the original intent of Fabry and collaborators was to determine the standard meter in terms of fundamental units, at that time the wavelength of the red cadmium line, the spectroscopic properties of the device were quickly appreciated, described and used (Fabry and Perot, 1898a, 1898b, 1898c, 1899, 1900a, 1900b, 1901a, 1901b, 1901c, 1901d, 1902a, 1902b; Perot and Fabry, 1898a, 1898b, 1898c, 1899a, 1899b, 1900a, 1900b, 1901a, 1901b, 1904a, 1904b, 1904c, 1904d; Fabry, 1904a, 1904b, 1905a, 1905b; Fabry and Buisson, 1908a, 1908b, 1910a, 1910b, 1911, 1914, 1919; Buisson and Fabry, 1908, 1910, 1912, 1913, 1921; Benoit, Fabry and Perot, 1913; Lummer, 1901a). The first published report on the instrument built for Fabry was given by Jobin (1898) and diagrams of the device have been shown by Fabry and Perot (1899) and Benoit, Fabry and Perot (1913). Because of the high spectral resolution of the multiple-beam-interferometer, it was promptly employed in remeasuring the wavelengths of the solar spectrum (Perot and Fabry, 1900b, 1901a, 1904a; Fabry and Perot, 1901b, 1902a, 1902b; Fabry, 1905b; Fabry and Buisson, 1910a, 1910b; Buisson and Fabry, 1910), the wavelengths of the iron arc (Fabry and Perot, 1901b, 1901d, 1902a, 1902b; Pfund, 1908; Fabry and Buisson, 1910a) and rare gases (Buisson and Fabry, 1913; Merrill, 1917; Burns *et al.*, 1918; Meggers, 1921). Other contemporary users of the device were Lummer and Gehrcke (1902), Barnes (1904), Gehrcke (1906), Hamy (1906), Lord Rayleigh (1906, 1908), Pfund (1908), Zeeman (1908), Nagaoka and Takamine (1915), Meissner (1916), Nagaoka (1917), Strutt (later to become Lord Rayleigh, 1919) and Merton (1920). Lord Rayleigh's contribution was to recognize the Fabry-Perot interferometer as a (light) resonating cavity.

About this time, a variant of the Fabry-Perot, based on the Herschel reflection prism, was introduced by Lummer and Gehrcke (1903) and bears the name of these authors. Although this Lummer-Gehrcke plate was a strong contender with the Fabry-Perot (Lummer and Gehrcke, 1904; Perot and Fabry, 1904c; Gehrcke, 1905; Gehrcke and von Baeyer, 1906; Nagaoka and Takamine, 1912; Bogros, 1930), technical difficulties in its construction, as well as its inflexibility, soon placed this device on the shelf of historically interesting optical instruments.

After this aside, we return to the extension of the applicability of the interferometer by Fabry and coworkers. In 1912, Buisson and Fabry made use of this device to test the kinetic theory of gases by detecting the temperature broadening of the emission lines of gases, Doppler broadening, and then measuring the wavelength shift of the bulk motion of the emitters, or Doppler shift (Fabry and Buisson, 1914). Then, they combined the above into the astrophysical measurement of the Doppler shifts and widths in the Orion Nebula (Buisson, Fabry and Bourget, 1914a, 1914b), while at the same time they measured very accurately the wavelengths of the 'nebulium', now known to be lines from the O II $\left({}^4S{}^0_{3/2} - {}^2D{}^0_{5/2,3/2} \right)$ forbidden transitions.

Further uses of the Fabry-Perot include measurements of the viscosity effects of air (Fabry and Perot, 1898d), an electrostatic voltmeter (Perot and Fabry, 1898d, 1898e), the Zeeman effect (Zeeman, 1908, Nagaoka and Takamine, 1915), the index of refraction of gases (Meggers and Peters, 1918) and isotopic and fine structure investigations (Nagaoka, 1917; Strutt, 1919; Merton, 1920).

By 1920 the Fabry-Perot was well-entrenched in spectroscopic circles and spreading into other disciplines, such as its first geophysical application in the wavelength determination of the green line of the night sky (Babcock, 1923) at 557.7345 nm and now known to be the $\left({}^1D_2 - {}^1S_0 \right)$ transition of atomic oxygen.

The maturity of the instrument and its varied uses led to the usual mark of approval of extended reviews and books (Fabry, 1923; Childs, 1926; Hansen, 1928; Williams, 1930).

Developments of the late 1920s and 1930s include the measurement and determination of reference wavelengths (Babcock, 1927; Humphreys, 1930, 1931; C.V. Jackson, 1931, 1932, 1933, 1936), and fine structure of emission lines (Houston, 1926, 1927; D.A. Jackson, 1934; D.A. Jackson and Kuhn, 1935, 1936, 1937, 1938a, 1938b) . Some of these investigations required further capabilities than a single-etalon Fabry-Perot interferometer could deliver, thus, the use of two or more interferometers in tandem was introduced (Nagaoka, 1917; Houston, 1927; Gehrcke and Lau, 1927, 1930; Lau, 1930, 1932; Lau and Ritter, 1932; Pauls, 1932). However, note that Perot and Fabry (1899b) had already used a double interferometer, although not as a spectroscopic device. As the capabilities of the Fabry-Perot device were further developed into the realm of line-breadth measurements, the need for a quantitative mathematical expression for a practical Fabry-Perot interferometer became increasingly apparent (Minkowsky and Bruck, 1935a, 1935b, 1935c). Duffieux (1935, 1939) succeeded in this endeavor, and his work can be considered the highlight of this period.

The two-part review by Meissner (1941, 1942) consolidated the spectroscopic aspect of the Fabry-Perot up to that time, and this review, coupled with the appearance of Tolansky's (1947) book on high-resolution instruments, their construction, alignment and care, set the stage for the next advances in the field of high-resolution spectroscopy. The first development was the production of multilayer dielectric coatings (Banning, 1947a, 1947b; Dufour, 1948; and the review by Heavens, 1960) which allows the attainment of high-reflectivity mirrors with small absorption and/or scattering. These mirror coatings are vastly superior to the previously used, mainly silver, mirror coatings, as they allow the use of much higher reflectivities without the concomitant flux losses associated with absorption by metallic coatings. As will be discussed later, the use of higher reflectivity means, other things being equal, higher resolving power, but the price of this advance was to emphasize the surface defects of the substrates where these multilayer coatings are deposited. This lack of complete flatness of the substrates (as well as the distortions introduced by the multilayer coatings) limits the improvements that can be achieved with higher and higher reflectivity coatings (Dufour and Picca, 1945; Dufour, 1951). The next advance was made by Jacquinot and Dufour (1948) when they moved away from the previous recording methods used with Fabry-Perot interferometers, namely the human eye and the ever-present photographic plate, into quantum detectors (e.g. photomultipliers). Although the ultimate gains of the Jacquinot and Dufour scheme were the efficient use of light and the higher effective quantum efficiency of the detectors (coupled with their inherent linearity and wide dynamic range), the milestone was the realization and accomplishment by Jacquinot and Dufour of the use of the radial symmetry of the Fabry-Perot and the necessary scanning (and its many varieties) of the instrument in order to take full advantage of quantum detectors. In addition to the above, these authors also developed a criterion for the best performance of the instrument, which is called the luminosity-resolution-product (LRP) criterion. Chabbal (1953) fully developed the LRP criterion for the many varied and sundry uses of the instrument.

The impact of Jacquinot and Dufour's (1948) work slowly reached the field (see, for instance, Barrell, 1949, Candler, 1951, and Kuhn, 1951, where in the last, only a few lines are given to photoelectric recording), but applications soon were suggested (Giacomo, 1952; Armstrong, 1953, 1956; Biondi, 1956). Jacquinot (1954) showed that the Fabry-Perot is a more luminous instrument than either grating or prism instruments at the same conditions. P. Connes (1956) further showed that the luminosity of the Fabry-Perot could be increased with his spherical Fabry-Perot, i.e., a compensated instrument. This gain in luminosity is most useful at very high resolving powers where the normal Fabry-Perot starts showing the effects of its limited practical size and fixed LRP. As mentioned earlier, two or more interferometers had been used to increase the the capabilities of a single Fabry-Perot interferometer, and Chabbal (1957, 1958a, b, c) began his studies of multiple etalon devices which eventually led to the very useful PEPSIOS interferometer (Mack *et al.*, 1963). This device is widely used where the highest resolving power coupled with selectivity is required, yet preserving the high luminosity inherent in the Fabry-Perot device.

Tolansky's book (1955), the CRNS Colloquia (1958, 1967), the NPL Symposium (1960) and Jacquinot's review (1960) on interference spectroscopy have served as springboards for most of the recent developments in the field. Besides the PEPSIOS device previously mentioned, we find the proposal to use the Fabry-Perot as the cavity for an optical maser (Dicke, 1958; Prokhorov, 1958; Schawlow and Townes,1958) culminating with the first laser (Maiman, 1960), photoelectric spatial scanning (Armstrong, 1958; Bradley, 1962a; Shepherd *et al.*, 1965; Hirschberg and Platz, 1965; Katzenstein, 1965; Shepherd and Paffrath, 1967; Hirschberg and Cooke, 1970; Hirschberg and Fried, 1970; Hirschberg *et al.*, 1971), dynamic alignment (Ramsay, 1962, 1966; Gadsden and Williams, 1966; Hernandez and Mills, 1973), and the use of insect-eye lenses (Courtès *et al.*, 1966; Courtès and Georgelin, 1967). The 1939 Duffieux approach to the analytical description of a Fabry-Perot interferometer was continued by Krebs and Sauer (1953), Bayer-Helms (1963a, 1963b, 1963c, 1963d, 1964a, 1964b), Tako and Ohi (1965), Del Piano and Quesada (1965), Ballik (1966), Best (1967) and these results have been extended and collected by Hernandez (1966, 1970). The classical treatment of active material in the etalon cavity (Kastler, 1962), illustrates the hitherto unexplored capabilities of the Fabry-Perot. The reviews of Girard and Jacquinot (1967), Herriott (1967), Steel (1967), Shepherd (1967, 1969a, b, 1972), Vaughan (1967), Jacquinot (1969), Koppelmann (1969) and Dyson (1970) show in great detail the contemporary status of high resolution spectroscopy, as well as geophysical and astrophysical applications. In the following years there was a consolidation of techniques and applications in varied fields and the highlights are extension of measurements into the near (Chantrel *et al.*, 1964) and vacuum ultraviolet (Abjean and Johannin-Gilles, 1970; Guern *et al.*, 1974), the proof that an interference pattern exists even at extremely low photon fluxes (Reynolds *et al.*, 1969; Bozec *et al.*, 1970), the matrix treatment of the instrumental function by Neuhaus and Nylén (1970), and the response of the etalon to rapid changes in optical length (Gerardo *et al.*, 1965; Dangor and Fielding, 1970). Also the deconvolution work of the measured spectra by Roig and collaborators (Roig *et al.*, 1967; Fourcade *et al.*, 1968; Fourcade and Roig, 1969), Cooper (1971), Hays and Roble (1971), the optimization of the instrument by Velichko *et al.* (1971) and Pátek's (1967) book on lasers deserve mention. A new technique for automatic alignment , based on Jones and Richards' (1973) report on capacitive micrometers, was first used by Hicks *et al.* (1974) and is now the basis of the more-commonly used automatic-alignment scheme. The tandem etalon system of Perot and Fabry (1899b) was simplified by the double-passed etalon arrangement (Dufour, 1951; Hariharan and Sen, 1961; Müller and Winkler, 1968). Multi-passing an etalon has been described by Sandercock (1971), and this arrangement is used to great advantage in Raman and Brillouin spectroscopy. The Fabry-Perot has also been used as a comb filter in Raman studies (Barrett and Myers, 1971). The analytical treatment of the Fabry-Perot was extended to the off-axis case (Hernandez, 1974), which includes the behavior of interference filters when they are used to scan a spectrum at low resolving power.

The reviews by Roesler (1974), Fabelinskii and Chistyi (1976), Meaburn (1976) and Genzel and Sakai (1977) are excellent references on the PEPSIOS multi-etalon devices and interferometric techniques. In recent years, studies on the fundamental limitations of the signal, with its inherent noise, in the retrieval of information from Fabry-Perot measurements have been made (Hernandez, 1978, 1979; Jahn *et al.*, 1982), and the results provide an optimum set of operational points based on minimum uncertainty criteria. Thus far, such derivations have been made for the determination of Doppler widths (i.e., temperatures) and shifts.

When quantum detectors are employed, typically only part of the central order of the fringe pattern is utilized and many schemes have been attempted to use a larger fraction of the flux gathered by the Fabry-Perot interferometer. Most of these reported results take advantage of the Fabry-Perot multiple orders by using multi-slit or multi-annular masks based on the original suggestions by Jacquinot and Dufour (1948) (Allard, 1958; Meaburn, 1976; Sipler and Biondi, 1978; Dupoisot and Prat, 1979; Okano *et al.*, 1980), by extension into multiple detectors and/or multiple anode detectors (Shepherd *et al.*, 1965; Chaux *et al.*, 1976, Chaux and Boquillon, 1979; Abreu *et al.*, 1981; Rees *et al.*, 1981a; Killeen *et al.*, 1983) which include television camera tubes (Sivjee *et al.*, 1980), and by multiplexing techniques (Shepherd *et al.*, 1965; Hoey *et al.*, 1970; Hirschberg *et al.*, 1971; Neo and Shepherd, 1972; Shepherd *et al.*, 1978). The radiation-modulation concept of the two-etalon TESS (Hernandez *et al.*, 1981, Hernandez, 1982a) preserves the spectral information in the incoming light and thus the (theoretical) gain in luminosity is very large and dependent on the resolving power. This gain is limited in practice to values less than about 1000-fold.

The range of utilization of the Fabry-Perot has been enlarged into the vacuum ultraviolet down to 138 nm (Bideau-Mehu *et al.*, 1976, 1980), and X-ray operation of the device has been proposed (Steyerl and Steinhauser, 1979). The time-measurement method to determine the Fabry-Perot maxima by Pole and collaborators (1978, 1980) and the matched-etalon camera concept of Young and Clark (1980) deserve mention as some of the newer developments.

At this point, this historical overview should end by referring to the reviews of Heilig and Steudel (1978), Hernandez and Roble (1979), Hernandez (1980), Meriwether (1983), Atherton *et al.* (1981), Pismis (1982) and Stenholm (1984) to provide the reader with an idea of the present status of the use of Fabry-Perot devices in the laboratory, and their applications in geophysics, astronomy and laser spectroscopy, while the reviews of Baker and Walker (1982), Chantry (1982) and

Clarke and Rosenberg (1982) cover the millimeter, microwave and infrared regions.

FIGURE 1.2.   A. Perot (1863–1925).

Chapter 2

Mathematical development

1. Ideal interferometer

In this section the inherent behavior and properties of a Fabry-Perot will be reviewed in the ideal classical sense, i.e., where the effects of the limited size of the mirrors, diffraction effects, etc. are considered to be negligible. By taking this (arbitrary) approach, it is possible to consider the main properties of the device in an ideal sense and, then later, to treat the practical limitations as corrections of the idealized picture.

Although almost every book on optical interference phenomena (Fabry, 1923; Born and Wolf, 1964; Steel, 1967 ; Cook, 1971) shows the basic behavior of the multiple interference device known as the Fabry-Perot interferometer, it will be repeated here for the sake of completeness and notation. This ray-path approach is suitable for the conventional Fabry-Perot, where diffraction effects are negligible. Diffraction, and its effects, is discussed in Chapter 6, Section 4.

Consider a plane-parallel slab of refractive index μ with surfaces that partly reflect and partly transmit light, and upon which a plane wave of monochromatic radiation is incident at an angle Θ to the normal as indicated in Figure 2.1. At the first surface, the wave is divided into two plane waves, one reflected, as indicated by the dashed line leading to a , and the other transmitted, and the division process of the wave remaining inside the slab continues as shown in the figure. The phase lag,

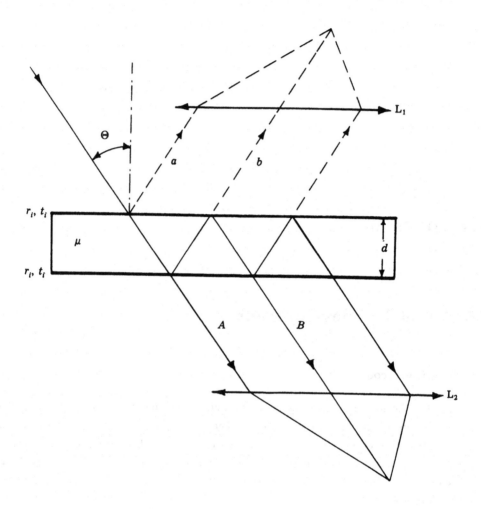

FIGURE 2.1. Multiple-beam interference in a Fabry-Perot etalon.

Φ , between two emerging waves, say, A and B , or a and b , in the figure is given by:

$$\Phi = 4\pi \, \nu \mu \, d \, c^{-1} \cos\Theta \, , \qquad\qquad 2.1.1a$$

$$\Phi = 2\pi \, \nu t \, \cos\Theta \, . \qquad\qquad 2.1.1b$$

In the above equations ν is the frequency of the radiation in vacuum, μ the index of refraction of the slab, d is the thickness of the same, c is the speed of light, and t is the transit time for each double passage normal to the surfaces, i.e., $t = 2 \, \mu \, dc^{-1}$ (Steel, 1967). Consider each reflective surface to be described by its complex reflection and transmission amplitude coefficients r_i and t_i. r_1 (t_1) is defined to be the reflection (transmission) coefficient for a wave traveling from the surrounding medium towards the slab, while r_2 (t_2) applies to a wave traveling from the slab towards the surrounding medium. For a unit amplitude incident wave the amplitude of the successive waves reflected from the plate is:

$$r_1 \, , \, t_1^2 r_2 e^{i\Phi} \, , \, t_1^2 r_2^3 e^{2i\Phi} \, , \, \ldots\ldots\ldots \, . \qquad 2.1.2$$

The transmitted amplitudes are given by:

$$t_1 t_2 \, , \, t_1 t_2 r_2^2 e^{i\Phi} \, , \, t_1 t_2 r_2^4 e^{2i\Phi} \, , \, \ldots\ldots \, . \qquad 2.1.3$$

When the rays are brought to focus, by lenses L_1 and L_2 of Figure 2.1, the total reflected and transmitted amplitudes are then the sum of the amplitudes given in Equations 2.1.2) and 2.1.3) , i.e,

$$A_r(\Phi) = r_1 + t_1^2 r_2 e^{i\Phi} + \cdots$$

$$= \left[\left(1 - e^{i\Phi}(R+\tau) \right) R^{1/2} \right] \left(1 - R e^{i\Phi} \right)^{-1} , \qquad 2.1.4$$

$$A_t(\Phi) = t_1 t_2 + t_1 t_2 r_2^2 e^{i\Phi} + \cdots = \tau \left(1 - R e^{i\Phi} \right)^{-1} , \qquad 2.1.5$$

where the r and t subscripts indicate reflection or transmission. In the above, use has been made of the following relations (Born and Wolf, 1964):

$$t_1 t_2 = \tau \, ,$$

$$r_1 = -r_2 \, ,$$

$$r_1^2 = r_2^2 = R \, ,$$

$$\delta = \Phi + 2\chi \, ,$$

where χ represents the phase changes upon reflection on each of the surfaces, e.g., the sum of the arguments of the complex amplitude reflections. The resultant reflected and transmitted intensities are then given by the product of the amplitude and its complex conjugate:

$$Y_r(\delta) = A_r(\delta) \, A_r^*(\delta)$$

$$= R \left[1 - 2(R+\tau)\cos\delta + (R+\tau)^2 \right] \left[1 + R^2 - 2R\cos\delta \right]^{-1} , \qquad 2.1.6a$$

$$Y_t(\delta) = A_t(\delta) \, A_t^*(\delta)$$

$$= \tau^2 \left[1 + R^2 - 2R\cos\delta \right]^{-1} . \qquad 2.1.7a$$

Note that the above functions can be expressed in terms of the absorption/scattering coefficient A, since $R + \tau + A = 1$, and the results are given by :

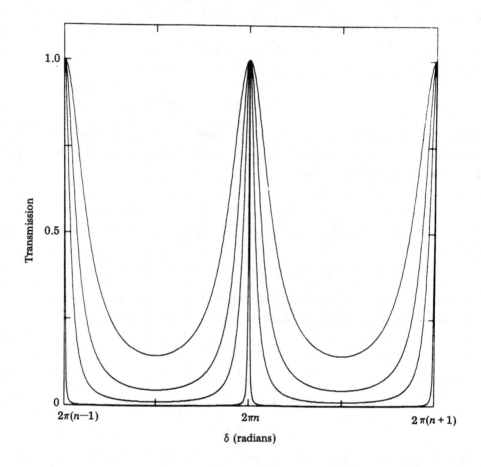

FIGURE 2.2. Transmitted radiation of a Fabry-Perot etalon as a function of the
phase retardation of the beams for various reflectivities. The values of the latter are:
0.98, 0.81, 0.65 and 0.45 for the narrowest to the widest profile.

$$Y_r\,(\delta) \;=\; R\;\left[1+(1-A\,)^2-2(1-A\,)\;\cos\delta\right]\left[1+R^2-2R\;\cos\delta\right]^{-1}\,, \qquad 2.1.6b$$

$$Y_t\,(\delta) \;=\; [1-A\,(1-R\,)^{-1}]^2(1-R\,)^2\Bigl(1+R^2-2R\;\cos\delta\Bigr)^{-1}\,. \qquad 2.1.7b$$

This last equation is often found described by its equivalent formulations:

$$Y_t\,(\delta) \;=\; [1-A\,(1-R\,)^{-1}]^2(1-R\,)(1+R\,)^{-1}\left(1+2\sum_{m=1}^{m=\infty} R^{\,m}\cos(m\,\delta)\right)\,, \qquad 2.1.7c$$

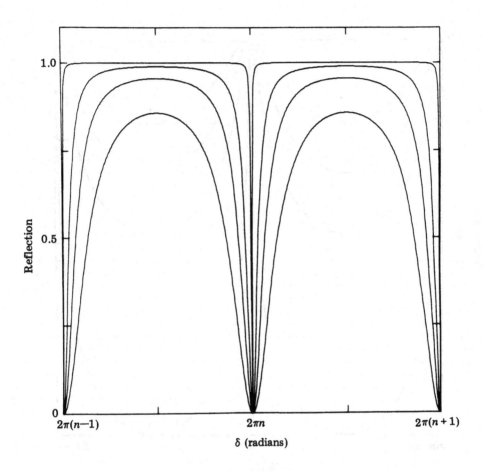

FIGURE 2.3. Reflected radiation of a Fabry-Perot etalon as a function of the phase retardation of the beams. Same reflectivities as Figure 2.2.

$$Y_t(\delta) = [1 - A (1-R)^{-1}]^2 \left[1 + 4R (1-R)^{-2} \sin^2 \frac{\delta}{2} \right]^{-1} , \qquad 2.1.7d$$

$$Y_t(\delta) = [1 - A (1-R)^{-1}]^2 (1-R)(1+R)^{-1} 2 \ln(R^{-1})$$

$$\times \sum_{n=-\infty}^{n=\infty} \left\{ [\ln(R)^{-1}]^2 + (\delta - 2\pi n)^2 \right\}^{-1} . \qquad 2.1.7e$$

The last term of this Equation shows the transmission intensity to be composed of the sum of dispersion, or Lorentzian, profiles of width $l = \ln(R^{-1})$ and transmission

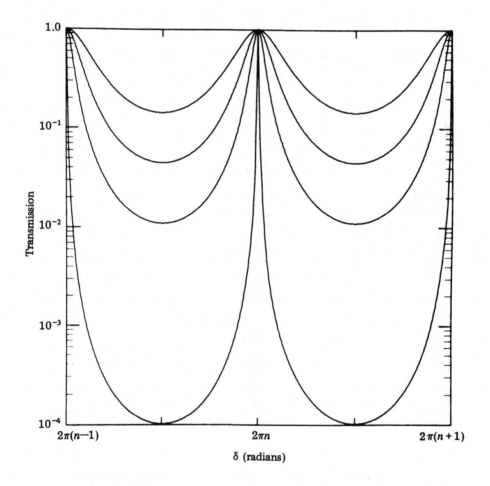

FIGURE 2.4. Transmitted radiation of a Fabry-Perot etalon as a function of the phase retardation of the beams. This is the same data as in Figure 2.2, except it is shown in logarithmic scale.

less than unity.

For the case when $A = 0$, the sum of Equations 2.1.6b) and 2.1.7b), or their variants, is equal to unity; that is, the incoming wave energy has been conserved. The behavior of the Fabry-Perot etalon as a function of the reflectivity R is shown in Figures 2.2, 2.3 and 2.4. Note in particular the increase in instrumental resolution, or ability to separate closely spaced lines, as the reflectivity increases. In this example the absorptivity/scattering coefficient A has been set to zero.

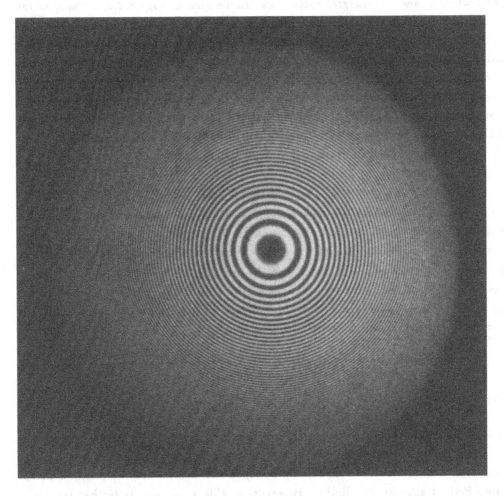

FIGURE 2.5. Transmitted radiation of a Fabry-Perot etalon. Mercury discharge lamp source, 1.5 cm spacer etalon with 0.85 reflectivity.

Consider many plane waves of equal intensity, incident over a range of angles to the etalon of Figure 2.1, where the resultant reflected and transmitted beams are collected by two separate lenses and thus made to interfere, as given in the figure. The following results are then observed. For the transmission case it is found that there exist maxima of intensity, or constructive interference, when δ has integral values of 2π; that is, when the order of interference, n, as defined below, has integer values:

$$n = \delta(2\pi)^{-1} = 2\mu d\sigma \cos\Theta .\qquad\qquad 2.1.8$$

In the above equation, σ is defined as $\sigma = \nu\,c^{-1} = \lambda^{-1}$, or the inverse wavelength in vacuum, which will be called wavenumber, as is usual in spectroscopic notation. The wavenumber unit adopted here is the Kayser (K) = cm^{-1}. However, note that Kayser in his tables (1925) used m^{-1}. Minima, or destructive interference, will be found when n has half-integral values. As can be seen in Figure 2.5, because of the circular symmetry of the device, the resultant pattern is a set of bright circular concentric rings, or fringes, on a dark background. For the reflected radiation, the resultant is found to be the complement of the transmitted case, i.e., a bright field crossed by narrow dark fringes. The orders of interference are, in the reflection case, equal to integer values for the dark fringes and one-half integers for the maxima. For simplicity, the central or operating order, of interference for transmitted radiation is defined as that occurring perpendicular to the reflecting surfaces:

$$n_0 = 2\,\mu\,\sigma\,d\ .$$

$$2.1.9$$

The set of observed fringes shows the fundamental periodic behavior of the Fabry-Perot interferometer, namely, a single wavenumber appears a multiple number of times at specific angles given by Equation 2.1.8). These angles are a function of the wavenumber (inverse wavelength) and the optical spacer $\mu\,d$. The distance between the mth and the $(m^{th}+1)$ order can be considered in terms of a change in wavenumber of the radiation, or a change in the optical spacing. If this optical spacing is fixed, the necessary change in wavenumber to vary the order by one unit, say from m to $(m+1)$, is called the free spectral range, $\Delta\sigma$, and which is defined by:

$$\Delta\sigma\ =\ (2\mu d)^{-1}\ =\ \frac{\sigma}{n_0}\ .$$

$$2.1.10a.$$

The name free spectral range arises from the early use of this instrument to resolve fine structure components in spectra. Imagine two superimposed, equal-intensity, monochromatic lines whose wavenumber separation can slowly be changed by shifting the wavenumber of one of the lines while the other is held constant. Eventually the separation between the two lines will be apparent as two sets of maxima in the Fabry-Perot transmitted radiation. However, as this separation is further increased, the second maxima will blend and disappear into the next maxima of the original unchanged wavenumber line, and eventually reappear at the other side. When the two lines blend exactly, it is then said there has occurred an overlap of orders and the true separation in wavenumber between the two lines is ambiguous by one order, and in the general case, by any number of orders. Indeed, it is not possible to say that there exist two lines in the transmitted pattern unless a separate measurement is made. As the previous example demonstrates, the separation between two lines is free of ambiguity over the wavenumber spectral range defined by one order, thus the name free spectral range.

Again referring to Equation 2.1.8), for a fixed wavenumber σ, a change in the optical spacing, $\mu\,d$, forces the wavenumber to appear at any arbitrary angle Θ, actually a set of Θs. This change can be accomplished by either changing the value of d, i.e., physical spacing, or the value of the index of refraction μ. In the latter case, it is perhaps more correct to say that the effective wavenumber $\mu\sigma$ has been changed by varying μ. To avoid this slight ambiguity, the nominal free spectral range is defined in vacuum, i.e., $\mu \equiv 1$, or:

$$\Delta\sigma = (2d)^{-1} .$$ 2.1.10b

As seen earlier in Figure 2.2, the width of the bright fringes observed in transmission (and their complementary reflection dark fringes) is governed by the value of R in Equations 2.1.6b) and 2.1.7b), and for $R \leqslant 1$, their width is finite and given, as a half-width at half-height (HWHH), by:

$$a = 2 \sin^{-1}\left[(1-R)(2R^{1/2})^{-1} \right] .$$ 2.1.11a

The above can be approximated for small values of the argument of the inverse sine, i.e., R near unity, by:

$$a \approx (1-R)R^{-1/2} .$$ 2.1.11b

The correct expression given in Equation 2.1.11a) is not defined for values of R less than $(3 - 8^{1/2})$. The above definitions of a are given for the 2π periodicity of the etalon, and in order to allow for changes in scale, the normalized HWHH is defined as the ratio of the width to the period, namely:

$$a^* = a (2\pi)^{-1} = \pi^{-1}\sin^{-1}\left[(1-R)(2R^{1/2})^{-1} \right] .$$ 2.1.12a

In the text that follows, normalized widths will be used unless otherwise specified. For the sake of completeness, the normalized HWHH, l^*, of the individual Lorentz profiles in the sum of Equation 2.1.7e) is related to the normalized HWHH a^* of Equation 2.1.12a) by:

$$\sin (\pi a^*) = \sinh (\pi l^*) .$$ 2.1.12b

The inverse of Equation 2.1.12a) is normally called the reflective finesse, N_R, of the etalon (Chabbal, 1953):

$$N_R = (2a^*)^{-1} = \pi \left\{ 2 \sin^{-1}\left[(1-R)(2R^{1/2})^{-1} \right] \right\}^{-1} ,$$ 2.1.13a

$$N_R \approx \pi R^{1/2}(1-R)^{-1} .$$ 2.1.13b

The $4R (1-R)^{-2}$ term in Equation 2.1.7d) is called here the **F** parameter, rather than the coefficient of reflective finesse, in order to avoid confusion with the finesse N_R of Equation 2.1.13). In terms of the (approximate) finesse, defined above, this **F** parameter can be described by:

$$\mathbf{F} = 4R(1-R)^{-2} = (2N_R \pi^{-1})^2 = 4a^{-2} = (\pi a^*)^{-2} . \qquad 2.1.13c$$

Equations 2.1.12a) and 2.1.13a) show that the normalized width, and thus the reflective finesse, depend only on the reflectivity R, and therefore both are constants of the instrument. The width of the fringes decreases as R approaches the limiting value of unity, at which point the fringes become a Dirac comb, a result which will be used later. In the absence of absorption and/or scattering, i.e., $R + r = 1$, inspection of Equations 2.1.7a), 2.1.7b), 2.1.7c), 2.1.7d), and 2.1.7e) reveals that the transmission at the maximum of the fringes is equal to unity, while the minimum value is given by $(1-R)^2(1+R)^{-2}$. This result indicates there exists a finite leakage of the line emission near and at the minimum. A quantitative expression of this effect is given by the contrast, C:

$$\mathbf{C} = Y_t(\delta)_{\max} \left[Y_t(\delta)_{\min} \right]^{-1} = (1+R)^2(1-R)^{-2} \simeq 1 + (2N_R \pi^{-1})^2 , \qquad 2.1.14$$

where N_R is the approximate finesse given in Equation 2.1.13b). The contrast can be interpreted as a measure of the ability of a Fabry-Perot interferometer to differentiate between a strong and a weak line in the same spectrum. Figure 2.6 shows the contrast effects for a transmission Fabry-Perot. In the figure, two emission lines are given where the weaker of the two lines has been arbitrarily set to be 1/100 of the stronger one. Note both the decrease of line leakage at the minimum and the ability to discriminate between the two lines as the etalon reflectivity increases.

The finite width of the fringes gives the Fabry-Perot a finite resolving power, or ability to distinguish the existence of two closely spaced, equal-intensity, lines in a spectrum. This resolving power can be defined by the usual criteria (Rayleigh, 1879; Schuster, 1904; Sparrow, 1916; Houston, 1927; Ditchburn, 1930; Buxton, 1937). The most commonly used criterion for Fabry-Perot interferometers is the Houston criterion (Ramsay *et al.*, 1941 ; Chabbal, 1953); that is, two equal-intensity lines are considered to be resolved when their maxima are separated by their full-width at half-height (FWHH). Thus, assuming that these two lines give rise to a true HWHH a, and its associated normalized HWHH a^*, then the resolving power is:

$$\mathbf{R} = \sigma(2a)^{-1} = n_0(2a^*)^{-1} = n_0 N_R . \qquad 2.1.15$$

In the above equation, σ is the mean wavenumber of the radiation of the two lines. For the sake of completeness, the resolving limit is defined as the inverse of the resolving power times the wavenumber, which in terms of Equation 2.1.15) is given by:

$$\mathbf{r} = \frac{\sigma}{\mathbf{R}} = 2a = 2a^* \Delta\sigma = \Delta\sigma N_R^{-1} . \qquad 2.1.16$$

At this point it should be noted that the classical approach used thus far is not the only method employed to obtain the derivations. Indeed, as Connes (1961) and

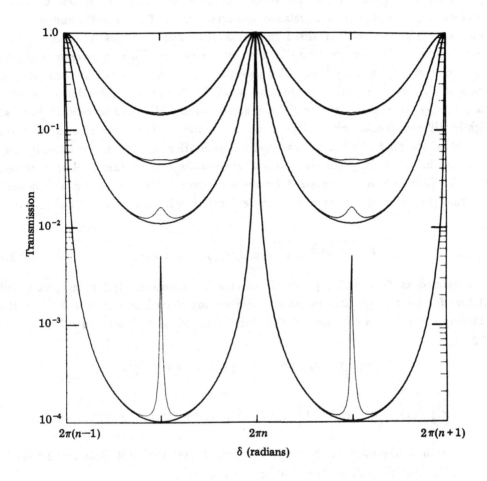

FIGURE 2.6. Contrast effects on a transmission etalon. Another emission line with 1 % of the emission rate of the main emission line has been superimposed at the minimum transmission point of the latter. Same conditions as Figure 2.4.

Stoner (1966) have shown, the same results can be obtained by using the elegant optical transient response method on the instrument, just as in electrical filter theory, to determine the instrumental function. This topic will be discussed in Chapter 5, Section 3.2. The classical approach has been used here in order to provide continuity with the earlier literature on the subject.

2. Real sources

In order to operate in the real world, the rather arbitrary requirement of monochromatic light sources must be relaxed and line sources of finite width and shape be employed. This can be visualized in terms of a series of closely spaced monochromatic lines with a given intensity distribution as a function of wavenumber . As has been shown in Equation 2.1.8), this wavenumber distribution can be considered in terms of the etalon order n and/or the phase δ. When this collection of lines is examined by an etalon, there will be a superposition of etalon-broadened lines at slightly different angles which depend on the center wavenumber of each of these lines; thus, the final result is a broadened representation of the original intensity distribution. In the limit, when these monochromatic lines are infinitely close together, their distribution can be expressed by an analytical expression. Using a unit area Gaussian profile as an example, the source function with a true HWHH dg is given by:

$$\mathbf{S} = \left(\pi^{1/2} G \right)^{-1} \exp \left[- (\sigma - \sigma_0)^2 G^{-2} \right] . \qquad 2.2.1$$

G is defined as $G = dg\,[\ln(2)]^{-1/2}$. Employing Equations 2.1.8) for the phase and 2.1.10a) for the free spectral range, and introducing them into Equation 2.1.7b), the following expression is obtained for the etalon response to the assumed source profile of 2.2.1):

$$Y(\sigma_0) = (\pi^{1/2}G)^{-1}[1-A(1-R)^{-1}]^2(1-R)^2$$

$$\times \int_{-\infty}^{\infty} \left\{ \exp - \left[(\sigma - \sigma_0)^2 G^{-2} \right] \left[1 + R^2 - 2R\,\cos(2\pi\sigma\,\Delta\sigma^{-1}\cos\Theta) \right]^{-1} \right\} d\sigma . \quad 2.2.2a$$

The solution to Equation 2.2.2a) can be obtained analytically if Equation 2.1.7c) is used for the etalon function. The solution is given below:

$$Y(\sigma_0) = [1-A(1-R)^{-1}]^2 \,(1-R)(1+R)^{-1}$$

$$\times \left\{ 1 + 2 \sum_{k=1}^{k=\infty} R^k \exp - [\pi\,\Delta\sigma^{-1}G\,k]^2 \,\cos(2\pi k\sigma_0\,\Delta\sigma^{-1}\,\cos\Theta) \right\} . \qquad 2.2.2b$$

The argument of the exponential is found to be equal to $k^2\,\pi^2 dg^2 [\Delta\sigma^2\,\ln(2)]^{-1}$, which in our notation of normalized widths reduces to $k^2\,\pi^2 (\,dg^{\,*})^2 [\ln(2)]^{-1}$. Note, in particular, that these results indicate that the important width is the normalized width of the source, i.e., the ratio of the source width to the etalon period, or free spectral range, and not the absolute width. Referring to Equation 2.2.2a), the mathematical operation that describes the integral is the cross-correlation integral (Bracewell, 1965). For symmetrical functions, such as the presently assumed Gaussian function, i.e., when $F(-x) \equiv F(x)$, it is possible to change the order of the argument of the exponential without affecting the results. Thus, if in Equation 2.2.2a) one would have used $(\sigma_0 - \sigma)$ in the argument rather than $(\sigma - \sigma_0)$, the operation resulting from this change would have been the convolution integral (Bracewell, 1965). Since convolution is commutative, associative, and distributive, it is very convenient to use; however, the correct results will be obtained only when the functions under consideration are symmetric, such as the Gaussian source function of Equation 2.2.1.). In the following text the convolution operational calculus will be used for convenience; however, the reader should keep in mind that cross-correlation for symmetrical functions is being used. When the need arises, the proper cross-correlation calculus must be used to obtain the correct answer.

Another example of a source shape is the Lorentz, or dispersion, profile, assumed in this case to have unit area and true HWHH l, i.e.,

$$S = (l\,\pi)^{-1}\left\{1 + [(\sigma - \sigma_0)^2 l^{-2}]\right\}^{-1} .$$

2.2.3

The etalon response to this profile is:

$$Y(\sigma_0) = [1 - A\,(1-R)^{-1}]^2 (1-R)(1+R)^{-1}$$

$$\times \left\{1 + 2\sum_{k=1}^{k=\infty} R^k \exp-[2\pi\,\Delta\sigma^{-1} lk]\cos\left(2\pi\,k\,\sigma_0\,\Delta\sigma^{-1}\cos\Theta\right)\right\} .$$

2.2.4a

Again, in this expression, it is found that the important width is the width of the source relative to the etalon periodicity. Note that the reflectivity and the exponential terms can be combined into an effective reflectivity given by $R_E = R\exp-(2\pi l^*)$. Since for l^* greater than zero the exponential is then less than unity, it is found that $R_E \leqslant R$. This example has been used specifically to show the effects of the etalon on the original source. For instance, introducing the value of R_E in Equation 2.1.7b) the latter becomes:

$$Y_t(\delta) = [1 - A\,(1-R)^{-1}]^2 (1-R)^2 [1 + R_E^2 - 2R_E\cos\delta]^{-1} .$$

2.2.4b

If, for convenience, A is arbitrarily set equal to zero, then the results are obvious from Equation 2.2.4b). First, the resultant maximum transmission has decreased from unity to $(1-R)(1+R_E)/[(1+R)(1-R_E)] \leqslant 1$ and correspondingly the minimum has changed from $(1-R)^2(1+R)^{-2}$ to $(1-R)(1-R_E)/[(1+R)(1+R_E)]$ and the latter is numerically larger than the former. Thus the contrast, or ratio of maximum to minimum, has also decreased. Second, the HWHH of the resultant profile has increased as can be seen by introducing R_E in Equation 2.1.12a). Also, since the area under the profile (for one period of the interferometer) has been kept constant, the loss of transmission at the maximum is made up by a wider profile with higher tails, i.e., higher minimum. The above results show, in a simplified manner,

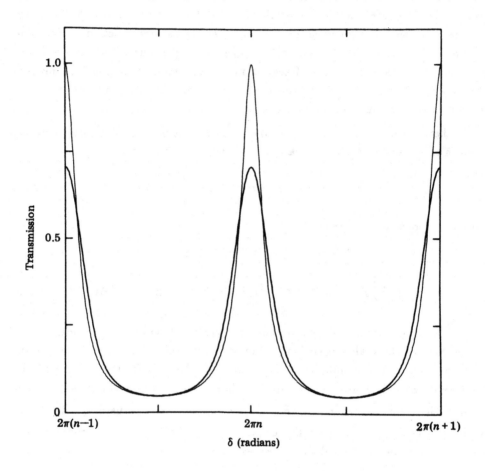

FIGURE 2.7. Broadening effects of an etalon on a Gaussian emission profile. The monochromatic line response is shown for comparison.

the basic effects of the broadening and consequent loss of transmission on a source profile when examined by an etalon. For illustration purposes, the broadening effects of a Gaussian line on the resultant profile are given in Figure 2.7. The breadth of this Gaussian line profile has been set equal to the breadth obtained from the reflective coatings alone. The monochromatic line, or ideal etalon, behavior is included for comparison.

As it will be seen later, it is convenient to define the general source coefficient in equations such as 2.2.2b) and 2.2.4a) to be equal to the exponentials in these equations. If this coefficient is called s_k, the equation that expresses the etalon transmission to a given arbitrary source is:

$$Y(\sigma_0) = [1 - A (1-R)^{-1}]^2 (1-R)(1+R)^{-1}$$

$$\times \left\{ 1 + 2 \sum_{k=1}^{k=\infty} R^k s_k \cos(2\pi k \sigma_0 \Delta\sigma^{-1} \cos\Theta) \right\}. \qquad 2.2.5$$

In the above definition of the source coefficient, it is understood that s_k stands for any given arbitrary profile or a combination of profiles, as is usually the case.

When the line widths, such as l or dg of Equations 2.2.2b) and 2.2.4a), tend to the limit of zero width, the just-mentioned equations revert to the case of a mono-chromatic line derived earlier when discussing the behavior of the basic instrument, given in Equations 2.1.7a) → 2.1.7e). On the other extreme, when the line width becomes very large, i.e., l (or dg) $>> \Delta\sigma$, the value of the exponentials in either 2.2.2b) or 2.2.4a) [or s_k of 2.2.5)] goes to zero and the summation term also goes to zero. The etalon response to a very wide line is then:

$$Y(\sigma_0) = [1 - A (1-R)^{-1}]^2 (1-R)(1+R)^{-1} . \qquad 2.2.6$$

The response to a very wide line is, in the limit, the response of the etalon to white, or continuum, light and is only dependent on the reflectivity and the absorption/scattering properties of the coatings. This behavior was the basis of the proposal by Giacomo (1952) for a measurement method of the reflectivity of etalon coatings. In particular, note that Equation 2.2.6) indicates that an etalon discrim-inates against a continuous background, a very useful property when measuring an emission line in the presence of such a background.

3. Imperfect etalons

With this mathematical framework, it is now possible to examine the effects of having reflective surfaces which are not absolutely flat or held perfectly parallel. For the purposes of this discussion, it does not matter whether or not the surface defects are inherent to the substrates where the reflective coatings are deposited, or the reflective coatings themselves. Further, it is possible to presume these defects are associated with only one of the reflective surfaces while the other surface is assumed to be ideally perfect. As a first example, it is assumed that the imperfect surfaces are microscopically inhomogeneous in a random fashion, with a root mean square varia-tion which is very small in relation to the size and spacing of the etalon. In the classi-cal sense this arrangement can be viewed as being equivalent to a large collection of elemental etalons of varying spacing, with a distribution given by the microscopic inhomogeneities. Furthermore, since these elemental etalons only redistribute the radiation, the functions that describe them must have unit area (Chabbal, 1953). In general, the imperfect-surface etalon is thus expressed as a sum of elemental etalons weighed by a distribution function. This classical approach presumes incoherent illumination, where the final result is obtained by summing the resultant intensities from the elemental etalons. The more appropriate summation of the amplitudes of the elemental etalons, and the differences with the classical approach, is discussed in Chapter 8, Section 3. For the present random-microscopic inhomogeneities case, the distribution function of the phase is a normal, or Gaussian, distribution of unit area:

$$Y(\delta_0) = \tau^2 \, (\pi^{1/2}\Delta\delta)^{-1}$$

$$\times \left\{ \int_{-\infty}^{\infty} \exp-[(\delta_i - \delta_0)^2 \Delta\delta^{-2}][1 + R^2 - 2R \, \cos\delta_i]^{-1} \, d\delta_i \right\}$$

$$= [1 - A \, (1-R)^{-1}]^2 (1-R)(1+R)^{-1}$$

$$\times [1 + 2 \sum_{k=1}^{k=\infty} R^k \exp-(2^{-1}\Delta\delta \, k)^2 \cos(k \, \delta_0)] \, .   \qquad 2.3.1$$

In the above equation, $\Delta\delta$ is a measure of the dispersion in phase caused by the varying length of the elemental etalons. This dispersion can also, by similarity, be associated with a change in wavenumber of the source such that its dispersion appears as a phase dispersion represented by $\Delta\delta$. Thus, by relating the changes in frequency with those in etalon spacing length and finally reducing the latter into terms of wavelength, after some manipulation, it is found that the equivalent HWHH, expressed in terms of fractions of wavelength (λ/m) is given by (Chabbal, 1953):

$$df = 2.35 \, \Delta\sigma \, m^{-1} \, .   \qquad 2.3.2a$$

In terms of normalized HWHH the above equation becomes:

$$df^* = 2.35 \, m^{-1} \, .   \qquad 2.3.2b$$

The results derived for the surface defects in the above equations demonstrate that these defects can be treated as a constant of the etalon, since the function that describes them is found to be independent of the spacing of the etalon and the value of the reflectivity. By arguments similar to those used in the reflectivity case discussed earlier, a surface finesse is defined as (Chabbal, 1953):

$$N_D = (2 \, df^*)^{-1} = (4.71)^{-1} \, m \, .   \qquad 2.3.3$$

When this value of df^* is reintroduced into the argument of the exponential of Equation 2.3.1), and after converting it into a phase, we obtain the result that $(2^{-1}\Delta\delta)^2 = (\pi \, df^*)^2 [\ln(2)]^{-1}$. This is, indeed, the same result obtained for the argument of the exponential of Equation 2.2.2b) for a Doppler-broadened source emission profile. Thus, the normalized HWHH of these microscopic-surface-inhomogeneities will be distinguished by the label df_G^* .

For an etalon that has errors of curvature, usually associated with polishing defects of the substrate, or is misaligned in the sense that the flats form a very small wedge, the probability distribution is found to be constant over a given phase interval (Duffieux, 1939; Chabbal, 1953). The associated wavenumber distribution is then also constant over a small wavenumber range and zero outside this interval. The implicit assumptions in the above are circular flats for the curvature case and either very large area or square-shaped flats for the misalignment case. For other examples see the reports of Duffieux (1939) and Hill (1963). The result for the errors of curvature and misalignment, when these errors are very small compared with the spacing and the area of the flats, can then be given in terms of fractions of a wavelength m (Duffieux, 1939; Chabbal, 1953):

$$df = \Delta\sigma \; m^{-1} . \qquad\qquad 2.3.4\text{a}$$

In terms of the normalized widths, the result is:

$$df^{\;*} = m^{-1} . \qquad\qquad 2.3.4\text{b}$$

Then by definition the surface finesse becomes:

$$N_D = (2 \; df^{\;*})^{-1} = 2^{-1} m . \qquad\qquad 2.3.5$$

The normalized-curvature defects and misalignment HWHH will be distinguished by the label $df^{\;*}_C$. The analytical representation of an etalon with curvature-polishing defects, or with a slight wedge misalignment, analyzing a monochromatic line is then:

$$Y(\sigma_0) = [1 - A \; (1 - R)^{-1}]^2 (1 - R)(1 + R)^{-1}$$

$$\times \left\{ 1 + 2 \sum_{k=1}^{k=\infty} R^k \text{sinc}(2 \; k \; df^{\;*}) \cos(2\pi \; k \; \sigma_0 \; \Delta\sigma^{-1} \cos\Theta) \right\} . \qquad 2.3.6$$

In the above equation, and subsequently, the sinc function is defined as (Bracewell, 1965):

$$\text{sinc}(x) = (\pi \; x)^{-1}\sin(\pi \; x) .$$

The effects of less-than-perfect etalon flats (and misalignment) are seen from Equations 2.3.1) and 2.3.6), to result in instrumental broadening relative to an ideally perfect etalon with the same reflectivity as well as its concomitant decrease in peak transmission. As done previously, the assumption of microscopic inhomogeneities distributed with a Lorentz population provides the simpler visualization of the effects of other-than-perfect surfaces of an etalon. Since the results obtained are identical with those of Equations 2.2.4a) and 2.2.4b), they will not be repeated here. The main result is that the etalon can be described by an effective reflectivity, R_E , which is the product of the original reflectivity, R, and a function related to the assumed Lorentzian defects, i.e.,

$$R_E = R \, \exp-(2\pi \, l^{\,*}) \, . \qquad\qquad 2.3.7$$

Again, in this equation it is possible to define the finesse and HWHH in terms of fractions of a wavelength to be (approximately):

$$df_L^{\,*} = (2N_D)^{-1} \approx 2 \, m^{-1} \, . \qquad\qquad 2.3.8$$

Therefore, using equation 2.3.7) as a general representation of the effects of less-than-perfect surfaces of an etalon, it is found that the loss in reflectivity, with the associated loss in peak transmission and consequent broadening of the profile, is the main effect. For non-zero values of the surface defect the effective reflectivity has a value less than unity even though the actual reflectivity R can be made as close to unity as desired. This behavior of a maximum effective reflectivity leads to the concept of a limiting finesse for an etalon (Chabbal, 1953), that is, that finesse (or normalized HWHH) an etalon can only asymptotically approach. For instance, it is possible to define a finesse, N_{R_E}, by inserting R_E of Equation 2.3.7) into Equation 2.1.13b), as follows:

$$N_{R_E} = \pi \, R_E^{1/2} \, (\, 1 - R_E \,)^{-1}$$

$$= \pi \, R^{1/2} \, [\exp(-2\pi l^{\,*}) \, (1-R) + (1-\exp(-2\pi l^{\,*})) \,]^{-1} \, . \qquad 2.3.9$$

The normalized etalon width, $e_E^{\,*}$, can be found, with the help of Equation 2.3.9), to be equal to:

$$e_E^{\,*} = (\, 2 \, N_{R_E} \,)^{-1} \, , \qquad\qquad 2.3.10$$

and the limiting width, $e_l^{\,*}$, can be obtained as follows:

$$e_l^{\,*} = \lim_{R \to 1} e_E^{\,*} = [1-\exp(-2\pi l^{\,*}) \,] \, [2\pi \, \exp(-\pi l^{\,*}) \,]^{-1} \, . \qquad 2.3.11$$

This limiting width has a value greater than zero (for $l^{\,*}$ greater than zero), unlike the value of $a^{\,*}$ of Equation 2.1.12a) which goes to zero as the limit $R \to 1$ is approached. Therefore, the limiting finesse can be re-written from Equation 2.3.11), or:

$$N_l = (2e_l^{\,*})^{-1} = [\pi \, \exp(-\pi l^{\,*})] \, [1-\exp(-2\pi l^{\,*}) \,]^{-1} \, . \qquad 2.3.12$$

The peak transmission of an etalon with surface defects, in the absence of absorption, is given by:

$$Y_t(0) = (1-R) \, (1+R)^{-1}(1+R_E) \, (1-R_E)^{-1} \quad < \quad 1 \, , \qquad 2.3.13a$$

since $R_E < R$. For $l^{\,*} << 1$ and R approaching unity, Equation 2.3.13a) can be approximated by:

$$Y_t(0) \approx 1 - \pi \, l^* .$$ 2.3.13b

This result illustrates (very approximately) the loss of transmission due to the existence of a limiting finesse caused by surface defects. The minimum transmission is also affected by the presence of surface defects. This is better illustrated in terms of the contrast, $\mathbf{C}_E$, i.e.:

$$\mathbf{C}_E = (1+R_E)^2 \, (1-R_E)^{-2} < (1+R)^2 \, (1-R)^{-2} .$$ 2.3.14

For very small values of l^* , a limiting contrast can be found to exist:

$$\mathbf{C}_l = \lim_{R \to 1} \mathbf{C}_E \approx 1 + (2 \, N_l \, \pi^{-1})^2 \simeq (\pi \, l^*)^{-2} .$$ 2.3.15

In this expression, the approximate relationship between contrast and finesse given in Equation 1.1.14) has been used.

Thus, it becomes necessary to find the best compromise for the reflective coatings to be used with an etalon known to be imperfect. Chabbal (1953) has addressed this problem and his results, based on the Luminosity Resolution Product (LRP) assumption that a user requires the highest product of peak etalon transmission times the smallest reasonable etalon width (or HWHH), show that the best compromise occurs for values of the ratio df^*/a^* near unity, that is, finesse ratios $N_R \, (N_D)^{-1}$ near unity. These results must be tempered in the presence of absorption/scattering in the coatings, in particular for small values of a^* (or high R), since the absorption effect enters as $[1-A \, (1-R)^{-1}]^2$.

Since both a^* and df^* , i.e., the etalon width associated with the reflective coatings and the width associated with the lack of flatness of the etalon, are independent of the etalon spacing, the total normalized width of the etalon, e^* , is also a constant of the instrument. As shown earlier, the reflective and surface-defect widths 'add' in a cross-correlation operation form (or convolution sense for symmetric profiles), and the results show that the total normalized etalon width e^* is no less-narrow than the widest of the profiles being cross-correlated. Therefore, as done in Equations 2.1.13), 2.3.3) and 2.3.5), it is possible to define the etalon finesse (Chabbal, 1953):

$$N_e = (2 \, e^*)^{-1} = (2 \, a^* \oplus df^*)^{-1} ,$$ 2.3.16

where the $\oplus$ operator indicates the effect on the widths when two symmetric functions are cross-correlated. The actual width of the etalon function must be extracted from expressions such as Equation 2.3.6). For some special cases the Fourier series representation expressions can be reduced to simple analytical expressions [for instance, see Khashan (1979)] but, as a rule, the Fourier series treatment is the only available solution. In the present text the Fourier representation will be employed for the sake of generality.

Although, as has been discussed earlier, the cross-correlation integral operation represents the correct resultant function of an ideal etalon combined with an arbitrary source function, or surface defects (Wilksch, 1985), the convolution integral operation provides the same answer for symmetrical functions. Since symmetric functions are normally used in discussing Fabry - Perot spectrometers (Chabbal, 1953; Hernandez, 1970), the convolution operation notation is widely employed in this context. In the following text a rather brief overview of the definitions and operations associated with convolution, as used in Fabry - Perot spectrometer literature, will be presented.

The convolution of two functions $\mathbf{f}(x)$ and $\mathbf{g}(x)$, giving rise to a new function $\mathbf{h}(x)$, is defined as follows (Bracewell, 1965):

$$\mathbf{h}(x) = \mathbf{f}(x) * \mathbf{g}(x) = \int_{-\infty}^{\infty} \mathbf{f}(u)\, \mathbf{g}(x - u)\, du \ . \qquad 2.3.17$$

Note that the area under $\mathbf{h}(x)$ is equal to the product of the areas of $\mathbf{f}(x)$ and $\mathbf{g}(x)$. The commutative, associative and distributive properties of the convolution operation are (Bracewell, 1965):

$$\mathbf{f} * \mathbf{g} = \mathbf{g} * \mathbf{f} \ , \qquad 2.3.18$$

$$\mathbf{f} * (\mathbf{g} * \mathbf{h}) = (\mathbf{f} * \mathbf{g}) * \mathbf{h} \ , \qquad 2.3.19$$

$$\mathbf{f} * (\mathbf{g} + \mathbf{h}) = \mathbf{f} * \mathbf{g} + \mathbf{f} * \mathbf{h} \ . \qquad 2.3.20$$

The abbreviated asterisk notation is convenient, since the asterisks behave like multiplication signs (Bracewell, 1965).

Further, if the functions $\mathbf{f}$, $\mathbf{g}$, and $\mathbf{h}$ have widths (either HWHH or FWHH) f, g, and h, it is customary to state the effects of the convolution operation on the function widths as (Chabbal, 1953):

$$h = f \oplus g \ . \qquad 2.3.21$$

Thus, in general terms, the resultant profile obtained from a Fabry - Perot device which has a width y, is expressed as:

$$y = s \oplus e \oplus f \ . \qquad 2.3.22$$

In Equation 2.3.22) s represents the source width, e the etalon width, and f the aperture width. Note that these widths may themselves be the result of further convolution operations. As would be expected, the commutative, associative and distributive properties of the parent function operations are reflected in the widths.

In order to present general usage graphs (Chabbal, 1953) and tabulations (Hernandez, 1970) on the properties of Fabry - Perot spectrometers, a non-dimensional method is normally used. For instance, to show the broadening properties of surface defects of width f on an etalon of width a, the quantity

$a \ /(\ a \ \oplus \ f \ )$ as a function of $f \ /a$ is given. The limiting property of the former ratio is:

$$\lim_{f \to 0} [\ a \ /(\ a \ \oplus \ f \ )] = 1 \ . \qquad \qquad 2.3.23a$$

The above result is based on the property of a function becoming a Dirac delta (δ) function as the width becomes negligibly small, i.e.:

$$\lim_{f \to 0} \mathbf{f}(x) = \delta(x) \ . \qquad \qquad 2.3.24$$

Similar operations to those given above for the widths can also be defined for the peak value, or transmission τ, of the functions. A useful transmission operation to remember is:

$$\lim_{f \to 0} [\ \tau(a) \ / \tau(\ a \ \oplus \ f \ )] = 1 \ . \qquad \qquad 2.3.23b$$

A most useful property of the convolution operation is its connection with the Fourier transform (Bracewell, 1965). In the following discussion, it is explicitly assumed that the transforms exist in the limit. Thus, if a function $\mathbf{f}(x)$ has a transform $F(s)$, this is expressed in shorthand as:

$$\mathbf{f}(x) \supset F(s) \ . \qquad \qquad 2.3.25$$

For real functions, the following property applies:

$$\mathbf{f}(-x \ ) \supset F(-s \ ) \ . \qquad \qquad 2.3.26$$

The relations between transform pairs and convolution operations are:

$$\mathbf{f} * \mathbf{g} \supset F \bullet G \ , \qquad \qquad 2.3.27$$

$$\mathbf{f} \bullet \mathbf{g} \supset F * G \ . \qquad \qquad 2.3.28$$

In the last two equations the $\bullet$ symbol indicates simple multiplication. Extension of these equations to the associative, commutative, and distributive properties of convolution is trivial.

For completeness, the relation between convolution and cross-correlation should be given. The cross-correlation integral (Bracewell, 1965) is:

$$\mathbf{h}(x) = \mathbf{f}(x) \ \blacksquare \ \mathbf{g}(x) = \int_{-\infty}^{\infty} \mathbf{f}(u - x) \ \mathbf{g}(u) \ du \ , \qquad \qquad 2.3.29$$

where the symbol $\blacksquare$ indicates the cross-correlation operation. From the above definition, both the properties of cross-correlation and its relation with convolution and the Fourier transform can be found:

$$\mathbf{f}(x) \ \blacksquare \ \mathbf{g}(x) \neq \mathbf{g}(x) \ \blacksquare \ \mathbf{f}(x) \ , \qquad \qquad 2.3.30$$

$$\mathbf{f}(x) \ \blacksquare \ \mathbf{g}(x) = \mathbf{f}(-x) * \mathbf{g}(x) \ , \qquad \qquad 2.3.31$$

$$\mathbf{f}(-x) \ \blacksquare \ \mathbf{g}(x) = \mathbf{g}(-x) \ \blacksquare \ \mathbf{f}(x) = \mathbf{f}(x) * \mathbf{g}(-x) \ . \qquad 2.3.32$$

$$\mathbf{f}(x) \ \blacksquare \ \mathbf{g}(x) \supset F(-s) \bullet G(s) \ . \qquad 2.3.33$$

Therefore, for symmetrical functions, when $\mathbf{f}_s(x) \equiv \mathbf{f}_s(-x)$, the properties of cross-correlation become those of convolution described earlier, since:

$$\mathbf{f}_s(x) \ \blacksquare \ \mathbf{g}_s(x) = \mathbf{f}_s(x) * \mathbf{g}_s(x) \ . \qquad 2.3.34$$

This last equation shows the basis for employing the convolution approach, with its simpler rules of operation.

To recover a line profile from a measurement, the process of deconvolution is employed. When the measurement is described by $\mathbf{h}$, the instrument by $\mathbf{f}$ and the line profile by $\mathbf{g}$, the measurement is defined as:

$$\mathbf{h} = \mathbf{f} * \mathbf{g} \supset F \bullet G \ . \qquad 2.3.35$$

Therefore, a function $\mathbf{f}^{-1}$ can be defined such that the following operation applies:

$$\mathbf{f}^{-1} * \mathbf{h} = \mathbf{f}^{-1} * \mathbf{f} * \mathbf{g} \supset F^{-1} \bullet F \bullet G \ . \qquad 2.3.36$$

When $F^{-1} \bullet F \equiv 1$, the convolution $\mathbf{f}^{-1} * \mathbf{f}$ is, by definition, a Dirac delta (δ) function, thus:

$$\mathbf{f}^{-1} * \mathbf{h} = \delta * \mathbf{g} = \mathbf{g} \supset G \ . \qquad 2.3.37$$

The function F^{-1}, associated with $\mathbf{f}^{-1}$, is found to be equal to:

$$F^{-1} = F^* \, [\, F^* \, F \,]^{-1} \ , \qquad 2.3.38$$

where F^* is the complex conjugate of F. Deconvolution is then easier to perform in the Fourier transform plane, since the operation is reduced to a simple division. In practice, the presence of noise in the measurement renders this approach nearly useless. A method to accomplish deconvolution, when noise is present in the measurement, will be discussed later in the text.

The usefulness of the cross-correlation, convolution and Fourier transform relations lies in the ability to recognize functions, via their transform, as the product (or convolution) of other basic functions, as well as to manipulate functions which are not themselves analytic in the real plane as analytic functions in the transform plane (i.e., Voigt profiles). For a complete and rigorous treatment of convolution, cross-correlation and Fourier transforms, the reader is referred to a book such as Bracewell's (1965).

After this aside on notation, it is now possible to present some of the consequences of surface defects on ideal etalons. Figures 2.8 and 2.9 illustrate the effects of the combination of the etalon reflectivity function with microscopic random defects of polishing and curvature of the flats respectively. Figures 2.8a and 2.9a show the broadening effects expressed as the ratio of the limiting function width to the resultant etalon width [a / ($a \oplus g$) in Figure 2.8] as a function of the ratio of the limiting function width to the reflective function width (g / a). Figures 2.8b and 2.9b show the product of the etalon peak transmission times the fraction of the limiting function width relative to the etalon width. The Chabbal (1953) LRP criterion suggests the best reflective coatings to be used are those associated with the maximum of the curves in Figures 2.8b and 2.9b. As will be seen later, this is not the only criterion available for determining the best compromise of operation for a Fabry-Perot interferometric spectrometer. These figures are based on the assumption of a high-finesse etalon where the periodic character of a Fabry-Perot can be neglected. For lower finesse, when the overlap of the tails of the neighboring profiles is significant, similar figures can be derived from the fixed-finesse tabulations of Hernandez (1970).

4. Real interferometric spectrometers

In a real instrument, it is necessary to observe over a finite solid angle in order to receive and collect some flux. The most obvious example is the human eye, where this solid angle is defined by the effective detectors (rods, cones, optical nerve) of the eye-lens combination, although as a rule, the imperfections of the eye set the basic limitation on this solid angle. For other detectors, the solid angle is defined by a lens and aperture combination, and the latter may be a silver halide grain in a photographic emulsion, the size of the electron beam of a television tube, the slit of a microdensitometer examining a photographic plate, an aperture projected onto a quantum detector, etc. Thus, it is possible to generalize that a measurement with a Fabry-Perot consists of the recording of the source's flux, after modification by the etalon, by a detector of finite angular spread. In the following discussion, the detector is considered to be ideal, i.e., it changes the signal in such a way that the overall result can be considered to be a simple change of scale of the analytical expressions.

The most obvious effect to be noticed, when a finite angular spread detector is used, is that the signal recorded will be changed from that transmitted by the etalon because of the spatial integration of the etalon signal over this finite angular spread of the detection device. Therefore, in general, the flux received by the detector can be defined as follows:

$$Y(\sigma) = \mathbf{I}\,\mathbf{A}\,\epsilon\,\tau_L\,\Omega\,\tau_{e,\Omega}\ .\qquad\qquad 2.4.1$$

In the above equation, $\mathbf{I}$ is the irradiance of the source, $\mathbf{A}$ is the useful area of the etalon, Ω is the solid angle projected on the etalon by the detector, ϵ is the quantum efficiency of detection (a simple proportionality constant in the present case; however, with a value less or equal to unity). τ_L includes losses in transmission not included in the etalon function $\tau_{e,\Omega}$. This last function is the combined etalon

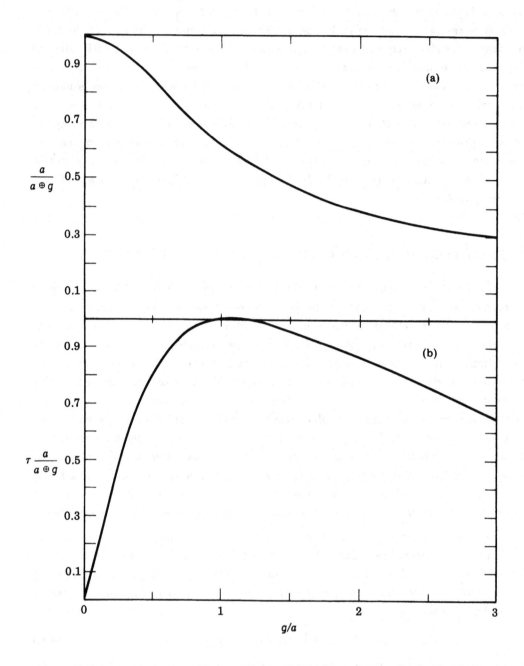

FIGURE 2.8. Effects of microscopic random surface defects on an ideal etalon. The results are shown as: a) the ratio of the ideal etalon width to the resultant width due to the surface defects and ideal etalon, and b) the product of the etalon peak transmission times the ratio given in a). Both are shown as a function of the ratio of the surface defects function width to the perfect etalon width. After Chabbal (1953).

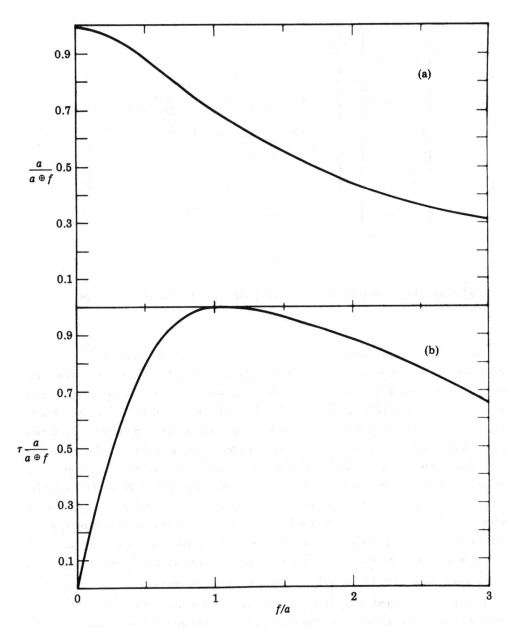

FIGURE 2.9. Same as Figure 2.8, except for curvature defects of the etalon surfaces. After Chabbal (1953).

modulating effect $\tau_e(\sigma)$ with the aperture smearing effects. Note that $\tau_e(\sigma)$ is the shorthand notation of a function such as that given in Equation 2.3.6).

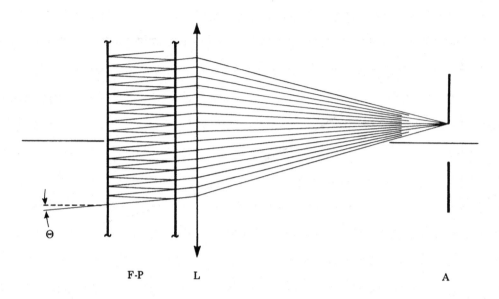

F-P   L   A

FIGURE 2.10. Schematic representation of the Jacquinot and Dufour (1948) spectrometer.

For ease of mathematical treatment, a finite aperture is considered to be centered about the etalon axis, as well as being symmetrical about this axis. Also, this aperture is constrained to have unity transmission where desired and zero everywhere else. The general case of an annulus centered about the optical axis at a plane where a lens has projected the etalon transmission fringes, and a linear detector behind this annulus, will satisfy the requirements described above. In fact, this is a description of the Jacquinot and Dufour (1948) spectrometer, shown in Figure 2.10, where they used an annulus of zero inside diameter, or circular aperture. When this circular aperture size is defined in terms of the angles it projects at the etalon, the necessity of dealing directly with a lens is dispensed, although this lens is a necessary part of the interference process. Therefore, a circular aperture, or the more general case of an annulus, will receive radiation which passes through the etalon between angles Θ_1 and Θ_2 associated with the annulus inner and outer radii respectively. Note that these angles are measured away from the normal to the etalon reflective surfaces and, because of the radial symmetry of the etalon and the annulus, for any angle Θ_i there exists a circle where the spectral information is constant.

The flux detected can be calculated easily by integrating over the solid angle range the annulus transmits:

$$Y(\sigma) = \mathbf{I} \, \mathbf{A} \, \epsilon \, \tau_L \int_0^{2\pi} \int_{\Theta_1}^{\Theta_2} \tau_e(\sigma) \sin\Theta \, d\Theta \, d\Phi \ . \qquad 2.4.2a$$

Because of the previously mentioned radial symmetry, the integration over Φ is straightforward and reduces to 2π. The function describing $\tau_e(\sigma)$ can be any of the results derived earlier, namely, a monochromatic source and a perfect etalon of Equations 2.1.7a) $\rightarrow$ 2.1.7e), a finite width source function and perfect etalon of Equations 2.2.2b) and 2.2.4b), a monochromatic source and an imperfect etalon of Equations 2.3.1) and 2.3.6), or for that matter any combination of the above (Hernandez, 1966). The etalon transmission function $\tau_e(\sigma)$ will be given by the general expression:

$$\tau_e(\sigma) = Y(\sigma_i) = [1-A(1-R)^{-1}]^2(1-R)(1+R)^{-1}$$

$$\times \left\{ 1+2\sum_{k=1}^{k=\infty} a_k \cos(2\pi k \; 2\mu d\sigma_i \cos\Theta) \right\}. \qquad 2.4.3$$

In the above equation, and throughout this text, the a_ks represent the kth coefficient of the etalon function consisting of the etalon reflectivity, surface defects, source function, etc. Although substitution of Equation 2.4.3) into Equation 2.4.2a) will provide a solution, it is more illustrative to remember the relationship between the angle Θ and wavenumbers given in Equation 2.1.8), that is, for a given wavenumber, any angle Θ_i can be associated with its related order n_i and the central order n_0, thus:

$$\cos \Theta_1 = n_1 n_0^{-1}. \qquad 2.4.4a$$

The angle associated with Θ_2 can be similarly defined; however, it is more convenient to define n_2 in terms of n_1 by the relation $n_2 = n_1 - f$, where f is the size of the aperture in orders, then:

$$\cos \Theta_2 = n_2 n_0^{-1} = (n_1 - f) n_0^{-1}. \qquad 2.4.5a$$

As the reader can discern, the definition of f in the previous equation is related to the normalized HWHH f^* by the relation:

$$f^* = f / 2. \qquad 2.4.6$$

The complete solution of Equation 2.4.2a), using the explicit form of Equation 2.4.3), as well as the relations defined in Equations 2.4.4) and 2.4.5), is found to be equal to:

$$Y(\sigma_i) = I \, A \, \epsilon \, \tau_L \, 4\pi f^* n_0^{-1} [1-A(1-R)^{-1}]^2(1-R)(1+R)^{-1}$$

$$\times \left\{ 1 + 2\sum_{k=1}^{k=\infty} a_k \, \text{sinc} \left[2k \left[f^* - f^*(\sigma_0-\sigma_i)/\sigma_0 \right] \right] \right.$$

$$\left. \times \cos \left[2\pi k \left[\sigma_i \, \Delta\sigma^{-1} - f^* - \Delta n_1 + (\sigma_0-\sigma_i)/\sigma_0 (\Delta n_1 + f^*) \right] \right] \right\}. \qquad 2.4.2b$$

In the above equation Δn_1 denotes the location of n_1 from the central order n_0, expressed in terms of orders. The term containing Δn_1, i.e., $(\sigma_0 - \sigma_i)/\sigma_0 (\Delta n_1 + f^*)$, or second order term, is quite small since $(\sigma_0 - \sigma_i)/\sigma_0$ is itself very small. Note also the existence of another second order term in the argument of the sinc function. These second order terms are usually neglected in high resolution studies.

Comparison of the results of Equation 2.4.2b) (neglecting second order terms) with Equation 2.4.1) identifies the solid angle Ω with $4\pi f^* n_0^{-1}$, thus showing the linear relationship between the solid angle and f^*.

Further investigation of Equation 2.4.2b) shows the properties of the resultant output $Y(\sigma)$ as a function of the value of f^*. First, for values of $2f^*$ equal to integer values, where the sinc function is equal to zero, the summation term becomes equal to zero and the output is constant, thus practical values of f^* are always less than 0.5. Second, for values of f^* where the sinc function has a finite value, there is a shift of the profile peak by f^* relative to the position of the original profile (Bruce, 1966; Hernandez, 1974). Third, as the value of f^* is increased monotonically from some arbitrarily small value, the flux increases because of the larger solid angle; however, this is coupled with a decreased depth of modulation by the cosine terms since the sinc function decreases in value as its argument gets larger and larger. This loss in the depth of modulation can be identified with broadening of the original etalon profile, as was shown in Equation 2.2.4b).

The relationship between the size of an aperture, in orders, and the angle it subtends, limits the amount of useful flux a Fabry-Perot device can collect. For the simple example of a circular aperture on the axis, i.e., $n_1 = n_0$, Equations 2.4.4a) and 2.4.5a) become:

$$\cos \Theta_1 = 1.0 \ , \qquad\qquad\qquad\qquad 2.4.4b$$

$$\cos \Theta_2 = 1.0 - f \ n_0^{-1} \ , \qquad\qquad\qquad 2.4.5b$$

or the description of an annulus of zero internal angular radius. Therefore, the solid angle associated with a given aperture f is:

$$\Omega = 2\pi \left(1 - \cos \Theta_2 \right) = 2\pi f \ n_0^{-1} \ . \qquad\qquad 2.4.7a$$

For small values of Θ_2, the previous expression can be approximated by (Zucker, 1964):

$$\Omega \simeq \pi \Theta_2^2 \qquad\qquad \Theta_2 << 1 \ . \qquad\qquad 2.4.7b$$

This result illustrates a Fabry-Perot spectrometer as a second order angular light gathering device.

When an ideal etalon of very high reflectivity examining a monochromatic line is considered, then the resolving power of Equation 2.1.15) is:

$$\mathbf{R}_0 = n_0\, f^{-1}\ , \qquad\qquad 2.4.8$$

which leads to the desired result (Jacquinot, 1954):

$$\Omega = 2\,\pi\, f\ n_0^{-1} = 2\,\pi\ \mathbf{R}_0^{-1}\ . \qquad\qquad 2.4.7c$$

Therefore, the product of the solid angle of acceptance and resolving power is fixed and equal to $2\,\pi$. However, for a real instrument examining a finite width line source, the actual resolving power, $\mathbf{R}$, is less than $\mathbf{R}_0$ of Equation 2.4.8), since the fringe width is greater than f [cf., Equation 2.3.22)]. Then, in general:

$$\Omega\ \mathbf{R} = 2\,\pi\ \mathbf{R}\,\mathbf{R}_0^{-1} \leqslant 2\,\pi\ . \qquad\qquad 2.4.9$$

This topic will be considered in more detail in Chapter 5, Section 1.

Thus, for a practical measurement, there exists a value of f^* where the product of flux and modulation (or resolving power × angle of acceptance) is highest, and this was found by Chabbal (1953) to occur when the ratio of the HWHH of the aperture f^* to the HWHH e^* of the etalon-source combination is near unity, as seen in Figure 2.11. However, as discussed earlier, the practical value of f^* must always be smaller than 0.5. This is the case because for those values of f^* when the sinc function is zero, i.e., when $2\,f^*$ is an integer or f^* is very large, the output tends to become constant and it is only dependent on the intrinsic source irradiance, the reflectivity and the absorption/scattering of the coatings. Giacomo (1952) achieved the same result in a practical manner by using a very broad source, that is, the a_k s of Equation 2.4.2b) are forced to very small values and, thus, the summation term tends to zero, and the measurement with and without the etalon in the system gives a measure of the reflectivity. These methods to obtain the reflectivity, by increasing either the aperture size or the source width, provide good results when the absorption/scattering coefficient A is small. A concurrent measurement of the optical flats' transmission is necessary when the absorption/scattering is not negligible.

There is a large similarity between Equations 2.4.2b) for the annulus, and Equation 2.3.6) where the latter referred to curvature defects of the etalon surfaces. Indeed, this similarity can be seen easily since both the curvature defects distribution function and the annulus itself can be represented by a rectangular function. Thus, the results of Equation 2.4.2b) could have been obtained by the cross-correlation of a rectangular function and the etalon-source function, although at the price of making *ad hoc* assumptions about the field of view. The physical treatment under which Equation 2.4.2b) was obtained is therefore to be preferred, as the results naturally follow.

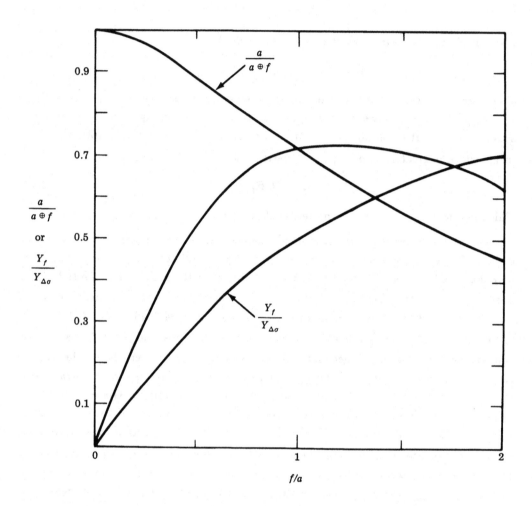

FIGURE 2.11. Effects of a finite aperture on the resultant profile width as a function
of the ratio of aperture width to etalon width. The width effects are given as the ratio of the ideal etalon width to the resultant width, and as the ratio of the flux
transmitted by the aperture to that flux transmitted over one free spectral range.
The product of these two ratios is shown as the unlabeled curve, which shows the existence of an optimum range of operation as indicated by the broad maximum. Note
that the ordinate of this curve has been increased by a factor of two. After Chabbal
(1953).

The Jacquinot and Dufour (1948) spectrometer is obtained for the solution of
Equation 2.4.2b) when $\Delta n_1 = 0$, as would be given for a circular aperture. As these
two authors found, the flux transmitted by the spectrometer can be considerably
increased, without any loss of resolution, by placing a number of concentric annuli

separated by one order from each other and all of these annuli are made with the same width f^*. However, note that the second order effects discussed in Equation 2.4.2.b) become important in this instance. When this multiple annuli arrangement is attempted by mechanical means, it is found to be limited by stringent mechanical tolerances. Reports in the literature show that about 10 annuli are the practical limit (Sipler and Biondi, 1978; Okano *et al.*, 1980) for purely mechanical construction and about 50 for photographic masks (Dupoisot and Prat, 1979). Note, however, that these multiple annuli masks are wavenumber-sensitive, as given by Equation 2.1.8). Although other systems using two etalons for flux selection have been reported (Jacquinot and Dufour, 1948), their details will be considered when multiple etalon systems are discussed.

Reduction of Equation 2.4.2b) to the Jacquinot and Dufour (1948) spectrometer in its simplest configuration of an on-axis circular aperture i.e., when Δn_1 is made equal to zero and second order terms are neglected, is given by:

$$Y(\sigma) = \mathbf{I}\,\mathbf{A}\,\epsilon\,\tau_L\,4\pi\,f^*\,n_0^{-1}\,[1-A\,(1-R)^{-1}]^2(1-R)(1+R)^{-1}$$

$$\times \left\{ 1 + 2 \sum_{k=1}^{k=\infty} a_k\,\operatorname{sinc}(2kf^*)\,\cos[2\pi k\,(\sigma\,\Delta\sigma^{-1} - f^*)] \right\}. \qquad 2.4.2c$$

For the purposes of the following discussion, it is only necessary to note that the argument of the cosine in the above equation consists of a variable, $2\pi k\sigma\,\Delta\sigma^{-1}$, and a constant, $2\pi kf^*$. The constant gives rise to the previously discussed shift, considered to be a constant of the instrument for a given f^*, and will not be discussed further. The varying part shows the possible choices available to affect the modulation of the signal. The available variables are the wavenumber, the index of refraction and the spacing of the etalon, with the latter two embedded in the free spectral range, as given in Equation 2.1.10a).

The approach used by Jacquinot and Dufour was to change the index of refraction, as was done earlier by Jackson and Kuhn (1938a), to tune the passband of their etalons. Note that this mode of operation is the inverse of that employed in the early days to determine the index of refraction of gases (see for instance Meggers and Peters, 1918). The change in this index of refraction can be considered to be a change in the optical thickness in the gap μd, or a change in the effective wavenumber $\mu\sigma$ undergoing multiple reflections in the etalon. Since either of these approaches is valid within its context, the change in the signal modulation effected by varying the index of refraction of the material between the two coated surfaces of the etalon hereafter will be called index of refraction scanning. As noted above, index of refraction scanning is usually achieved by changing the pressure of a gas within the etalon gap.

Historically, changing the etalon gap mechanically was used first by employing sliding ways (Fabry and Perot, 1899), or by having the etalon built from two wedge-shaped substrates which could be moved with respect to each other (Lummer, 1901b), and was supplanted by the more convenient and reliable index of refraction scanning. Mechanical scanning has received considerable attention recently because of the availability of new materials and techniques (Chantrel, 1958; Dupeyrat, 1958; Gobert, 1958; Greenler, 1957, 1958; Chabbal and Soulet, 1958; Roig, 1958; Shepherd, 1960; Ramsay, 1962; Slater *et al.*, 1965; Gadsden and Williams, 1966; Rozuvanova, 1971; Hernandez and Mills, 1973 and references therein) and it is nowadays commonly employed for the scanning of the Jacquinot and Dufour type spectrometer. Combined mechanical and index of refraction scanning has been achieved by heating/cooling a solid material etalon by Burger and Van Cittert (1935) and Auth (1968). Scanning, in the more general approach, includes the movement of apertures, or its equivalent, the presence of multiple apertures, while the etalon itself is held constant. Among the possibilities reported in the literature are the use of varying size apertures (Shepherd *et al.*, 1965; Hoey *et al.*, 1970) multiplexed apertures (Hirschberg and Fried, 1970; Hirschberg *et al.*, 1971; Neo and Shepherd, 1972; Shepherd *et al.*, 1978), axicon prisms (Katzenstein, 1965), as well as the use of multiple apertures with multiple detectors (Hirschberg and Platz, 1965; Shepherd *et al.*, 1965; Chaux *et al.*, 1976), multiple detectors and multiple anode detectors (Chaux and Boquillon, 1979; Sivjee *et al.*, 1980; Abreu *et al.*, 1981; Killeen *et al.*, 1983), to cite a few.

Thus far in the present discussion, it has been implicitly assumed that the detector and recording devices have unlimited bandwidth, that is, the ability to respond to all frequencies equally. When this assumption is not applicable, the resultant recorded profile will be affected, i.e., shifted and broadened. An analytical expression for this effect has been given (Hernandez, 1966):

$$Y(x) = \mathbf{I} \, \mathbf{A} \, \epsilon \, \tau_L \, 4\pi \, f^* \, n_0^{-1} \, [1 - A \, (1 - R)^{-1}]^2 (1 - R)(1 + R)^{-1}$$

$$\times \left\{ 1 + 2 \sum_{k=1}^{k=\infty} a_k \, [1 + (kt)^2] \, [\cos(kx) + kt \sin(kx)] \right\} . \qquad 2.4.10$$

In the above expression, x has been used to represent the variable causing the modulation by some arbitrary scanning method, presumed to be linear with time, and t is the time constant of the equivalent RC circuit of the bandwidth limiting circuit which must be expressed in the same units as x. As usual, the a_k represent the coefficients generated by the combination of source, etalon and aperture in a given experiment.

5. Off-axis systems

5.1. Etalon spectrometers

The mathematical treatment given thus far refers to the centered on-axis, or symmetrical, case which is useful for a Jacquinot and Dufour (1948) spectrometer. The off-axis, or asymmetric, behavior is necessary when the Fabry-Perot is used with a television-like detector, a discrete-area solid-state detector, such as a charge coupled device (CCD), etc., when the angle subtended by the smallest effective detector area is not negligible with, say, the free spectral range of the etalon. The above is also equivalent to a microdensitometer examining a photographic plate whose image is a set of Fabry-Perot rings.

This asymmetric behavior has been treated in some detail by Hernandez (1974) for the specific case of a circular aperture whose center is displaced from the Fabry-Perot axis of symmetry. Figure 2.12 shows the particular asymmetric example explored where $\dot{R}$ is the radius of the circular aperture, ρ is the radius of the projected Fabry-Perot rings, on the same plane as the aperture, and Φ is the angle about the axis of the etalon. ρ as well as $\dot{R}$ are better described in terms of angles, and will be treated this way in the following discussion. Note the equivalence of ρ with Θ of the earlier discussion. The displaced aperture then is described in polar coordinates as a circle of radius $\dot{R}$ whose center is displaced by α units from the origin:

$$\dot{R}^2 = \rho^2 + \alpha^2 - 2\,\alpha\rho\,\cos\Phi\ . \qquad\qquad 2.5.1$$

For transmission purposes, the useful values of ρ extend from ($\alpha - \dot{R}$) [or zero if ($\alpha - \dot{R}$) < 0] to ($\alpha + \dot{R}$) on the line joining the origin and the center of the aperture. The angle Φ_1, described by ρ_1 in the figure, encompasses a full circle, therefore defining unity transmission, while Φ_2 defined by ρ_2 being less than a full circle, is associated with less-than-unity transmission. Angles associated with values of ρ outside the aperture, such as ρ_3, have, by definition, zero transmission. A transmission can be defined, with unity as a maximum, by:

$$\tau(\rho) = \pi^{-1}\cos^{-1}[(\,\rho^2 + \alpha^2 - \dot{R}^2\,)/\,2\alpha\rho\,]\ . \qquad\qquad 2.5.2$$

Equation 2.5.2) has, in effect, reduced the two dimensional problem, given in Figure 2.12, into a one-dimensional problem as Φ has been defined in terms of ρ. The value of the argument of an inverse cosine is defined to lie within the values $+1$ and -1, yet close examination of the argument of the inverse cosine of Equation 2.5.2) shows it is possible to obtain values outside the defined range. The physical significance of these outside values can be ascertained by using selected values of ρ. For instance, for values like ρ_1, where a full circle is encompassed within the circular aperture, the value of the argument is less than -1, yet by definition, the transmission is unity. Values like ρ_3, found outside the aperture, lead to argument values greater than $+1$.

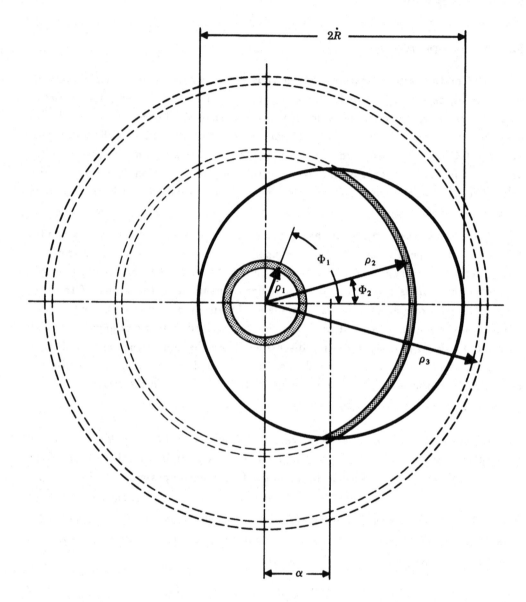

FIGURE 2.12. Off-axis aperture on projected Fabry-Perot etalon fringes. Hernandez
(1974).

Again, by definition, any ray outside the aperture must have zero transmission. On
the basis of the above, it can be generalized that values of the argument less than -1
represent unity transmission, while values greater than +1, being outside the aperture, are associated with zero transmission. Using the symbol z as a shorthand for
the argument of the inverse cosine of Equation 2.5.2), the transmission of the aperture can be expressed for all values of z by:

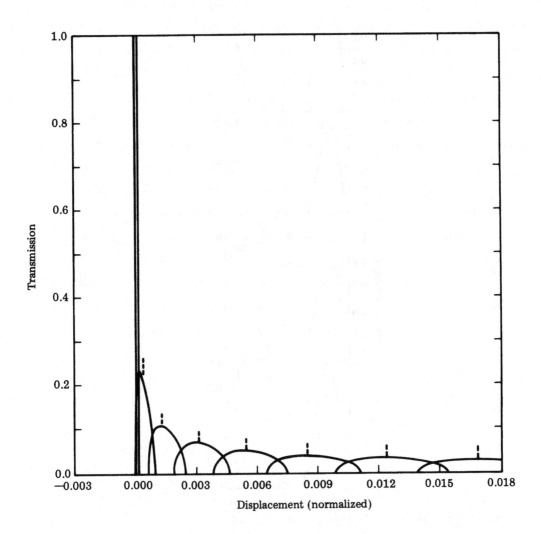

FIGURE 2.13. Deformation of the aperture in (normalized) wavenumber space as the center of the aperture is moved away from the optical axis. Tick marks indicate the aperture center. Hernandez (1974).

$$\tau(\rho) = 1 - \mathbf{H}(z + 1)\, \pi^{-1}\,[\,\Pi(z/2)\,\cos^{-1}(z)\,]\,. \qquad\qquad 2.5.3$$

The Heaviside step function, $\mathbf{H}(x)$, and the rectangular function, $\Pi(x)$, are defined (Bracewell, 1965) as:

$$\mathbf{H}(x) = \begin{cases} 0 & x < 0 \\ 1/2 & x = 0 \\ 1 & x > 0 \end{cases}, \qquad\qquad 2.5.4$$

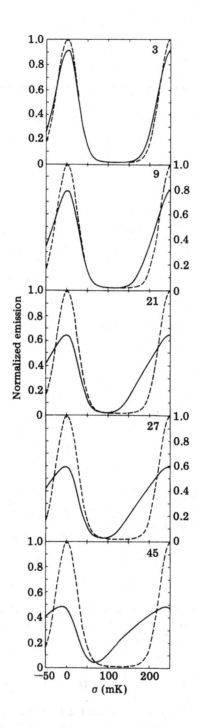

FIGURE 2.14. Effects of an off-axis aperture on a measured profile. The aperture has been shifted away from the optical axis as indicated in the frames. All units are in mK. Hernandez (1974).

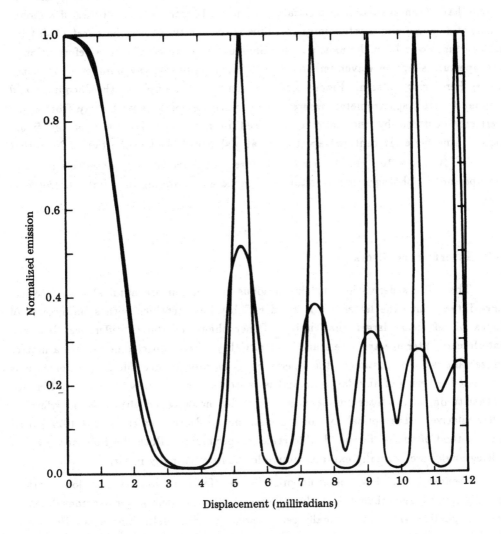

FIGURE 2.15. Finite aperture effects in photographic densitometry of a Fabry-Perot pattern. Hernandez (1974).

$$\Pi(x) = \begin{cases} 1 & \lfloor x \rfloor < 1/2 \\ 1/2 & \lfloor x \rfloor = 1/2 \\ 0 & \lfloor x \rfloor > 1/2 \end{cases} . \qquad 2.5.5$$

The total flux transmitted by the aperture then can be calculated by introducing $\tau(\rho)$ into Equation 2.4.2a), and remembering that ρ is identical to Θ, then:

$$Y(\sigma) = \mathbf{I}\,\mathbf{A}\,\epsilon\,\tau_L \int\limits_0^{2\pi} \int\limits_0^{\pi/2} \tau_e\,\tau(\Theta)\,\sin\Theta\,d\Theta\,d\Phi . \qquad 2.5.6$$

The inclusion of $\tau(\Theta)$ in Equation 2.5.6) makes the equation not analytically integrable, except for the case of the centered aperture [cf. Equation 2.4.2c)], and the results have been evaluated numerically. Figure 2.13, for the special case of a monochromatic source and a reflectivity R approaching unity, shows the effect of shifting an aperture away from the axis, and the obvious result observed is the deformation of the aperture shape in wavenumber space by the non-linear wavenumber scale of the Fabry-Perot ring pattern. Figure 2.14 is given as an example of the Jacquinot and Dufour (1948) spectrometer, where the aperture has been moved away from being perfectly centered by the amount indicated in the upper-right side of the figure panels. The perfectly centered aperture is also shown dashed in the figure. Note that $1\,\text{mK} = 10^{-3}\text{K} = 10^{-3}\text{cm}^{-1}$. Finally, Figure 2.15 demonstrates the effects of a finite aperture in photographic densitometry or vidicon scanning of a Fabry-Perot pattern.

5.2. Interference filters

The off-axis derivations in the previous sub-section are applicable to interference filters, since the latter are low-order Fabry-Perot etalons with solid spacers of index of refraction larger than unity. Other than the above, differences between interference filter spectrometers and typical Fabry - Perot spectrometers are a matter of semantics. For instance, the aperture of an interference filter spectrometer is expressed in terms of angular units and is called the field of view. Another difference is that tuning of the filter passband is effected by inclining, or tilting, the interference filter relative to the optical axis of the instrument. Note that Houston (1927) used this method of tuning for his double etalon experiments. Both the field of view and tilting angles are normally expressed outside the filter, usually in air.

Because of the low operating order of interference filters, the etalon width is usually much larger than typical line source widths, and the latter are considered to be of negligible width and formally represented by Dirac delta functions (Bracewell, 1965).

Numerical calculations using Equation 2.5.6) have been made for a first order interference filter 20 K wide (FWHH), centered at 15867 K and with an index of refraction equal to 2 (Hernandez, 1974). These results are given in Figure 2.16, where the left panel of the figure shows the results for an ideal etalon, while the right panel presents the results for an etalon with microscopic random surface defects, or a Gaussian-defect-limited case. In both cases, the width of the filter at normal incidence, at negligible field of view and zero tilt, has been made equal, and the transmission at this point has been normalized to unity. The different panels in the figure are for varying fields of view, as indicated, while in each panel the effects of tilting in 3 ° steps is shown. The asymmetrical behavior of the aperture, as given in

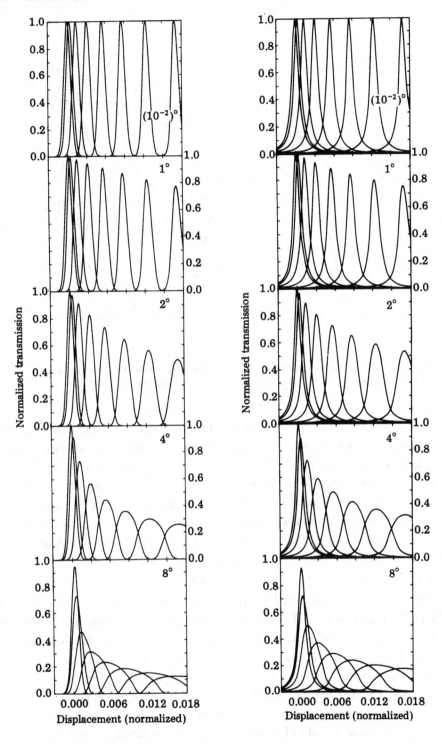

FIGURE 2.16. Calculated effects of tilting an interference filter at arbitrary fields of view, as indicated in the panels. The tilting is done in 3° steps. The left picture is the ideal etalon case, while the right picture is for a microscopic-surface-defect-limited etalon. Hernandez (1974).

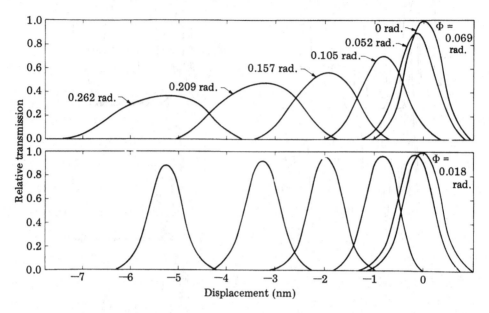

FIGURE 2.17. Laboratory measurement of an interference filter with similar characteristics as those used in the calculations of Figure 2.16. Hernandez (1974).

Figure 2.13, is quite evident. Note that the Gaussian-defect-limited case, shown in the right panel, does not decrease in transmission as fast as the ideal etalon case, in the left panel, as the filter is tilted. This effect can be explained in terms of the extended tails of the perfect etalon function, relative to the Gaussian-defect-limited function, causing a wider final filter profile with a smaller peak transmission. As a comparison, Figure 2.17 shows a laboratory measurement of a filter 20 K wide (FWHH) and centered at 15867 K (however in nanometer units), and is the filter whose characteristics were used in the calculations of Figure 2.16.

The quantitative comparison between the laboratory measurements and the calculations are given in Figure 2.18 for shift, relative transmission and relative width (FWHH). In this figure, both the ideal etalon and the Gaussian-surface-defect-limited cases are shown. In the upper two panels of the figure, the difference between the two calculated cases is not detectable, but on the relative width panel there exist significant differences between the assumed functions used to represent the filter in the calculations and the microscopic-defect-limited-function appears to be a better representation of the actual filter, as has been found elsewhere (Stoner, 1966). Although the agreement found in Figure 2.18 is good, the functions used to obtain the calculated values are not analytic and do not provide means to express the changes in the filter profile as a function of the tilt and the field of view.

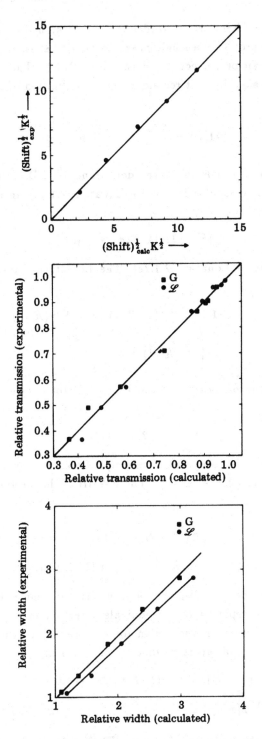

FIGURE 2.18. Comparison of measured and calculated filter properties. From top to bottom: shift, relative transmission, and relative width. Hernandez (1974).

Since interference filters are low-order etalons of high finesse, it is possible to approximate the filter profile as a single peak function, that is, the periodicity of the Fabry - Perot etalon can be neglected (Hernandez, 1974). The formal approach to this approximation is taken by first remembering the definition of a Dirac comb function $III(x)$, namely:

$$III(x) = \sum_{n=-\infty}^{n=\infty} \delta\left(x - n\right) . \qquad 2.5.7$$

In this expression $\delta(x - n)$ is the Dirac delta function (Bracewell, 1965). With this definition it becomes possible to rewrite Equation 2.1.7e), in terms of the phase δ, as (Connes, 1961):

$$Y_t(\delta) = III(\delta / 2\pi) * L(\delta) , \qquad 2.5.8$$

which is a restatement of Equation 2.1.7e). The Lorentzian function $L(\delta)$ is then defined by:

$$L(\delta) = [1 - A(1-R)^{-1}]^2 (1-R)(1+R)^{-1} \ln(R^{-1}) \pi^{-1}$$

$$\times \left\{ [\ln(R^{-1})]^2 + \delta^2 \right\}^{-1} . \qquad 2.5.9$$

Therefore, neglecting the periodicity of the Fabry - Perot device, Equation 2.5.8) is reduced to:

$$Y_t(\delta) = 2\pi L(\delta) . \qquad 2.5.10$$

The above expression can be recognized as a simple Lorentz profile, with a peak value of slightly less than unity (for $A \equiv 0$). The FWHH of this profile, ω_0, and its peak value, τ_0, are found to be:

$$\omega_0 = 2\left[\ln(R^{-1})\right] , \qquad 2.5.11$$

$$\tau_0 = 2(1-R)\left[(1+R)\ln(R^{-1})\right]^{-1} = 4(1-R)\left[(1+R)\omega_0\right]^{-1} . \qquad 2.5.12$$

If it is further assumed that the displaced aperture is sufficiently small such that the off-axis solution can be replaced by an equivalent section of an annulus, then the effects of this finite aperture on the etalon function reduce to the convolution of a Lorentz function with a simple aperture function, $F(x)$, i.e.:

$$L(x) * F(x) = 2(1-R)\left[\Delta y (1+R)\right]^{-1}$$

$$\times \left[\tan^{-1}[2(x-y)\omega_0^{-1}] - \tan^{-1}[2(x-y-\Delta y)\omega_0^{-1}] \right] . \qquad 2.5.13$$

In the above equation, the function $F(x)$ has a FWHH Δy, is centered at $y + \Delta y /2$, and its area has been set to unity. The function given by Equation 2.5.13 has a maximum at $x = y + \Delta y /2$, with a peak value τ and FWHH ω given (relative to the unconvolved profile) by:

$$\omega \, \omega_0^{-1} = [\, 1 + (\, \Delta y \, \omega_0^{-1})^2]^{1/2} \, . \qquad\qquad \text{2.5.14a}$$

$$\mathbf{T} = \tau \, \tau_0^{-1} = (\, \omega_0 \, \Delta y^{-1}) \, \tan^{-1}(\, \Delta y \, \omega_0^{-1})$$

$$\approx [\, 1 + (\, \Delta y \, \omega_0^{-1})^2]^{-1} = (\, \omega_0 \, \omega^{-1})^2 \, . \qquad\qquad \text{2.5.15a}$$

The approximation for the $\tan^{-1}$ given in Equation 2.5.15a) is valid for values of $(\, \Delta y \, \omega_0^{-1})^2 \neq -1$, and is quite good for small values $\Delta y \, \omega_0^{-1}$ (Zucker, 1964). Equations 2.5.14a) and 2.5.15a) can be rearranged to show the explicit dependence of the width and transmission of the filter on Δy, namely:

$$(\, \omega \, \omega_0^{-1})^2 - 1 = \mathbf{T}^{-1} - 1 = (\, \Delta y \, \omega_0^{-1})^2 \, . \qquad\qquad \text{2.5.16a}$$

The microscopic-defect-limited etalon function has the same shift dependence as the ideal etalon, namely $y + \Delta y$, while the relative FWHH and peak values are approximately given by:

$$(\, \omega \, \omega_0^{-1})_g = [\, 1 + \ln 2 \, (\, \Delta y \, \omega_0^{-1})^2 \,]^{1/2} \, , \qquad\qquad \text{2.5.14b}$$

$$\mathbf{T}_g = (\, \tau \, \tau_0^{-1})_g \approx [\, 1 + \ln 2 \, (\, \Delta y \, \omega_0^{-1})^2 \,]^{-1} = (\, \omega_0 \, \omega^{-1})_g^2 \, . \qquad\qquad \text{2.5.15b}$$

These two expressions show the existence of a smaller broadening effect, and associated higher transmission, for the microscopic-surface-defect-limited function and are in reasonable agreement with the off-axis aperture results given in Figure 2.17. Because of the approximations already involved in the present derivations, it suffices to say that the microscopic-surface-defects-limited function filter suffers from smaller broadening effects than the ideal etalon. Therefore a factor $k^2 \leqslant 1$ will be introduced in lieu of the logarithmic factor in Equations 2.5.14b) and 2.5.15b), and thus in general, a filter's broadening properties will be described by:

$$(\, \omega \, \omega_0^{-1})^2 - 1 = \mathbf{T}^{-1} - 1 = (\, k \, \Delta y \, \omega_0^{-1})^2 \, . \qquad\qquad \text{2.5.16b}$$

Although the expressions derived thus far are useful, they still need to be transformed to the notation of tilt and field of view angles, called Θ and Φ respectively. When these angles are defined at the spacer, and distinguished by the subscript s, they have the following relationships with y and Δy:

$$y + \Delta y \, 2^{-1} = \pi \, (\, \Theta_s^2 + \Phi_s^2 \, 4^{-1}) \, , \qquad\qquad \text{2.5.17}$$

$$\Delta y = 2 \, \pi \, \Theta_s \, \Phi_s \, . \qquad\qquad \text{2.5.18}$$

The above are valid for $\Theta \geqslant \Phi \, 2^{-1}$ but, within the approximations carried thus far, they can be made valid for all cases if the following redefinition is used:

$$\Theta_s = \Phi_s \, 2^{-1} = |\, (\Theta_s - \Phi_s \,) \, 2^{-1} | \qquad \text{if } \Phi_s \, 2^{-1} > \Theta_s \, . \qquad\qquad \text{2.5.19}$$

The result given in Equation 2.5.17) has been theoretically predicted from thin film

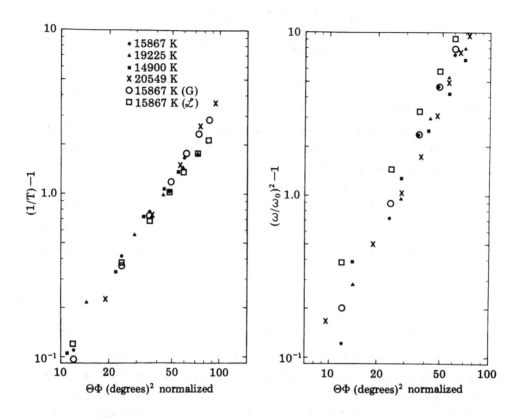

FIGURE 2.19. Transmission and broadening properties of a number of interference filters given by the approximate theory. The complete off-axis calculations (□ and O) are included for comparison. Hernández (1974).

theory (Lissberger and Wilcock, 1959), and experimentally confirmed (Blifford, 1966; Eather and Reasoner, 1969). Note the separation of the effects of the field of view and tilt changes in Equation 2.5.17). Replacing the value of Δy given in Equation 2.5.18) into Equation 2.5.16b), as well as remembering that the angular width of the filter ω_0 can be expressed in terms of the operating order as $2\,\pi\,n_0\,\delta\sigma\,\sigma_0^{-1}$, then, after some re-ordering, the following expression is obtained:

$$(\omega\,\omega_0^{-1})^2 - 1 = \mathbf{T}^{-1} - 1 = [\,k\,\Theta_s\,\Phi_s\,\sigma_0\,(\,n_0\,d\sigma\,)^{-1}\,]^2\,. \qquad 2.5.16c$$

In this equation σ_0 is the peak wavenumber of the unconvolved filter. Finally, the tilt and field of view angles, thus far defined at the spacer, must be converted into their air equivalents by using Snell's law. For simplicity, it will be assumed that for small angles the sine can be expressed by its argument, therefore:

$$(\omega\,\omega_0^{-1})^2 - 1 = \mathbf{T}^{-1} - 1 = [\,k\,\Theta\,\Phi\,\sigma_0\,(\,\mu^2\,n_0\,d\sigma\,)^{-1}\,]^2\,. \qquad 2.5.16d$$

The value of the index of refraction of the spacer μ can be experimentally determined using the following expression:

$$\mu = \sin(\Theta_{air}) \times \left\{ \sin\left[\cos^{-1}\left[(\sigma_0 - d\sigma)\sigma_0\right]\right] \right\}^{-1}. \qquad 2.5.20$$

The dependence of both transmission and width relative to the filter uncon-volved width, center wavenumber, order and index of refraction given in Equation 2.5.16d) can be removed altogether, since all of the above are measurable parameters. Therefore filters of widely varying widths and center wavenumber position can be treated together (Hernandez, 1974). Figure 2.19 shows the transmission and broadening properties of a number of filters, normalized to the 20 K wide filter at 15867 K used earlier, as well as the calculated properties already given in Figure 2.16. Again, it is found here that the perfect etalon case is not as good a description of interference filters as is the microscopic-surface-defects-limited model, as indeed is found in practice (Stoner, 1966).

The approximate results given in Equation 2.5.16c) and 2.5.17) show the basic properties of an interference filter to be: a) The wider a filter is, the smaller the effects of tilt and field of view; b) Two filters alike in all respects, except index of refraction of the spacer, behave equally for a given wavenumber shift; c) Higher order filters are better behaved than lower order filters, since the higher order filters are, in effect, broader filters.

Although the details of practical interference filters are more complicated (Dobrowolski, 1978), the treatment presented here provides a basic understanding of their behavior and properties as low-order Fabry-Perot devices.

6. Selectivity and continuous spectra

In Section 2, the response of an etalon to white light was shown to be, in the limit, the response to a very wide line. The main result presented there was that an etalon would discriminate against this white, or continuous emission. If this continu-ous emission being examined by an etalon is further examined, or crossed, with another dispersing element, the following will be observed. When the resolving limit of the second dispersing element, assumed to be a spectrograph, is smaller than the free spectral range of the etalon, the result is a spectrum composed of bright narrow bands which are separated by dark intervals (Fabry and Perot, 1897; Edser and Butler, 1898; Fabry, 1905b). For the assumed spectrograph, the narrow bright fringes are of near-parabolic shape (Meissner, 1942; Cook, 1971), i.e., the mapping of constant order fringes on the spectrograph wavenumber scale.

When the continuum radiation contains absorption line features, the appearance of the bright fringes will be interrupted by dark regions, whose positions correspond to the wavenumbers of the absorption lines. As the resolution of the crossed dispersing element is decreased by opening the slit of the spectrograph, the parabolic bright fringes, or channeled spectra, are broadened until they coalesce and disappear. At this stage, the spectrum has the appearance of a continuous spectrum containing the 'dark' interference fringes of the absorption lines, or the negative of an emission line spectrum (Meissner, 1942). This crossed dispersive element technique has been used by Fabry (1905b), Fabry and Buisson (1910a, 1910b) and many others thereafter, to measure the wavelengths of the solar Fraunhofer spectrum. Therefore, the study of absorption lines can be treated similarly to that of emission lines, except for the instrumental need to increase the wavenumber selectivity of the Fabry-Perot by external means with another dispersive device, such as the spectrograph used in the earlier example. The need to increase the selectivity of a Fabry-Perot is not limited to absorption spectra, but to dense emission spectra as well, when neighboring order lines will interfere in the desired measurements and filtering with colored glass filters is no longer sufficient.

Although a spectrograph has been used as an example of a crossed dispersive element to increase the selectivity of the Fabry-Perot, such a scheme is fairly inefficient in the use of the available radiation, since a linear dispersion device, or spectrograph, is being matched with a radially symmetric disperser and, at best, it can be hoped that a reasonable part of an etalon fringe can be utilized (Hirschberg, 1958). Therefore, the most efficient matching of a dispersive element with a Fabry-Perot etalon is reached when the former has radial dispersion symmetry. Although the obvious matching device is another Fabry-Perot, there are other dispersive devices with radial dispersion symmetry, such as the Michelson (1881, 1882) interferometer, which can be used as well (Langley and Ford, 1969).

Chapter 3

Optimum operation

1. Practical considerations

The mathematical structure developed in Chapter 2 for an idealized Fabry-Perot spectrometer now can be used to deal with finite width spectral lines. A single emission line can be described uniquely by its location in the wavenumber spectrum, by its profile shape and by its emission rate, or area under the profile.

Historically, the absolute location of a line has been of great metrologic and spectroscopic interest, as it has been used to define the standard of length, i.e., the original reason for the work of Fabry and Perot, as well as to understand the structure of matter. In principle, the determination of the absolute location of an emission line is quite simple. In present terms, all that is required is a number of nearby reference lines (or secondary standards) against which the unknown is compared. This measurement is usually carried out at several etalon spacings to account correctly for order overlap effects. In practice, there are several serious difficulties in the above idealized procedures, and the two most important drawbacks are phase dispersion and selectivity (actually the lack of) by the Fabry-Perot device. The last drawback can be overcome for most circumstances by crossing the Fabry-Perot with another dispersive device such as a spectrograph (Meissner, 1941, 1942 and references therein). Then the isolated Fabry-Perot patterns can be measured by determining the center of gravity of the lines.

Phase dispersion can arise from several causes, and the simplest cause can be envisaged when an etalon with severe spherical curvature surface defects, or a poorly aligned etalon, is used in the measurement. If different areas of the etalon are used for different wavenumber lines, as will be the case when the etalon is used after the dispersing element of the crossed lower dispersion apparatus, then it is found that, in effect, a number of etalons with different spacing are being used. This problem can be alleviated if the etalon is placed ahead of the input slit and the Fabry-Perot pattern is projected by an achromatic lens onto the slit, thus ensuring all sections of the etalon contribute equally to the formation of the fringes (Meissner, 1941, 1942). A more fundamental cause of phase dispersion is obtained when the phase changes upon reflection from the coated surfaces are a function of wavenumber, i.e., χ of Chapter 2, Section 2.1 is not constant (Fabry and Perot, 1897; Perot and Fabry, 1901b). As would be expected, the effect of this phase change will increase with the separation in wavenumber among lines being compared. This phase change arises from the properties of thin metal films (Bauer, 1934) or, when multilayer mirror coatings are used, from the change in the 'equivalent' depth in the coatings at which the reflections occur. This phase change can be directly measured by setting the mirrors as a thin wedge (Fabry and Buisson, 1908b), or by making the desired measurement with the same etalon at two or more spacings (Meissner, 1941, 1942). For those measurements when only the relative line location is desired, the simplest case being given by the determination of Doppler shifts where the shift itself is very small relative to the wavenumber of the line being measured (Fabry and Buisson, 1914; Buisson *et al.*, 1914a, 1914b), the phase dispersion is negligible, as would be expected. Unless otherwise stated, in this text it is presumed that the effects of phase dispersion have been taken into account.

1.1. Noise and precision of measurement

Since emission lines have a finite width and a shape associated with them, a basic limitation on the measurement of their location is given by this line width. Up to this point, the emission of the line source and the flux received by the detector have been implicitly assumed to be described by noise-free continuous functions. In practice, a photon field is fluctuating and by its very nature is discontinuous . For the purposes of the present discussion, it is necessary only to consider these two properties by stating that a photon field follows a discrete Poisson distribution (Parratt, 1961; Bevington, 1969) and the signal measured by the detector also follows this distribution, albeit scaled by the quantum efficiency ϵ. This fundamental limitation of having noise inherent in the signal, sets finite precision limits on a given determination. A quantitative measure of this limiting precision is given by the variance, which for the simple case with a quantum detector of N events, or photons at unit quantum efficiency, is equal to N. Another useful quantity to define precision is the signal-to-noise ratio (SNR) which is equal to the ratio of the signal to the one-half power of the variance. For the example given previously, where N events were measured, the SNR is equal to $N^{1/2}$. Consider the case where an emission line is being measured, and the measurement consists of a spectral scan at equidistant steps in wavenumber with an arbitrary spectrometer; then, the center of gravity of the emission line is given by (Gagnè *et al.*, 1974):

$$\bar{\sigma} = \left\{ \sum_{i=-\infty}^{i=\infty} \sigma_i S_i \right\} \left\{ \sum_{i=-\infty}^{i=\infty} S_i \right\}^{-1} . \qquad 3.1.1$$

In the above equation, σ_i is the wavenumber at the ith step and S_i is the number of events measured at that step. The variance of determination of the center of gravity of the line is then (Gagnè *et al.*, 1974):

$$\sigma_{\bar{\sigma}}^2 = \left\{ \sum_{i=-\infty}^{i=\infty} (\sigma_i)^2 S_i \right\} \left\{ \sum_{i=-\infty}^{i=\infty} S_i \right\}^{-2} . \qquad 3.1.2$$

Note that in the previous equation the symbol σ^2 is used to denote the variance. This has been done in order to be consistent with statistical usage (Bevington, 1969), while powers of wavenumbers σ will always be indicated by parenthesis, such as $(\sigma_i)^2$ in Equation 3.1.2). For the case of a Doppler-broadened emission line of true HWHH dg, Gagnè *et al.* (1974) have shown that the variance of determination is given by the following expression:

$$\sigma_{\bar{\sigma}}^2 = dg^2 \left[4 N \ln(2) \right]^{-1} . \qquad 3.1.3.$$

The meaning of N in this equation is the area under the profile, or total number of events measured, i.e.,

$$N = \left\{ \sum_{i=-\infty}^{i=\infty} S_i \right\} . \qquad 3.1.4$$

Since the result found in Equation 3.1.3) shows the precision to be proportional to the ratio of the HWHH squared to the total number of events measured, it indicates that higher precision of measurement is more efficiently approached by reducing the line width by, say, cooling the source, than by increasing the number of events counted. In the latter case, it makes no difference whether the source emission rate is increased or a longer time is spent in each scan step. A quantitative example of Equation 3.1.3) can be given for the measurement of a line 20 mK wide, such that 10^6 events are counted, and the result is a standard deviation (the one-half power of the variance) of about $1.2 \ 10^{-2}$ mK.

The results shown above indicate that any instrumental broadening of the measured profile will decrease the precision of the measurement; however, the use of an instrument with an infinitely narrow bandwidth will transmit an infinitely small flux, thus negating the resolving power obtained by having a negligible bandwidth. Therefore a compromise between instrumental broadening and luminous flux must be made, and indeed that is the LRP criterion (Chabbal, 1953), already mentioned in Chapter 2. This criterion is strictly applicable when a line is the sole occupant in the spectrum or when there are two well-separated lines. When the spectrum is contaminated by a continuous background, or when two or more lines of different intensity share the spectrum, other considerations, like the contrast, enter into play and the simple compromise offered by the LRP method is not sufficient. Under such conditions, changes of the etalon reflectivity, gap spacing, and aperture size may be necessary to fulfill the requirements of spectral separation. Therefore, other criteria become necessary to obtain the most reasonable compromise of operation..

2. Optimization of operation

In general, the flux detected in a Fabry-Perot interferometer can be described by a Fourier series such as that given in Equations 2.4.2b) or 2.4.6). For the sake of completeness, the presence of light leakage in the detector from sources other than the etalon, thermionic emission of the detector and a background, or continuous light, contribution of the source through the etalon will be considered. All these sources are considered under the same label $\mathbf{B}$. Thus, using the previous definitions of Equations 2.4.1) through 2.4.3), a general expression for the total flux detected is:

$$P(x) = \mathbf{I} \, \mathbf{A} \, \epsilon \, r_L \, 4\pi f^* n_0^{-1} \, [1 - A \, (1-R)^{-1}]^2 (1-R)(1+R)^{-1}$$

$$\times \left\{ 1 + \mathbf{B} \, \mathbf{I}^{-1} + 2 \sum_{k=1}^{k=\infty} a_k \, \cos[2\pi k \, (x - x_0) \mathbf{T}^{-1}] \right\} . \qquad 3.2.1$$

In the above equation, a number of simplifications have been made, including the use of x as the variable that describes the scan of the device, thus removing from consideration the particular type of scan adopted in a given experiment. Further, the use of $\mathbf{T}$ for the periodicity of the device allows the experimenter to employ either instrumental variables, such as pressure, or absolute units for the description of the measurement variable. Last, but not least, it has been assumed that the a_ks are the coefficients of the recorded profile and are related to the instrumental profile by a simple multiplicative operation, i.e., $a_k = d_k \times s_k$. The d_ks belong to the instrumental function and the s_ks are, by definition, the source function coefficients, as given in Equation 2.2.4). Thus, the actual measurements obtained with a Fabry-Perot spectrometer can be restated to be:

$$P(x) = b_0 + 2 \sum_{k=1}^{k=\infty} b_k \, \cos[2\pi k \, (x - x_0) \mathbf{T}^{-1}] . \qquad 3.2.2$$

The relations between the b_k s of Equation 3.2.2) and the a_k s of Equation 3.2.1) are obvious; however, they will be stated here using the following additional relations:

$$Q = \mathbf{I} \, \mathbf{A} \, \epsilon \, \tau_L \, 4\pi \, f^* \, n_0^{-1} \, [1 - A \, (1-R)^{-1}]^2 (1-R)(1+R)^{-1} \,, \qquad 3.2.3$$

$$J = \mathbf{B} \, \mathbf{I}^{-1} \,, \qquad 3.2.4$$

$$W = \int_0^{\mathbf{T}} P(x) \, dx \; = \; Q \, (1+J) \, \mathbf{T} \,. \qquad 3.2.5$$

With the help of the above definitions the coefficients of Equation 3.2.2) are given by:

$$b_0 = Q \, (1+J) = W/\mathbf{T} \,. \qquad 3.2.6$$

$$b_k = Q \, a_k = W \, a_k \, [\mathbf{T}(1+J)]^{-1} \,. \qquad 3.2.7a$$

Note that the b_k s are, in turn, obtained from the separate sine and cosine coefficients, i.e.,

$$s_k = Q \, a_k \, \sin(2\pi \, k x_0 \mathbf{T}^{-1}) = W \, a_k \, [\mathbf{T}(1+J)]^{-1} \, \sin(2\pi \, k x_0 \mathbf{T}^{-1}) \,, \qquad 3.2.8$$

$$c_k = Q \, a_k \, \cos(2\pi \, k x_0 \mathbf{T}^{-1}) = W \, a_k \, [\mathbf{T}(1+J)]^{-1} \, \cos(2\pi \, k x_0 \mathbf{T}^{-1}) \,, \qquad 3.2.9$$

$$b_k = (s_k^2 + c_k^2)^{1/2} = Q \, a_k = W \, a_k \, [\mathbf{T}(1+J)]^{-1} \,, \qquad 3.2.7b$$

$$x_{0_k} = (2\pi \, k)^{-1} \tan^{-1} (s_k \, c_k^{-1}) \equiv x_0 \, \mathbf{T}^{-1} \,. \qquad 3.2.10$$

Thus in principle, using the above equations, the coefficients b_k associated with the measurement can be obtained with infinite precision and the desired s_k s associated with the source are then extracted from the b_k s. However, in practice, this is not the case.

2.1. Equidistant equal-time sampling

Earlier in this discussion, the presence of noise inherent in the signal was mentioned and this, in turn, leads to finite uncertainties of determination of quantities derived from such inherently noisy measurements. In order to treat conveniently this limitation, it is assumed that $P(x)$ is sampled at $\mathbf{T}$ equidistant steps of x for each Fabry-Perot period, and each step is sampled for a fixed period of time. This equidistant sampling is presumed to be so fine that the Nyquist theorem limitations (Bracewell, 1965) are of no importance. Implicit in the last statement is that the function given by Equation 3.2.1) is bandlimited (Bracewell, 1965); that is, beyond some value of k the value of the coefficients a_k is zero. Note that the above describes a Jacquinot and Dufour (1948) photoelectric spectrometer being scanned linearly with time in evenly spaced steps. Under the presently described conditions, each value of

$P(x)$ belongs to a Poisson distribution with variance $P(x)$ (Bevington, 1969), and thus I, of Equations 3.2.1) and 3.2.3), must be expressed in units of photons (unit time)$^{-1}$ (unit area)$^{-1}$ (steradian)$^{-1}$. The uncertainties associated with the determination of the coefficients are given by Bevington (1969) to be :

$$\sigma_{b_k}^2 = \sum_{i=0}^{i=T} \left\{ \sigma_{P(x_i)}^2 \left[\frac{\partial b_k}{\partial P(x_i)} \right]^2 \right\} . \qquad 3.2.11$$

The above expression is not strictly applicable to Poisson distributions but, appealing to the central theorem, it can be used in the limit where the Poisson distribution approaches a normal distribution. As a rule, when the value of the mean of the Poisson distribution is ≥ 20, the above approximation can be considered to be satisfied. This approach, using the normal distribution, has been taken because of the difficulties associated with solving maximum likelihood equations associated with Poisson distributions (Bevington, 1969), and it does provide a reasonable measure of the uncertainties involved in the determination of spectroscopic quantities from a Fabry-Perot spectrometer measurement. Hernandez (1978, 1979, 1982b) has used the above technique to estimate the uncertainties and the results are:

$$\sigma_{b_0}^2 = Q(1+J)T^{-1} = WT^{-2} , \qquad 3.2.12$$

$$\sigma_{s_k}^2 = Q(2T)^{-1}[\, 1+J - a_{2k} \cos(4\pi k z_0 T^{-1}) \,]$$

$$= W[2T^2(1+J)]^{-1}[1+J - a_{2k} \cos(4\pi k z_0 T^{-1}) \,] , \qquad 3.2.13$$

$$\sigma_{c_k}^2 = Q(2T)^{-1}[\, 1+J + a_{2k} \cos(4\pi k z_0 T^{-1}) \,]$$

$$= W[2T^2(1+J)]^{-1}[\, 1+J + a_{2k} \cos(4\pi k z_0 T^{-1}) \,] , \qquad 3.2.14$$

$$\sigma_{b_k}^2 = Q(2T)^{-1}[1+J + a_{2k} \cos^2(4\pi k z_0 T^{-1})]$$

$$= W[2T^2(1+J)]^{-1}[1+J + a_{2k} \cos^2(4\pi k z_0 T^{-1})] , \qquad 3.2.15a$$

$$\sigma_{z_{0_k}}^2 = [\, Q2T(2\pi k a_k)^2]^{-1}[1+J - a_{2k} \cos^2(4\pi k z_0 T^{-1})]$$

$$= [2W(2\pi k a_k)^2]^{-1}(1+J)[1+J - a_{2k} \cos^2(4\pi k z_0 T^{-1})] . \qquad 3.2.16$$

The results given in Equations 3.2.12) through 3.2.16) provide the desired measure of the uncertainties caused by the inherently noisy measurements. For instance, Equation 3.2.15a) shows there is an irreducible value for the uncertainty associated with the determination of the coefficients b_k from a given measurement and this uncertainty is given in the limit of $a_{2k} \to 0$, as:

$$\sigma_{b_k}^2 = Q(1+J)(2T)^{-1} = W(2T^2)^{-1} . \qquad 3.2.15b$$

Another way to state the above is by using the ratio of the coefficient b_k to the standard deviation σ_{b_k}, and again, it is given for the limit when $a_{2k} \to 0$:

$$b_k(\sigma_{b_k})^{-1} = a_k \, (2QT)^{1/2}(1+J)^{-1/2} = a_k \, (2W)^{1/2}(1+J)^{-1} \, . \qquad 3.2.17$$

The above expression shows that the relative precision of b_k increases as the one-half power of the total flux and is inversely proportional to flux leakage and background. In turn, Equation 3.2.17) shows that at some point, the value obtained for b_k is not statistically significant. If the number of points used to derive b_k is very large then Student's-t test (1908) can be used as a criterion for deciding whether or not a given b_k is statistically meaningful. Assuming there existed 100 points, the t-test specifies that if the following equation is true, the coefficient b_k is not statistically significant at the 95% confidence limit for the null hypothesis $b_k = 0$, i.e.:

$$\left| \, b_k \, \sigma_{b_k}^{-1} \, \right| \leq 2.0 \, . \qquad 3.2.18$$

The above result is not a hard and fast rule and it must be used with caution as there exist cases where the value of a given b_k must be close to zero, such as when the coefficients contain sinc functions associated with circular apertures, and the test must then be used in the opposite sense intended in Equation 3.2.18). The usefulness of the significance test is that it shows the existence of a limited number of coefficients that can be extracted from a given measurement; that is, one of the constraints imposed by an inherently noisy signal has been found.

The determination of the position of a line can now be expressed in terms of the x_{0_k} of Equation 3.2.10), and its uncertainty of Equation 3.2.16), by defining the following expressions:

$$\mathbf{x_0} = \left[\sum_{k=1}^{k=N} x_{0_k}(\sigma_{x_{0_k}}^{-2}) \right] \left[\sum_{k=1}^{k=N} \sigma_{x_{0_k}}^{-2} \right]^{-1} , \qquad 3.2.19$$

$$\sigma_{x_0}^2 = \left[\sum_{k=1}^{k=N} \sigma_{x_{0_k}}^{-2} \right]^{-1} . \qquad 3.2.20$$

In the above two equations the summation is carried over the N useful coefficients obtained in the measurements, since the $\sigma_{x_{0_k}}$s depend on the statistically meaningful coefficients obtained, i.e.,

$$\lim_{a_{2k} \to 0} \sigma_{x_{0_k}}^2 = \sigma_{b_k}^2 (2\pi k b_k)^{-2} \, . \qquad 3.2.21$$

The above result can be seen, after some manipulation, from Equations 3.2.7b), 3.2.15) and 3.2.16). Equation 3.2.20) becomes, after the introduction of Equation 3.2.16), the following expression:

$$\sigma_{x_0}^2 = \left\{ 8\pi^2 Q \, \mathrm{T} \left[\sum_{k=1}^{k=N} (a_k k)^2 [1+J-a_{2k} \, \cos^2(4\pi k x_0 \mathrm{T}^{-1})]^{-1} \right] \right\}^{-1}$$

$$= \left\{ 8\pi^2 W(1+J)^{-1} \left[\sum_{k=1}^{k=N} (a_k k)^2 [1+J-a_{2k} \cos^2(4\pi k x_0 \mathbf{T}^{-1})]^{-1} \right] \right\}^{-1} . \quad 3.2.22$$

Equation 3.2.22), above, can be easily converted into wavenumber or Doppler shift (i.e., wind), since x_0 and σ_{x_0} have been expressed in units of orders [cf. Equation 3.2.10)]. In terms of wind, Equation 3.2.22) is easily transformed by the following:

$$\sigma_v^2 = \sigma_{x_0}^2 \ c^2 \ n_0^{-2} \ .$$

The units of the above equation are fixed by the units chosen to express the speed of light c. It is possible to compare the results of Equation 3.2.22) with those obtained from Equation 3.1.2). This has been done (Hernandez, 1979) under the assumption that the sums of Equation 3.1.2) can be carried for one free spectral range, rather than from $-\infty$ to ∞, and the line shape is a Fabry-Perot ideal profile. Using as a specific example for the line profile shape that is given by Equation 3.2.1), the result of the comparison, given as the ratio of the center of gravity determination uncertainties over the uncertainties of Equation 3.2.22), is equal to:

$$\sigma_{\sigma}^2 \ \sigma_{x_0}^{-2} = 8 \ \pi^2 \left[\sum_{k=1}^{k=N} (a_k k)^2 (1-a_{2k})^{-1} \right]$$

$$\times \left[12^{-1} + \pi^{-2} \sum_{k=1}^{k=N} (-1)^k a_k k^{-2} \right] . \quad 3.2.23$$

In the above comparison N is equal to ∞ for an ideal theoretical comparison, or in a practical case, equal to the number of statistically significant number of coefficients as described earlier. Also, note that the cosine term in Equation 3.2.22) has been set to unity. The comparison shown in 3.2.23) indicates that the center of gravity approach gives more weight to the lower harmonic coefficients, i.e., $a_k k^{-2}$, while the method of Equation 3.2.22) gives more weight to the higher frequencies by using ($a_k k$)^2. Therefore, the center of gravity method will tend to overestimate the uncertainty of determination of narrow profiles relative to the method of Equation 3.2.22). This is illustrated in Figure 3.1 as the one-half power of Equation 3.2.23), i.e., the ratio of the standard deviations, where the subscript G stands for the center of gravity method. As expected, the figure shows the overestimation of the uncertainty for the narrow profiles and an underestimation for wider profiles. Thus, the use of a deriva-tion of general applicability, such as that of Gagnè et al., (1974) given in Equation 3.1.2), will not provide as reliable an estimate of the variance as a specifically tailored derivation is capable of doing because it does not avail itself of all of the information in the measurement.

The procedure used thus far shows that the precision to which the position of an emission line can be determined is finite, and this precision is inversely proportional to both the flux received per free spectral range, W, and to the width of the emission

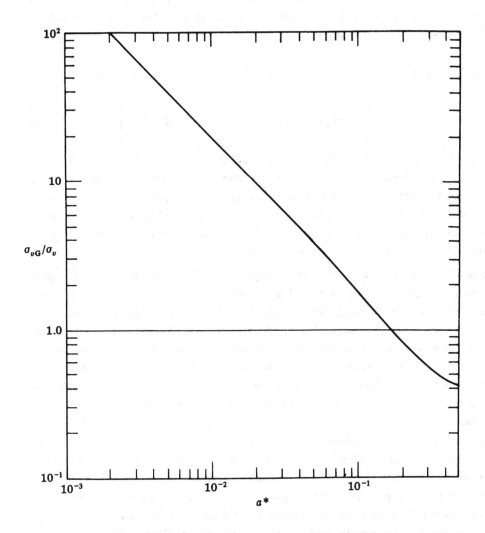

FIGURE 3.1. Ratio of the uncertainties of determination of the peak of a Fabry-Perot profile as a function of its width. This ratio is given as the uncertainties of a general method of determination over the uncertainties of a method specifically designed for the Fabry-Perot spectrometer. Hernandez (1979).

line being measured. The latter effect simply depends on the fact that a narrow line requires a larger number of coefficients for its description. Thus the summation of Equation 3.2.22) has a larger value than a broad line with a smaller number of coefficients, and the inverse of the summation will then be a smaller number.

When Equation 3.2.22) is carefully examined, it is found that the variance is not only a function of the source irradiance, but a function of the source line width and the instrumental parameters as well. Therefore it would be a reasonable expectation to presume there exists a combination of the above which will provide the lowest uncertainty for a given experiment. Following Hernandez (1979), Equation 3.2.22) can be rearranged in the following fashion:

$$\sigma_v^2\, n_0\,\mathbf{I}\,\mathbf{A}\,\mathbf{T}\,\epsilon\,\tau_L(dg^*)^{-2} = \left\{ 32\pi^3 c^{-2}(1-R)(1+R)^{-1}[1-A(1-R)^{-1}]^2 \right.$$

$$\left. \times\; f^*\,(dg^*)^2\left[\sum_{k=1}^{k=N}(a_k k)^2[1+J-a_{2k}\cos^2(4\pi kx_0 \mathbf{T}^{-1})]^{-1}\right]^{-1}\right\}^{-1}. \qquad 3.2.24$$

The right-side of this equation represents the variance of determination of a Doppler shift v, for a unit irradiance source, being examined by a unit area lossless etalon operating at the first order, with a scanning aperture f^*, and a unity quantum efficiency detector. This variance is, for convenience, made relative to the line width of the source. An expression such as Equation 3.2.24) is therefore of general applicability since it can be used with any real instrument by the proper scaling.

Equation 3.2.24) has been numerically solved for the case of a Doppler-broadened source profile being examined by a Fabry-Perot spectrometer consisting of a perfect etalon and a scanning circular aperture, as a function of their normalized HWHH's dg^*, a^*, and f^*, respectively. The results from the calculations are shown in Figure 3.2 in a series of panels, each a function of constant f^*. The upper left panel shows the limiting case when $f^* \to 0$ and $\mathbf{I}\mathbf{A}\epsilon\,\tau_L \to \infty$, or an infinitely small aperture. The contours shown in the figure are in terms of the standard deviations of the Doppler shift in units of 10^9 m/s, that is, the one-half power of Equation 3.2.24). The contour lines are spaced by a factor of $2^{1/2}$ beginning at the labeled contour. The figure shows there is a minimum uncertainty of determination near values of f^* = 0.128 when the values of a^* and dg^* are 0.070 and 0.140, respectively. These findings essentially state that, as a line becomes narrower, it becomes more efficient to increase the resolving power of the apparatus by decreasing the free spectral $\Delta\sigma$ of the etalon, rather than by simply increasing the reflectivity of the coatings to obtain a narrower instrumental profile. Since the results given in Figure 3.2 are in relative HWHH, it immediately follows that the reflective finesse, defined in Equation 2.1.13a), is fixed to be a value near 7 or a reflectivity near 0.65. This value coincidentally agrees with the value of reflectivity found for reasonably transparent, lightly silvered coatings in the early Fabry-Perot measurements of wavelengths. The low reflective finesse indirectly implies that the use of a cross-dispersing instrument in conjunction with the spectrometer is necessary to separate overlapping emission lines. With photoelectric spectrometers this drawback can be overcome by separating the emission lines into different channels (Hernandez and Mills, 1973), alternately chopping the sources by mirrors and/or filters, etc.

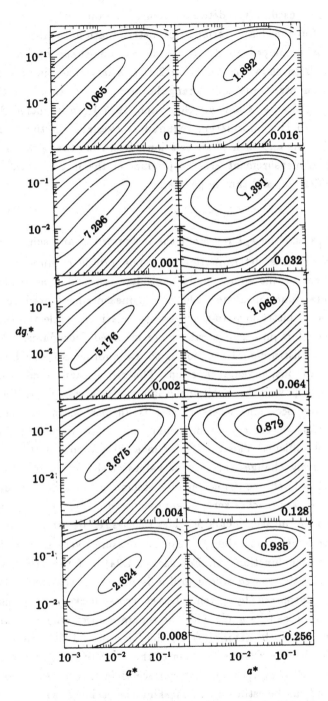

FIGURE 3.2. Uncertainties of determination of the peak of a Fabry-Perot profile, expressed as contours of the standard deviation of the equivalent wind (10^9 m/s units), as a function of the normalized source and etalon widths with the normalized aperture as a parameter. The contours are spaced by a factor of $2^{1/2}$. The uncertainties are given for a unit area lossless etalon examining a unit irradiance source in the first order with a unity quantum efficiency detector. From Hernandez (1979, 1982c).

The same process used in the determination of the optimum source and instrumental parameters for the least variance of determination of Doppler shifts, or wavenumber, can be employed to determine source line-shape optimum-operation parameters. As has been shown in Equation 2.4.3), a large combination of primary line shapes such as Lorentz, Gaussian, rectangular, etc., easily can be accommodated. Even not-so-simple profiles like Galatry (Goorvitch, 1977) can be included. In effect, as discussed earlier, all the information required is available in the Fourier series representation coefficients and their phase relations. Restating the problem, the coefficients derived from a measurement are the product of the instrumental coefficients with the source coefficients, and thus in general:

$$a_k = d_k \times s_k .$$

3.2.25

In the above, the d_ks and the s_ks represent the instrumental and source coefficients, respectively. Therefore, the determination of a line profile requires two measurements: a measurement of the source under consideration and a measurement that gives the instrumental function alone. The source measurement need consider only the best compromise of instrument-to-source width relations, while the instrumental parameter measurement requires, as a rule, more involved procedures. The obvious way to determine the instrumental function is to use another source whose width is negligible relative to the instrumental function. This requirement is met easily at low and medium resolving powers by available single-mode lasers, whose line width is fairly small. On those occasions where a laser is unsuitable, inconvenient, or not available, the procedure necessary to determine the instrumental profile is more cumbersome, but not difficult. This consists of measuring a source at various etalon spacers, which effectively changes the relative values of the d and s parameters. As the etalon gap is closed, the contribution of d increases relative to s until in the limit, as the etalon gap approaches zero, a approaches d. Thus extrapolation to zero spacer of the various gap measurements provides the desired determination. Note that if a circular aperture is used, its value f^* can be measured very accurately by finding the zeros of the sinc function [cf. Equation 2.4.2a)]. If the multiple gap procedure is followed and the source whose profile is desired is the source being used, then the answer is automatically given. Direct cross-checks on an instrumental parameter determination can be made with the separate spectrometer elements. For instance, the reflectivity can be determined by the Giacomo (1952) method, and the aperture size is given by the relative position of the zeros of the sinc function previously described, etc. However, note that surface defects are, as a rule, etalon-surface-dependent and can be estimated by masking the etalon down.

Thus, in principle, knowing the instrumental function (which may be wavenumber-dependent) in turn permits the direct determination of the source line profile. However, since the information being sought is usually embedded in the line profile, further manipulation (Hernandez, 1966, 1970, 1978; Hays and Roble, 1971;

Bennett, 1978) is necessary to make this information available. Using an extreme example, such as the pure Doppler-broadened profile of Equation 2.2.2a), where the source coefficients are of the form $\exp[-\pi^2(dg^*)^2 \, (\ln 2)^{-1} k^2]$, it is found that the logarithm of the coefficients as a function of k^2 forms a straight line whose slope is proportional to $(dg^*)^2$. For an emitting gas of negligible optical thickness and in statistical equilibrium with its surrounding medium, the Doppler line width is proportional to the kinetic temperature τ by the relation:

$$\gamma \, \tau = \pi^2 \, (\, dg^* \,)^2 \, (\, \ln 2 \,)^{-1} \, . \qquad\qquad 3.2.26$$

In this equation $\gamma = 2\pi^2 \, n_0^2 \, k \, A \, (c^2 \, M)^{-1}$, k is Boltzmann's constant, A is Avogadro's number, c is the speed of light and M is the molar mass of the emitting species.

The least squares determination of the temperature and the associated uncertainties from the available statistically useful coefficients is obtained by the following method. From an earlier definition in Equations 3.2.7a), 3.2.7b) the coefficients b_k are:

$$b_k = Q \, a_k = Q \, d_k \, \exp(-k^2 \gamma \, \tau) \, , \qquad\qquad 3.2.27a$$

$$b_k = W \, a_k [\mathbf{T}(1+J)]^{-1} = W[\mathbf{T}(1+J)]^{-1} d_k \, \exp(-k^2 \gamma \, \tau) \, . \qquad 3.2.27b$$

In the above equations the instrumental function is called d_k and the source function is denoted by the exponential. The uncertainties of the b_k are given by Equations 3.2.15a) and 3.2.15b) and need not be repeated here. From the above it is then possible to define a new quantity U_k, as follows (Hernandez, 1978):

$$U_k = \ln(b_k \, d_k^{-1}) = \ln(Q) - k^2 \gamma \, \tau \, , \qquad\qquad 3.2.28a$$

$$U_k = \ln \left\{ W[\mathbf{T}(1+J)]^{-1} \right\} - k^2 \gamma \tau \, . \qquad\qquad 3.2.28b$$

The uncertainty associated with U_k is found to be:

$$\sigma_{U_k}^2 = \sigma_{b_k}^2 \, b_k^{-2} + \sigma_{d_k}^2 \, d_k^{-2}$$

$$= [1 + J + a_{2k} \, \cos^2(4\pi k x_0 \mathbf{T}^{-1})](2 \, \mathbf{T} \, Q \, a_k^2)^{-1} \, , \qquad 3.2.29a$$

$$= (1+J)[1 + J + a_{2k} \, \cos^2(4\pi k x_0 \mathbf{T}^{-1})](2 \, W \, a_k^2)^{-1} \, . \qquad 3.2.29b$$

In the above result and in the discussion thereafter, it is assumed that the contribution of $\sigma_{d_k}^2 \, d_k^{-2}$ is negligibly small in comparison with the first term; that is, the instrumental coefficients are known with great precision. The best estimate of the temperature is given by the following, where $w_k = \sigma_{U_k}^{-2}$:

$$\tau = \gamma^{-1} \left[\left(\sum_{k=1}^{k=N} U_k w_k \sum_{k=1}^{k=N} k^2 w_k \right) \left(\sum_{k=1}^{k=N} w_k \right) - \sum_{k=1}^{k=N} U_k k^2 w_k \right]$$

$$\times \left[\sum_{k=1}^{k=N} k^4 w_k - \left(\sum_{k=1}^{k=N} k^2 w_k \right)^2 \left(\sum_{k=1}^{k=N} w_k \right)^{-1} \right]^{-1} . \qquad 3.2.30$$

The uncertainty for the best estimate of the temperature, given as the variance, is found to be :

$$\sigma_\tau^2 = \gamma^{-2} \left[\sum_{k=1}^{k=N} k^4 w_k - \left(\sum_{k=1}^{k=N} k^2 w_k \right)^2 \left(\sum_{k=1}^{k=N} w_k \right)^{-1} \right]^{-1} \qquad 3.2.31a$$

$$= (2Q\ T\ \gamma^2)^{-1} \left\{ \sum_{k=1}^{k=N} k^4 a_k^2 (1+J+a_{2k})^{-1} \right.$$

$$- \left[\sum_{k=1}^{k=N} k^2 a_k^2 (1+J+a_{2k})^{-1} \right]^2 \left[\sum_{k=1}^{k=N} a_k^2 (1+J+a_{2k})^{-1} \right]^{-1} \right\}^{-1} \qquad 3.2.31b$$

$$= [2\ W\ \gamma^2 (1+J)^{-1}]^{-1} \left\{ \sum_{k=1}^{k=N} k^4 a_k^2 (1+J+a_{2k})^{-1} \right.$$

$$- \left[\sum_{k=1}^{k=N} k^2 a_k^2 (1+J+a_{2k})^{-1} \right]^2 \left[\sum_{k=1}^{k=N} a_k^2 (1+J+a_{2k})^{-1} \right]^{-1} \right\}^{-1} . \qquad 3.2.31c$$

For the sake of convenience, the cosine term, i.e., $\cos^2(4\pi k z_0 T^{-1})$, associated with the a_{2k}s of Equations 3.2.31b) and 3.2.31c) has been arbitrarily set to unity. Examination of the above equations shows the sensitivity of the uncertainty on the emission rate and the background leakage. This is seen easily for the limiting case of a very high finesse etalon, i.e., $d_k \approx 1$, or $a_k \approx s_k$, where it can be shown (Hernandez, 1978) that for large N the uncertainty of the temperature can approximately be given by:

$$\sigma_\tau^2 \approx \alpha\ \gamma^{1/2} \tau^{5/2} (Q\ T)^{-1} = \alpha\ \gamma^{1/2} \tau^{5/2} W^{-1} (1+J) . \qquad 3.2.32$$

In the above, α represents a proportionality constant. Equation 3.2.32) shows the expected results that the uncertainties of determination increase with increasing line profile breadth (temperature) and are inversely proportional to the flux under the measured profile.

As the reader is aware, Equation 3.2.30) is implicitly a restricted (least squares) deconvolution of the profile where the answer is conveniently given in terms of the temperature. In the same spirit as the temperature determination, it is possible to extract other profile shapes and widths, such as a Lorentz profile . As shown in Equations 2.2.2b) and 2.2.4a) the only difference between a Doppler-profile expression and that of a Lorentz profile is that the logarithm of the coefficients of the latter is a linear function of the ordinal, or harmonic, number k.

Practical real sources are a mixture, i.e., cross-correlation, of various shape profiles with the most common being a combination of Doppler and Lorentz profiles. This combination is usually referred to as a Voigt profile and happens not to be amenable to analytical treatment. However, as shown in Equation 2.2.4b), a Lorentzian profile appears as a modifier of the reflectivity of a Fabry-Perot etalon or as mentioned earlier, as a first power of the harmonic number k of the Fourier representation. Therefore it is possible to extract a (suspected) Lorentzian width from a Voigt profile by the same methods used in Equations 3.2.28a) through 3.2.31c), except that a second order equation has to be solved.

Inspection of Equations 3.2.31b) or 3.2.31c) indicates, because of their similarity with Equation 3.2.22), there must exist a best source and instrumental width combination that leads to minimum uncertainties of determination of temperature, as was the case for the uncertainties of position. This has been done for a Doppler-broadened source, which is appropriate for measurements of the highly forbidden transitions in thermospheric physics problems (Hernandez, 1979). For the purposes of general applicability, as was done in Equation 3.2.24), the solution is better stated again as a ratio of the uncertainty of the temperature to temperature itself, however expressed for a unit area lossless etalon operating at the first order with an arbitrary scanning aperture f^*, examining a unit irradiance Doppler-broadened source and using a unity quantum-efficiency detector. The resultant expression is:

$$\sigma_\tau^2 \, (\tau^2 \, n_0)^{-1} \mathbf{I} \, \mathbf{A} \, \mathbf{T} \, \epsilon \, \tau_L \, = \, \left\{ 8\pi^5 [\ln(2)]^{-2} [1 - A(1-R)^{-1}]^2 (1-R)(1+R)^{-1} \right.$$

$$\left. \times \, f^* \, (dg^*)^4 \left[\sum_{k=1}^{k=N} k^4 w_k - \left(\sum_{k=1}^{k=N} k^2 w_k \right)^2 \left(\sum_{k=1}^{k=N} w_k \right)^{-1} \right]^{-1} \right\}^{-1} . \qquad 3.2.33$$

In this equation the w_ks are defined as:

$$w_k = a_k^2 \left[1 + J + a_{2k} \cos^2(4\pi k x_0 \mathbf{T}^{-1}) \right]^{-1} . \qquad 3.2.34$$

For most practical uses, the cosine term of the above equation is set to unity since it provides an upper limit for the estimate of the uncertainties (Hernandez, 1979).

The same limitations found in Equation 3.2.24) are found to be applicable to Equation 3.2.33), namely the summation terms are carried over to the N statistically useful coefficients and which are, in turn, derived from a profile with sufficiently fine sampling to satisfy the Nyquist theorem. Although it is possible to introduce surface defects in the equation, such as microscopic surface inhomogeneities of Equation 2.3.1), the interpretation of the results of Equation 3.2.33) are not changed significantly to warrant their introduction. Thus, for the purposes of the following discussion, it is considered that there are only three functions entering into the equation, namely an ideal reflective etalon, a circular scanning aperture and a Doppler-broadened emission source, which are described by their normalized HWHHs a^*,

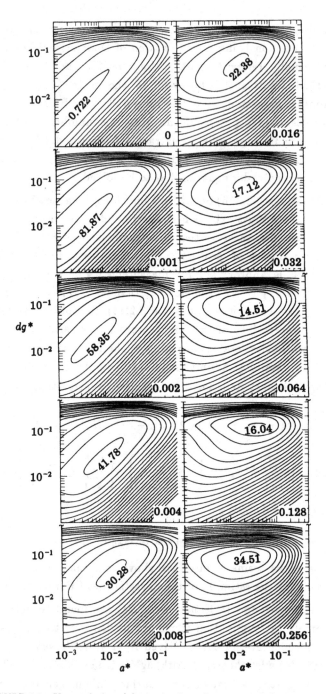

FIGURE 3.3. Uncertainties of determination of temperature from a measured Fabry-Perot profile. The uncertainties are expressed as the ratio of the uncertainty of determination to the temperature. Same conditions as Figure 3.2. From Hernandez (1979, 1982c).

f^*, and dg^* respectively.

Numerical results for Equation 3.2.33) are given in Figure 3.3 in the same manner as the results for the wind uncertainties have been shown in Figure 3.2. The panels in the figure illustrate the solution for the uncertainties of determination of the temperature, expressed as standard deviations, for the equivalent width dg^* as a function of the etalon width a^* with f^* held constant for each panel. The upper left side panel again shows the limiting case for an infinitely small aperture. In the same fashion as Figure 3.2, the present figure shows the existence of least minimum uncertainty for a set of values of a^*, dg^* and f^* at 0.035, 0.11, and 0.065, respectively. The 0.035 value for a^* corresponds to a reflective finesse [cf. Equation 2.1.13a)] of about 14, or a reflectivity of about 0.80. This result is similar to that found for the case of the Γoppler shifts, given earlier, in that the method shows that higher resolving power is more efficiently approached by increasing the etalon gap than by increasing the reflectivity of the coatings. This result also means that the requirements for both surface quality of the etalon and etalon alignment are also relaxed, which is cost-efficient.

The instrumental parameters at the point of minimum uncertainty for temperature of Figure 3.3 are not the same as the parameters found for the minimum uncertainty point for the winds of Figure 3.2. Examination of these two figures indicates that the change in the uncertainty for either, when the operation point is chosen to be the optimum for the other, is an increase in the corresponding uncertainty not exceeding a factor of 1.1. Because of the stronger sensitivity in the uncertainty of temperature, it is usually safe to adopt the parameters suitable for temperature determinations and live with the 10 % increase in the uncertainty of the winds.

The optimum parameter values found for the operation of the spectrometer under Equation 3.2.33) are different from the values obtained by the LRP method (Chabbal, 1953). For instance, the optimum ratio of the scanning aperture to the combined etalon-source width is found to be near 0.5 in the present method against a nominal value of 1.15 derived from the LRP criterion. The LRP criterion has no special preference for any particular values of a^*, dg^*, and f^* as long as the 1.15 value for the ratio is obeyed, while the least uncertainty derivations show that the choice of any values for the parameters, other than those obtained for minimum uncertainty, will give rise to increased uncertainties. Even under the best of conditions, i.e, when the values of a^*, dg^* and f^* are very near their optimum least uncertainty values yet still obeying the LRP criterion for a ratio of 1.15, the uncertainty of determination is about 1.5 times larger than the least uncertainty determined by the least uncertainty of determination criterion. As would be expected, operation of a Fabry-Perot spectrometer with source, etalon and aperture width parameters far-removed from their optimum values can increase the uncertainties of determination dramatically, even though the LRP criterion ratio of 1.15 is being

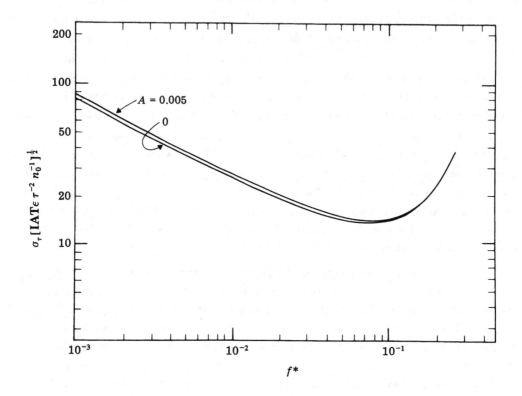

FIGURE 3.4. Minimum uncertainties of determination of temperature from a Fabry-Perot profile as a function of the normalized scanning aperture. From Hernandez (1979).

faithfully followed. Therefore the drawbacks of the LRP criterion when deriving source widths, Doppler in the present case, are that it does not provide unique values for optimum operation, nor does it attempt to give an estimate for the uncertainties of determination involved with such a choice.

The results of the optimization of parameters, given in Figures 3.2 and 3.3, indicate that for the limiting case of $f^* \to 0$, the optimum ratio of dg^* / a^* is near 2.5. This result, which is applicable for a reasonable range of values of dg^* and a^*, is to be compared with the results derived from the LRP criterion shown in Figure 2.8b. The LRP results in that figure show that a value of dg^* / a^* near unity is optimum. This example highlights the difference in the results obtainable using these two widely varying criteria to optimize Fabry-Perot measurements.

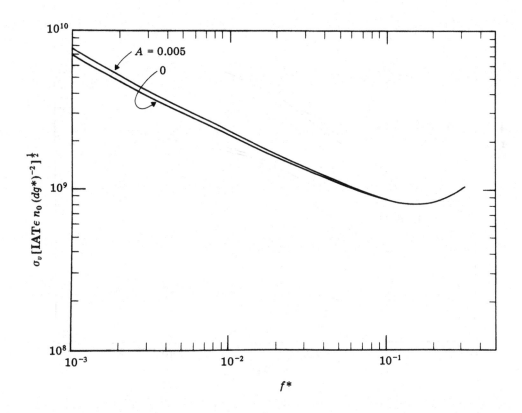

FIGURE 3.5. Minimum uncertainties of determination of the Doppler shift derived from a Fabry-Perot profile as a function of the scanning aperture. Same units as in Figure 3.2. From Hernandez (1979).

To summarize the results of Equations 3.2.24) and 3.2.33) given in Figures 3.2 and 3.3, the minimum uncertainties of determination for both Doppler breadth and Doppler shift as a function of the scanning aperture breadth are shown in Figures 3.4 and 3.5. Figures 3.6 and 3.7 indicate the optimum values of the source and etalon widths associated with minimum uncertainty also as a function of the scanning aperture width f^*. In these figures, allowance has been made for absorption/scattering of the etalon coatings for a practical value of $A = 0.005$, as indicated. The existence of this absorption/scattering effect is, as expected, more pronounced at the smaller values of f^*, where the loss in luminosity is largest because of the smaller values of a^* associated with these values of f^*.

In both Equations 3.2.24) and 3.2.33) it was presumed that the number of terms used in the summation terms was equal to N, the number of statistically useful coefficients. In the theoretical calculations, where the value of N goes to the limit of ∞, it

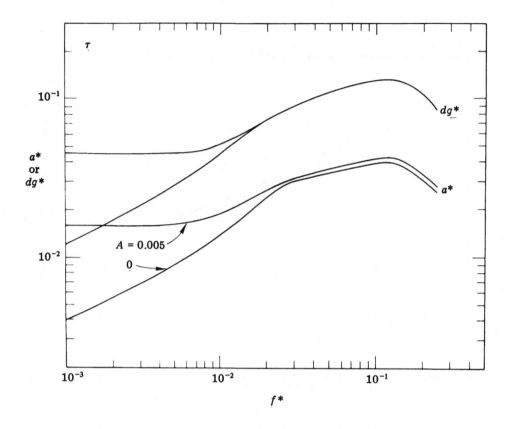

FIGURE 3.6. Optimum normalized widths for the etalon and source as a function of
the normalized aperture. (Temperature). Hernandez (1979).

was found (Hernandez, 1979, 1982b) that the results converge to a criterion of 1
ppm or less with a finite number of terms. The existence of a finite theoretical
number for **N**, although of different value for each calculation, carries with it the
implication of bandwidth limitation, thus the existence of a critical sampling interval.
These results indicate that to unambiguously determine the information contained in
a measured profile, a minimum of (equidistant) samples must be taken in the measurement. This number of samples is then given by twice the number of terms needed
to satisfy the minimum uncertainty conditions. That is, if the number of necessary
coefficients is **N**, then the critical sampling interval must be equal to $(2 N)^{-1}$ (Bracewell, 1965).

For the least uncertainty conditions of Figures 3.2. and 3.3. it is found that the
required minimum number of terms is 7 and 4 for temperature and Doppler-shift
measurements, respectively (Hernandez, 1982b). The required sampling spacing
with the above number of terms is then equivalent to at least 14 and 8 samples per

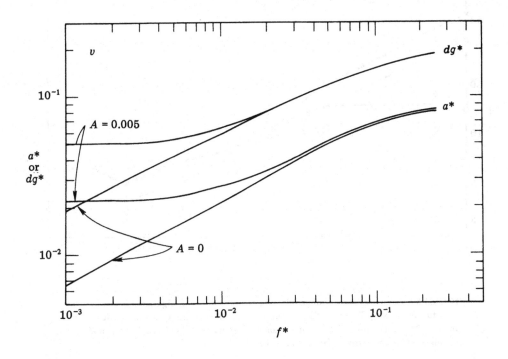

FIGURE 3.7. Optimum normalized widths for the etalon and source as a function of the normalized aperture. (Doppler shift). Hernandez (1979).

period, or order; since, as a rule, both temperature and wind determinations are desired simultaneously, the finer sampling must be adopted. Sampling coarser than this critical sampling will cause aliasing, i.e., the appearance of spurious amplitudes in the determined coefficients, in particular at the higher harmonic coefficients. Since, in the derivations thus far, use has been made of Doppler or Gaussian profiles with their associated fast-decreasing coefficient amplitudes, the effects of aliasing can be particularly severe. Oversampling, or taking a larger number of samples per free spectral range than required for critical sampling, has no adverse effects in the sense that all of the available information is preserved, or more correctly, no further information is acquired. On the practical side, gross oversampling has the disadvantage of more numerical manipulation because of the increase of raw data, and this is computationally inefficient since no further information can be retrieved which was not already available at the critical number of samples. For the sake of completeness, Figure 3.8 shows the critical number of coefficients required for the minima of uncertainty given in Figures 3.4 and 3.5.

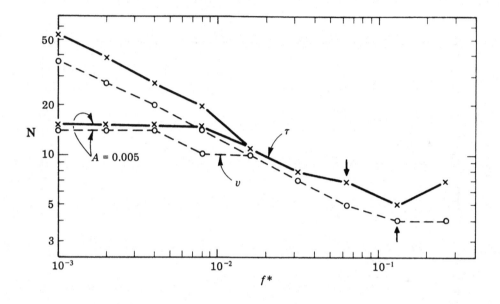

FIGURE 3.8. Critical number of coefficients required to unambiguously determine the properties of a Fabry-Perot profile for the minima of uncertainty given in Figures 3.4 and 3.5. The optimum points of operation for temperature and wind determinations are indicated by the arrows. Hernandez (1982b).

It is sometimes found that, using the criterion of Equation 3.2.18), the number of useful coefficients is less than the number required for the convergence of the uncertainty calculations, or critical sampling number. Presuming that the original sampling rate was at least as fine as, or finer than, the critical sampling rate, numerical calculations have been made to show how this lesser number of coefficients affects the final results (Hernandez, 1982b). This is given in Figure 3.9 as the ratio of the uncertainty obtained with the lesser number of coefficients to the converging uncertainty as a function of the number of terms used. This figure, calculated for the optimum uncertainty points for temperature and wind, illustrates that a value near the convergence value is reached within a few per cent relatively fast, i.e., 3 and 4 coefficients for the temperature and wind respectively against the 8 and 14 required to reach the convergence criterion. It is important to note that these results show the effects of the (unavoidable) noise in the measurement, and are not to be interpreted as a relaxation of the minimum number of terms necessary to define the profile unambiguously, but as the increase in uncertainty caused by the inability to use all of the information that would be otherwise available in the measurement. It still remains necessary to sample at least at the critical sampling interval to avoid aliasing of the results.

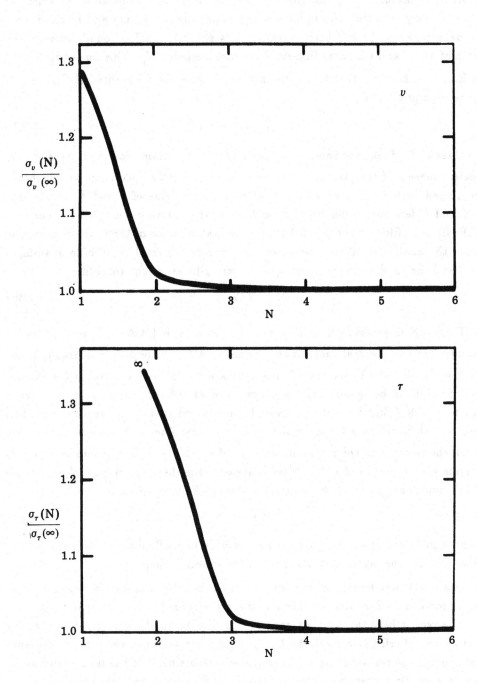

FIGURE 3.9. Increase in the uncertainty due to the inability to use all of the available coefficients in a profile determination. See text for details. Hernandez (1982b).

Another useful finding associated with the minimum number of samples, or terms, necessary to satisfy the minimum uncertainty requirements and to unambiguously determine the profile being measured, is the relationship found between the number of terms and the total finesse of the measurement N_t. This finesse is defined as in Equation 2.1.13a), that is, as the inverse of twice the normalized HWHH t^* of the measured profile, i.e.,

$$N_t = (2 t^*)^{-1} .$$

3.2.35

The values of N_t found for the least uncertainty, or optimum, points of operation for the determination of temperatures and winds are 3.5 and 2 respectively, which should be compared with their associated critical number of terms of 7 and 4. It also has been found (Hernandez, 1982b) that as long as the values of the HWHHs entering into the final profiles are separated by no more than a factor of three, the relationship between the number of terms (necessary to describe the resultant profile unambiguously) and the total finesse is given approximately by twice the total finesse:

$$N \approx 2 N_t .$$

3.2.36a

Therefore, in general, the critical sampling interval is $(4 N_t)^{-1}$, or the number of samples per free spectral range is then equal to $4 N_t$. The finesse is usually taken to be the number of effective interfering beams in a Fabry-Perot etalon (Meaburn, 1976), as well as being one-half the number of samples required to deconvolve a measured profile (Zipoy, 1979). It must be mentioned that Zipoy uses the reflective finesse N_R of Equation 2.1.13a), rather than the total finesse N_t. This relationship between the finesse and the minimum number of samples necessary to unambiguously determine the properties of a profile being measured is therefore of general applicability. For mnemonic purposes, Equation 3.2.36a) can be written as:

$$N = (t^*)^{-1} ,$$

3.2.36b

and this can be interpreted as the critical number of coefficients being equal to the number of times the profile HWHH fits in a free spectral range.

The equal-time treatment presented thus far is applicable to the classical measurement methods, where the information directly recorded is the spectrum as a function of wavenumber, the best example being the original Jacquinot and Dufour (1948) index of refraction method. In some applications, when the source irradiance varies over a wide range during the time of the measurement, it has been found useful to use a short-time scan and coherent addition measurement technique, since this approach tends to average out the source variations (Hernandez and Roble, 1976), and the final result is the equivalent of a (typical) slow scan of the time-averaged source irradiance. However, there are some Fabry-Perot spectrometers which, because of their mechanical construction or other physical limitations, can not be operated in this short-scan and coherent addition method. In the treatment that follows, a method is presented that allows not only the use of spectrometers with the above limitations, but it also is of general applicability for spectroscopic usage, since it permits attainment of fixed SNR determinations.

2.2. Equidistant equal-noise sampling

The equal-noise sampling approach consists of measuring a given spectral slice just long enough to obtain an arbitrary SNR, recording this time and then proceeding to the next slice, and so on (Hernandez *et al.*, 1984). This method has been described in detail, for use with a Fabry-Perot spectrometer operated in the pulse counting recording mode, by the above authors. Under this condition, the measurement method consists of counting the pulses to a preset number for each spectral slice and recording the time it takes to reach this number of counts. Assuming a constant irradiance source, the measurement at each spectral slice is described by (Hernandez *et al.*, 1984):

$$S = P(x_i) \, t_i \ , \qquad\qquad 3.2.37\text{a}$$

where $P(x)$ has already been defined in Equations 3.2.1) and 3.2.2) and t_i is the time necessary to satisfy the requirement imposed by S of Equation 3.2.37a). The recorded time t_i has an associated variance given by (Parratt, 1961):

$$\sigma_{t_i}^{2} = t_i^{\,2} \, S^{-1} \ . \qquad\qquad 3.2.38$$

The above is applicable to a photon field which, by its very nature, is fluctuating and discontinuous and thus obeys a discrete Poisson distribution (Parratt, 1961; Bevington, 1969).

The original flux in this equal-noise measurement can be recovered by carrying out the inverse operation of Equation 3.2.37a):

$$P(x_i) = S \, t_i^{\,-1} \ . \qquad\qquad 3.2.39$$

The variance of determination of $P(x_i)$ is related to the variance of determination of t_i in Equation 3.2.38), and can be deduced easily by appealing to the central theorem, or:

$$\sigma_{P(x_i)}^{2} = S \, t_i^{\,-2} = [P(x_i)]^2 \, S^{-1} \ . \qquad\qquad 3.2.40$$

The reconstructed profile of Equation 3.2.39) is now in the same form as that given in Equation 3.2.2). It is then straightforward to find the relationships between the b_k coefficients of Equation 3.2.39) and the a_k coefficients of Equation 3.2.1), which are the same results as those obtained for the equal-time sampling approach given in

Equations 3.2.6) through 3.2.10), and need not be repeated here. As expected, the uncertainties of determination for the equal-noise method will not be the same as those obtained for the equal-time approach, since the variance of determination of $P(x)$, as given in Equation 3.2.40), is not the same as that used earlier. The equal-noise uncertainties can be derived using Equation 3.2.11), and they are (Hernandez *et al.*, 1984):

$$\sigma_{b_0}^2 = Q^2 (TS)^{-1} \left[(1+J)^2 + \sum_{l=1}^{l=\infty} a_l^2 \right], \qquad 3.2.41$$

$$\sigma_{s_k}^2 = Q^2 (2TS)^{-1} \left[(1+J)^2 - 2(1+J)\, a_{2k}\, \cos(4\pi\, kx_0 T^{-1}) \right.$$

$$+ 2 \sum_{l=1}^{l=\infty} a_l^2 - \sum_{l=1}^{l=2k-1} a_{2k-l}\, a_l\, \cos(4\pi\, kx_0 T^{-1})$$

$$\left. - 2 \sum_{l=1}^{l=\infty} a_{2k+l}\, a_l\, \cos(4\pi\, kx_0 T^{-1}) \right], \qquad 3.2.42$$

$$\sigma_{c_k}^2 = Q^2 (2TS)^{-1} \left\{ (1+J)^2 + 2(1+J)\, a_{2k}\, \cos(4\pi\, kx_0 T^{-1}) \right.$$

$$+ 2 \sum_{l=1}^{l=\infty} a_l^2 + \sum_{l=1}^{l=2k-1} a_{2k-l}\, a_l\, \cos(4\pi\, kx_0 T^{-1})$$

$$\left. + 2 \sum_{l=1}^{l=\infty} a_{2k+l}\, a_l\, \cos(4\pi\, kx_0 T^{-1}) \right\}, \qquad 3.2.43$$

$$\sigma_{b_k}^2 = Q^2 (2TS)^{-1} \left\{ (1+J)^2 + 2 \sum_{l=0}^{l=\infty} a_l^2 \right.$$

$$+ \left[2(1+J)\, a_{2k}\, \cos^2(4\pi\, kx_0 T^{-1}) + 2 \sum_{l=1}^{l=\infty} a_{2k+l}\, a_l\, \cos^2(4\pi\, kx_0 T^{-1}) \right.$$

$$\left. \left. + \sum_{l=1}^{l=2k-1} a_{2k-l}\, a_l\, \cos^2(4\pi\, kx_0 T^{-1}) \right] \right\}, \qquad 3.2.44$$

$$\sigma_{x_{0_k}}^2 = [\, 2TS\, (2\pi\, a_k\, k)^2]^{-1} \times \left\{ (1+J)^2 + 2 \sum_{l=1}^{l=\infty} a_l^2 \right.$$

$$- \left[2(1+J)a_{2k}\, \cos^2(4\pi\, kx_0 T^{-1}) + 2 \sum_{l=1}^{l=\infty} a_{2k+l}\, a_l\, \cos^2(4\pi\, kx_0 T^{-1}) \right.$$

$$\left. \left. + \sum_{l=1}^{l=2k-1} a_{2k-l}\, a_l\, \cos^2(4\pi\, kx_0 T^{-1}) \right] \right\}. \qquad 3.2.45$$

Equations 3.2.41) through 3.2.45) are the counterparts of Equations 3.2.12) to 3.2.16) for the equal-time sampling approach. The basic limitation imposed by the noise in the measurement is the same as that already found earlier in Section 2.1, which is the existence of a limited number of coefficients which can be extracted from a given measurement. **T** is used in the above equations as both the periodicity of a Fabry-Perot profile and the number of spectral steps necessary to scan one free spectral range.

The determination of position of a measured line, and its corresponding uncertainty, can be obtained from the definitions of Equations 3.2.19) and 3.2.20), while the Doppler width and its uncertainty can be found by the same method given in Equations 3.2.26) through 3.2.31c). The uncertainties, expressed as Doppler shifts (winds) and temperatures are (Hernandez *et al.*, 1984):

$$\sigma_v^2 = c^2 n_0^{-2} \left\{ \sum_{k=1}^{k=N} \sigma_{z_{0_k}}^{-2} \right\}^{-1}$$

$$= \left\{ 8\pi^2 c^{-2} n_0^2 \ TS \ \sum_{k=1}^{k=N} e_k (a_k k)^2 \right\}^{-1} , \qquad \text{3.2.46a}$$

$$\sigma_\tau^2 \tau^{-2} = \left\{ 2\pi^4 \ [\ln(2)]^{-2} \ (dg^*)^4 \ T \ S \ \mathbf{W}_\tau \right\}^{-1} . \qquad \text{3.2.47a}$$

where e_k, $\mathbf{W}_\tau$ and its associated quantity w_k are defined, for convenience, as:

$$e_k = \left[(1+J)^2 - 2(1+J) \ a_{2k} + 2 \sum_{l=1}^{l=N} a_l^2 \right. $$

$$\left. - 2 \sum_{l=1}^{l=N} a_{2k+l} \ a_l - \sum_{l=1}^{l=2k-1} a_{2k-l} \ a_l \right]^{-1} , \qquad \text{3.2.48}$$

$$\mathbf{W}_\tau = \sum_{k=1}^{k=N} k^4 w_k - \left(\sum_{k=1}^{k=N} k^2 w_k \right)^2 \left(\sum_{k=1}^{k=N} w_k \right)^{-1} , \qquad \text{3.2.49}$$

$$w_k = a_k^2 \left\{ (1+J)^2 + 2(1+J) \ a_{2k} \right. $$

$$\left. + 2 \sum_{l=1}^{l=N} a_l^2 + \sum_{l=1}^{l=2k-1} a_{2k-l} \ a_l + 2 \sum_{l=1}^{l=N} a_{2k+l} \ a_l \right\}^{-1} . \qquad \text{3.2.50}$$

In the above equations, the $\cos(4\pi kz_0 \mathbf{T}^{-1})$ term has been set to unity in order to obtain an upper limit of the variances of the wind and the temperature and **N** is the minimum number of coefficients needed to describe the profile unambiguously. To study the properties of the general case of equal-noise sampling, it is useful to recast Equations 3.2.46a) and 3.2.47a) independent of the experimental variables:

$$\sigma_v^2 \ \mathbf{T} \ S \ n_0^2 \ (dg^*)^{-2} = \left\{ 8\pi^2 \ c^{-2} \ (dg^*)^2 \sum_{k=1}^{k=N} e_k (a_k k)^2 \right\}^{-1} , \qquad \text{3.2.46b}$$

$$\sigma_\tau^2 \tau^{-2} \ \mathbf{T} \ S = \left\{ 2\pi^4 \ [\ln(2)]^{-2} \ (dg^*)^4 \ \mathbf{W}_\tau \right\}^{-1} . \qquad \text{3.2.47b}$$

For the sake of consistency, Equation 3.2.46b) has been set in the same form as Equation 3.2.24), that is, the uncertainty is calculated relative to the Doppler line width of the source.

Equations 3.2.46b) and 3.2.47b) illustrate the fundamental advantage of the equal-noise approach; if a given uncertainty measurement is desired (with an arbitrary Fabry-Perot spectrometer), it can be obtained with only a simple assumption of the width of the line source. Note that the order, n_0, of Equation 3.2.46b) is fixed for a given line width at a given wavenumber. Once the source line width has been chosen, the only variable left is $\mathbf{T}\,S$. Since $\mathbf{T}$ is associated with $\mathbf{N}$, or the number of coefficients necessary to unambiguously describe the measured profile, and $\mathbf{N}$ itself is a function of the spectrometer and the source line width, then $\mathbf{T}$ for critical sampling defines the maximum value of S. The value found for S requires no knowledge of the source irradiance, since the measurement will be continued until the preset uncertainty is reached, regardless of the time it takes to make the measurements. This independence from the source irradiance, prior to the measurement, is of little consequence in laboratory experiments with a stable source, except possibly for the convenience factor. However, in other fields, such as auroral spectroscopy, where the source irradiance may vary by orders of magnitude over a short time, the equal-noise approach independence from the source irradiance is a definite advantage, since, regardless of this irradiance, the measurement will be carried to completion to the preset uncertainty. A direct corollary of this property is that the measurement process will slow down or speed up as the source irradiance changes, yet only the minimum amount of time will be spent on the measurement. The method prevents undersampling or oversampling. Under this condition of a varying source irradiance, normalization with a photometer device looking at the source becomes necessary, as is the case for the equal-time sampling method described earlier. When this sampling photometer measures the source for precisely the same time as the Fabry-Perot spectrometer, the ratio of the spectrometer measurements to the photometer measurements will give the equivalent result of a measurement of a steady source, as given in Equation 3.2.39), except for a constant factor which can be adjusted to be near unity.

The numerical solutions of Equations 3.2.46b) and 3.2.47b) have been made for the specific case of a Jacquinot and Dufour (1948) Fabry-Perot spectrometer consisting of an ideal etalon and a circular scanning aperture, examining a Doppler-broadened source profile, and they are given in Figures 3.10 and 3.11. The two figures show the common behavior of having the least uncertainty for the narrowest etalon width used, yet the source profile width at these least uncertainty points is rather substantial. This behavior can be deduced from the right side of both Equations 3.2.46b) and 3.2.47b), where maximum uncertainty is found for both very large and very small values of dg^{*}. At this point, the value of these graphs is limited to the determination of the necessary value for the product $\mathbf{T}\,S$, to attain a given uncertainty of measurement, without regard to the time required to do so.

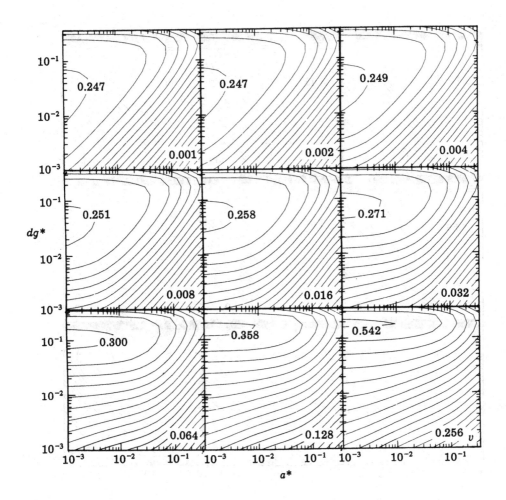

FIGURE 3.10. Uncertainties of determination of the peak of a Fabry-Perot profile, measured by the equal-noise method, expressed as contours of the standard deviation of the equivalent wind (10^9 m/s units), as a function of the normalized source and etalon widths with the normalized aperture as a parameter. The contours are separated by a factor of $2^{1/2}$. Hernandez (1985a).

The derivations and results presented thus far have provided no explicit information on the time requirements of this method, other than the time of measurement is the minimum necessary to reach an arbitrary uncertainty of measurement. Since the purpose of optimization is to obtain a given measurement in the least time for a given variance, time must be made explicit in the derivations. Equation 3.2.37a), which has time directly involved, can be rewritten as:

$$ S = P(x_i) \, t_i = P(x_0) \, t_0 = Q \, p(x_i) \, t_i = Q \, p(x_0) \, t_0 , \qquad 3.2.37b $$

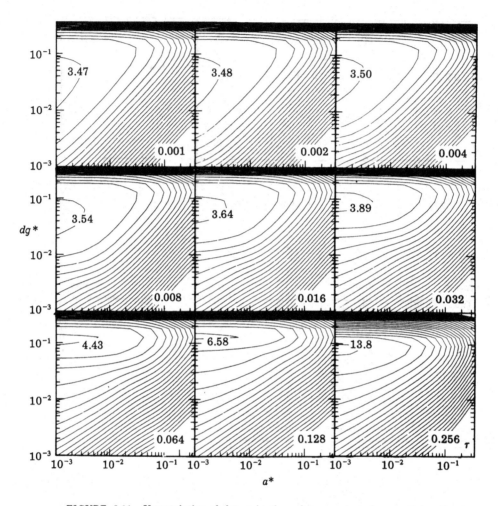

FIGURE 3.11. Uncertainties of determination of temperature from a Fabry-Perot profile measured by the equal-noise method. The uncertainties are expressed as the ratio of the standard deviation of determination to the temperature. Same conditions as Figure 3.10. Hernandez (1985a).

where Q has been defined previously in Equation 3.2.3) and $p(x)$ is given as:

$$p(x) = 1 + J + 2 \sum_{k=1}^{k=N} a_k \cos[2\pi k (x - x_0) \mathbf{T}^{-1}] . \qquad 3.2.51$$

An expression for t_i and the total time of measurement can be obtained from Equations 3.2.37a) and 3.2.37b), albeit referred to the relative peak transmission $p(x_0)$ and its associated time t_0:

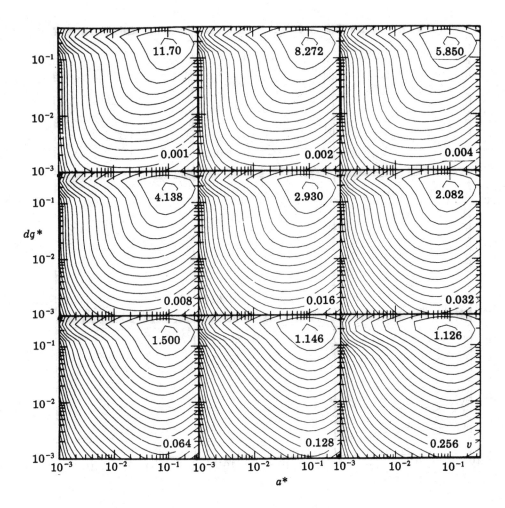

FIGURE 3.12. Uncertainties of determination of the peak of a Fabry-Perot profile measured with the equal-noise method. Same conditions as Figure 3.10, except the uncertainties are for unit time. Hernandez (1985a).

$$t_i = p(x_0) \, [\, p(x_i) \,]^{-1} \, t_0 \, , \qquad\qquad 3.2.52$$

$$\sum_{i=0}^{i=T} t_i = p(x_0) \, t_0 \sum_{i=0}^{i=T} [\, p(x_i) \,]^{-1} \, . \qquad\qquad 3.2.53$$

Replacing S in Equations 3.2.46b) and 3.2.47b) by its equivalent derived from 3.2.37b) and 3.2.53), or:

$$S = Q \, p(x_0) \, t_0 = Q \sum_{i=0}^{i=T} t_i \left[\sum_{i=0}^{i=T} [\, p(x_i) \,]^{-1} \right]^{-1} . \qquad\qquad 3.2.37c$$

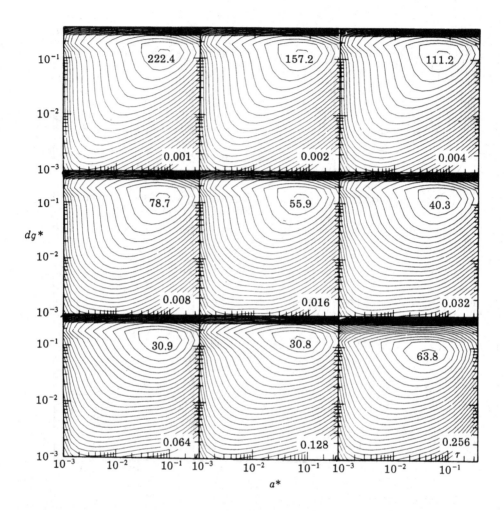

FIGURE 3.13. Uncertainties of determination of temperature from a Fabry-Perot profile measured by the equal-noise method. Same conditions as Figure 3.11, except the uncertainties are for unit time. Hernandez (1985a).

After replacing Q by its definition in Equation 3.2.3) and rearranging terms, the following expressions are obtained for the uncertainties:

$$\sigma_v^2 \, \mathbf{I} \, \mathbf{A} \, \tau_L \, \epsilon \, n_0 (dg^*)^{-2} \sum_{i=0}^{i=T} t_i \; = \; \mathbf{T}^{-1} \sum_{i=0}^{i=T} [\, p\,(x_i)\,]^{-1}$$

$$\times \left\{ 32\pi^3 \mathbf{c}^{-2} (dg^*)^2 f^* [1 - A\,(1-R\,)^{-1}]^2 \right.$$

$$\left. \times \; (1-R\,)(1+R\,)^{-1} \sum_{k=1}^{k=N} e_k \, (a_k\, k)^2 \right\}^{-1}, \qquad\qquad 3.2.46c$$

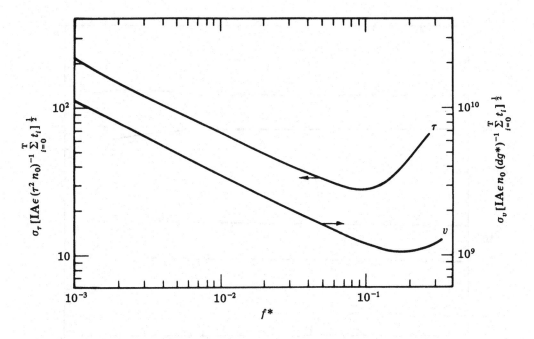

FIGURE 3.14. Minimum uncertainties of determination for wind and temperature for an equal-noise measurement of a Fabry-Perot profile, as a function of the normalized scanning aperture. Hernandez (1985a).

$$\sigma_\tau^2 \, \tau^{-2} \, \mathbf{I} \, \mathbf{A} \, \tau_L \, \epsilon \, n_0^{-1} \sum_{i=0}^{i=\mathbf{T}} t_i \;=\; \mathbf{T}^{-1} \sum_{i=0}^{i=\mathbf{T}} [\, p\,(x_i)\,]^{-1}$$

$$\times \left\{ 8\pi^5 \, [\ln(2)]^{-2} \, (dg^*)^4 \, f^* \, [1 - A\,(1-R\,)^{-1}]^2 \right.$$

$$\left. \times\; (\,1 - R\,)\,(\,1 + R\,)^{-1}\, \mathbf{W}_\tau \right\}^{-1} . \qquad\qquad 3.2.47\text{c}$$

When the background contribution, J, is negligible, it can be shown, with some manipulation, that:

$$\mathbf{T}^{-1} \sum_{i=0}^{i=\mathbf{T}} [\, p\,(x_i)\,]^{-1} = \mathbf{f}(a_k)\,, \qquad\qquad 3.2.54$$

that is, Equations 3.2.46c) and 3.2.47c) are independent of the periodicity $\mathbf{T}$ chosen, but only as long as $\mathbf{T}$ satisfies the Nyquist theorem requirements (Bracewell, 1965). Therefore, the right side of Equations 3.2.46c) and 3.2.47c) is the variance of determination of Doppler shifts and widths for a unit area etalon with losses τ_L, operating at unity order with a unity quantum efficiency detector for a unit time and examining a unit irradiance source profile. ·

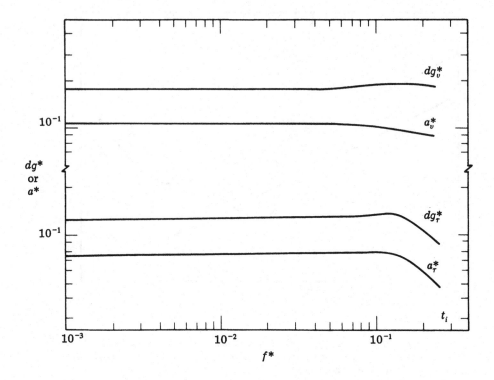

FIGURE 3.15. Optimum normalized widths for the etalon and source as a function of the normalized aperture. (Equal-noise method). Hernandez (1985a).

The (numerical) solutions for Equations 3.2.46c) and 3.2.47c) are given in Figures 3.12 and 3.13, which now show definite least uncertainty points. The optimum point of operation of the Fabry-Perot spectrometer is illustrated in Figures 3.14 and 3.15, and this point is located for the temperature (wind) determination at $a^* = 0.078$ (0.095), $dg^* = 0.13$ (0.19) and $f^* = 0.10$ (0.19). As illustrated in Figure 3.15, the optimum values of a^* and dg^* do not change except near the large values of f^*. The etalon finesse, for the optimum Doppler width (shift) uncertainty, is quite low, 6.4 (5.3) or a reflectivity, R, of 0.62 (0.56), when compared with the value obtained for the equal-time approach, i.e., a finesse of 14 (7), or a reflectivity of 0.80 (0.65). The number, N, of coefficients necessary to define a Fabry-Perot profile unambiguously at the least uncertainty points in Figures 3.12 and 3.13 is constant at 4 and 6 for position and temperature determinations respectively. Thus the critical number of samples, T, is 8 and 12 for these determinations.

The time required for a measurement with the equal-noise method can be obtained from Equations 3.2.37c) and 3.2.54), or:

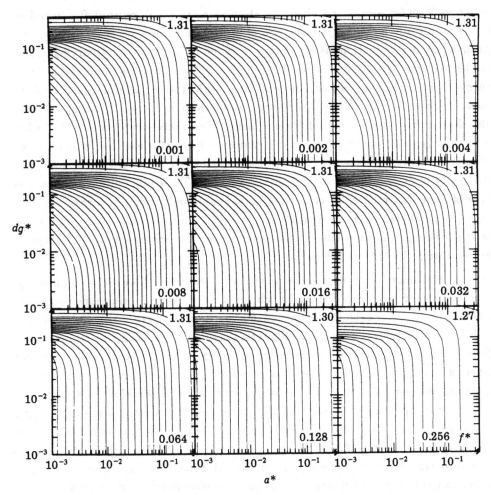

FIGURE 3.16. Values of the non-dimensional quantity $f(a_k)$ used to determine the time of measurement of an equal-noise experiment. The contours are separated by a factor of $2^{1/2}$. Hernandez (1985a).

$$\sum_{i=0}^{i=T} t_i = \mathbf{T}\, S\, Q^{-1}\, \mathbf{f}(a_k)\,, \qquad\qquad 3.2.55$$

where $\mathbf{f}(a_k)$ is given in Figure 3.16. The values of $\mathbf{T}\, S$ and Q are obtained from the combination of Equation 3.2.46b) and Figure 3.10 (or Equation 3.2.47b) and Figure 3.11) and Equation 3.2.3), respectively. Note that the units of $\mathbf{T}\, S$ must be in counts and Q in counts/(unit time).

Since Equations 3.2.46c) and 3.2.47c) have been cast in the same format as Equations 3.2.24) and 3.2.33), it is now possible to compare the relative efficiency of the equal-noise approach with the equal-time method. In terms of the standard deviations given in Figures 3.4, 3.5 and 3.14, it is found that the equal-noise approach has about twice the uncertainty of determination of the equal-time method for the determination of Doppler widths and about equal uncertainty for Doppler-shift determinations. For measurements of steady sources, this difference in the efficiency for Doppler-width determinations is significant, but for an unpredictably varying source it has been found (Hernandez *et al.*, 1984), this method provides more useful measurements than otherwise would be obtained with the equal-time measurement approach. These findings apply to a Fabry-Perot spectrometer which can not be used in the short-scan and coherent addition scheme. The short-scan and coherent addition method is to be preferred, since it is not only more efficient, but it requires no auxiliary photometer device for its operation and can be operated such that fixed uncertainties of determination can be obtained easily. The equal-noise method allows the use of some Fabry-Perot spectrometers in an environment for which they were not designed originally to operate, namely that of widely fluctuating source irradiances. Note that the equal-noise method is of general spectroscopic interest for absorption measurements (Hernandez, 1985a).

The least uncertainty approaches described thus far are limited by the assumptions made in the basic derivations, which are that Poisson distributions describing the photon field are nearly Gaussian and the uncertainties are reasonably approximated by the measured values. The present assumptions, using a Jaquinot and Dufour (1948) photoelectric spectrometer as an example, are applicable when the number of counts per sample bin exceed about 20. For extremely low count levels, the method outlined by Jahn *et al.* (1982), with its associated Monte Carlo computations, is more applicable. Needless to say, there are other approaches for extracting Doppler widths from Fabry-Perot recorded profiles. An excellent example is found in Turgeon and Shepherd (1962) for the use of precalculated graphs and tables to reduce into temperatures the width at fractional heights of experimental profiles. Another example is the visibility, or contrast, measurements of Buisson and Fabry (1912). The visibility measurements consist of the determination of the order, i.e., the spacing of the etalon, when the minima and maxima of the Fabry-Perot fringes can no longer be distinguished or the no-contrast point. As has been shown by Saha (1917), the no-contrast etalon gap is related to the wavenumber of the radiation and the one-half power of the ratio of the temperature to the molar mass of the emitting particle.

Chapter 4

Multiple-etalon devices

1. Filters

A quantitative description of a single-etalon interferometric spectrometer has been presented in Chapter 2. As discussed in that chapter, there are limitations to the usefulness of a single-etalon device. The most obvious limitation is that found in Equation 2.1.8); if a source rich in emission lines is being examined by a single-etalon device, these lines will appear together over a small range of the angle Θ, i.e., overlap has occurred. For a color-discriminating sensor, such as the human eye, the different wavelengths can be distinguished by their color (for lines sufficiently apart in the spectrum). When a detector does not discriminate color, such as a black-and-white photographic plate or a sensitive photomultiplier device, this overlap of many lines is no longer easily unscrambled. This is true, in particular, when the different wavelength lines have widely varying intensities. Under such circumstances, filtering is usually employed to separate the different lines.

Depending on the specific need, the filters used may range from broad-spectral-range types, such as colored glass, dyed gelatin, scattering filters and transparent solutions of selected chemicals, to higher-resolving-power devices such as spectrographs, spectrometers, interference filters, etc. As noted earlier, measurements of absorption spectra (embedded in a wide range continuous background) must include filtering, if results are to be obtained.

There are other occasions where simple filtering of a single Fabry-Perot spectrometer is not sufficient, since the experimental needs require not only the filtering but higher resolving power and high contrast as well. This is the case in hyperfine structure and Brillouin scattering experiments. For a Fabry-Perot device, the three simple solutions available to increase the resolving power are to decrease the free spectral range or increase the reflectivity, or both. The first choice (decreasing the free spectral range) leads to overlap, which would require further filtering, and does not increase the contrast because the reflectivity has not been changed [cf. Equation 2.1.14)]. Increasing the reflectivity will increase both the resolving power and the contrast; however, such gains are severely limited by the quality of the etalon substrate surfaces since they will limit the maximum etalon finesse attainable regardless of how large the reflectivity is made. As described in Chapter 2, Section 3, the increase in reflectivity to near the limiting finesse is also accompanied by drastic transmission losses.

2. Two-etalon systems

2.1. Identical etalons

For experimental problems, such as those described above, the more natural approach is to first consider the increase in the resolving power by means other than by increasing the reflectivity or the etalon spacer. The use of more than one etalon in spectroscopic work was reported by Nagaoka (1917) and further studies on this topic were continued by Houston (1927) and Gehrcke and Lau (1927). Although the etalons used by these workers were not identical, the actual use of multiple etalons can be dated from their pioneer work. However, it must be noted that Fabry and Perot (1897) used two etalons in series, but not for spectroscopic purposes, and Gehrcke (1905) used two Lummer-Gehrcke plates in series.

The two-identical-etalon problem has been studied in great detail by Dufour (1951). He has shown that in the limit of two widely spaced etalons, i.e., large separation relative to the etalon gaps, the function that describes their behavior is given by the product of the transmission functions of the two etalons. Using the terminology of Equation 2.1.7b) the result is (Fabry and Perot, 1897):

$$Y_{2e}(\delta) = [1 - A_1(1-R_1)^{-1}]^2(1-R_1)^2[1+R_1^2 - 2R_1\cos(\delta_1)]^{-1}$$
$$\times [1 - A_2(1-R_2)^{-1}]^2(1-R_2)^2[1+R_2^2 - 2R_2\cos(\delta_2)]^{-1} . \qquad 4.2.1a$$

In the above equation, the subscripts are used to distinguish between the two etalons. The example of two true identical etalons, reduces Equation 4.2.1a) into the following:

$$Y_{2e}(\delta) = [1-A(1-R)^{-1}]^4(1-R)^4[1+R^2-2R\cos(\delta)]^{-2}. \qquad 4.2.1b$$

Using the formulation of Equation 2.1.7d) to represent the etalon function gives the following result:

$$Y_{2e} = [1-A(1-R)^{-1}]^4[1+\mathbf{F}\sin^2(\delta/2)]^{-2}$$

$$= [1-A(1-R)^{-1}]^4\left\{1+\mathbf{F}\sin^2(\delta/2)[2+\mathbf{F}\sin^2(\delta/2)]\right\}^{-1}. \qquad 4.2.1c$$

In the above equation, the definition of $\mathbf{F}$ given in Equation 2.1.13c) has been used.

The last two equations show the obvious properties of the two-identical-etalon system, with the last equation [4.2.1c)] expressing the two-etalon combination as a single etalon with a more complex modulation term than usual. For a pair of ideal etalons with zero absorption/scattering coefficient, the value of maximum transmission is unity at $\delta = 0$. The minimum transmission, at $\delta = \pm\pi$, is now $(1-R)^4(1+R)^{-4}$, or the square of that obtained for a single etalon, that is, the contrast C_2 of the combination is the square of the single-etalon case of Equation 2.1.14). This behavior is shown in Figure 4.1 for the same etalons used in Figure 2.6. The increased contrast obtained with the tandem identical etalons makes it possible to distinguish clearly the presence of a weak line near the minimum transmission, even at low reflectivities. An experimental example is given in Figure 4.2.

The general case of n identical etalons then shows the contrast to be $C_n = C^n$, where C is the contrast of a single etalon. Again, in the general case of n identical etalons, the normalized HWHH of the profile a_n^* is given by:

$$a_n^* = \pi^{-1}\sin^{-1}[(2^{1/n}-1)^{1/2}(1-R)(2R^{1/2})^{-1}]. \qquad 4.2.2a$$

This equation can be simplified for values of R near unity, as was done with Equation 2.1.11a), to give the following result:

$$a_n^* \approx \pi^{-1}(2^{1/n}-1)^{1/2}(1-R)(2R^{1/2})^{-1},$$

$$a_n^* \approx (2^{1/n}-1)^{1/2}a^*. \qquad 4.2.2b$$

In the above, a^* is the normalized HWHH for a single etalon. Comparison of the above results with those given earlier in Equations 2.1.11a) and 2.1.11b), indicates the line width is indeed sharpened. It can also be easily shown from either Equations 4.2.2a) or 4.2.2b) that the overall profile becomes narrower and directly leads to the earlier result of an effective contrast $C_n = C^n$.

For combinations of etalons where the absorption/scattering coefficient is not negligible, the loss in the transparency becomes fairly large, i.e., the nth power of the single-etalon loss. In the early days of silvered etalon surfaces, with their concomitant absorption, the loss of transparency was a definite consideration in the use of

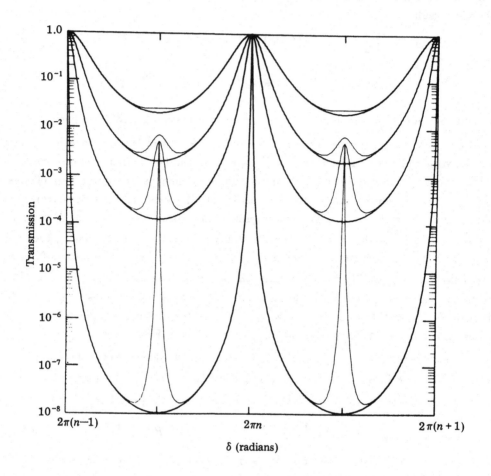

FIGURE 4.1. Increase of the contrast obtained by the use of two identical etalons in series. Monochromatic line case. Compare with Figure 2.6.

multiple-etalon systems. Losses in transparency also occur when the etalon surface defects are important, but, as described earlier, judicious choice of the reflectivity will minimize their effects. Although surface defects have further effects (Roesler, 1969) than those described in Chapter 2, they will be discussed in the poly-etalon section.

The use of a series of identical etalons does provide the desired properties of increased resolving power at a smaller loss in transmission, when compared with an unlimited increase in reflectivity using etalon substrates which are not ideally flat, yet the free spectral range is preserved. Although the solution for a series of identical etalons with differing reflectivities is not as amenable to analytical treatment, the results are sufficiently similar to the case just discussed to require further elaboration.

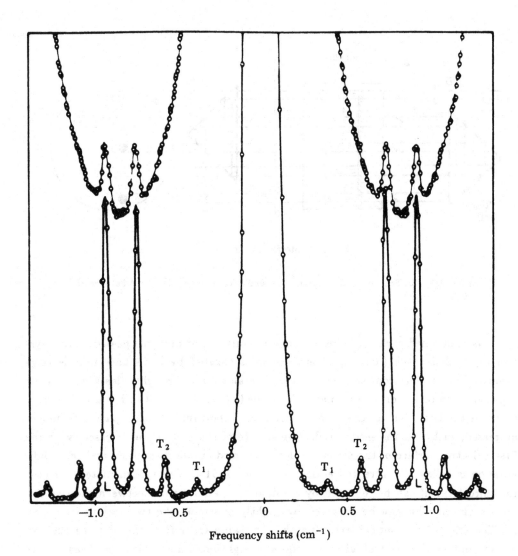

FIGURE 4.2. Single etalon and double-passed etalon spectrum of *SbSi*. (Sandercock, 1975).

Tandem identical etalons have been utilized to discriminate closely spaced emission lines of widely differing brightness (Bradley and Kuhn, 1948; Daehler and Roesler, 1968; Daehler, 1970; Cannel and Benedek, 1970).

Dufour (1951), Hariharan and Sen (1961), Müller and Winkler (1968), and Beysens (1973) obtained the two-identical-etalon behavior from a single etalon by using reflection to pass the light twice through the etalon. Sandercock (1970, 1975) further improved this system to accomplish multiple passes (≈ 5) with a single etalon, as shown in Figure 4.3.

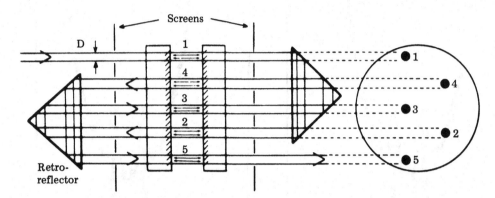

FIGURE 4.3. Schematic arrangement for single-etalon multipassing. (Sandercock, 1975).

As mentioned earlier, the results for the multiple, or multiple-passed, etalon systems apply when the (effective) etalons are separated by a distance that is large compared to the etalon spacing. As two etalons are brought closer and closer together, a third etalon is automatically formed. Actually, the third etalon has been there all the time, but its effects have not been important. As the spacing is further decreased, such that the second reflective surface of the first etalon coalesces with the first reflective surface of the second etalon, the third etalon disappears and this configuration is called a 'contracted etalon' (Dufour, 1951). For intermediate distances, the third 'etalon' can be removed by eliminating the reflections between the two original etalons. This can be obtained by slightly offsetting the two etalons from perpendicularity to the optical axis, and thus throwing the reflections between the two etalons out of the field of view. A more straightforward solution has been to use screens in a two-lens system (Dufour, 1951), or its equivalent of displaced axes (Bens *et al.*, 1965), such that the reflections from the second etalon do not return to the first etalon. This approach is shown in Figure 4.4. Other possible means to obtain a reduction or elimination of the third etalon include the use of polarizers, neutral density filters, etc. (Sturkey, 1940; Schwider, 1965). These approaches, just as the screen solution, involve a loss of light in the system.

2.2. Nearly identical etalons

For two nearly identical etalons, i.e., when δ_1 and δ_2 of Equation 4.2.1a) are not exactly the same, it is useful to express the separate etalons by their Fourier

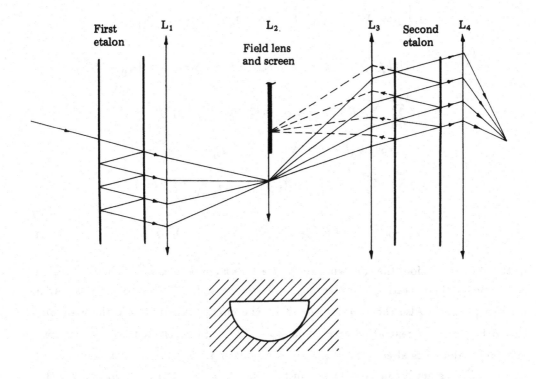

FIGURE 4.4. Elimination of reflections between two etalons using masks. The dashed lines indicate the offending reflections being eliminated. The shape of the mask is shown below the ray path drawing. After Dufour (1951).

decompositions as given by Equation 2.1.7c). Thus, consider an etalon examining a source of arbitrary line shape, with the response given by Equation 2.2.5), followed by an identical etalon behind it. Further, it is presumed that the fringes of the combination are projected onto a plane containing a centered circular aperture f orders wide or, in normalized HWHHs, $2 f^*$ wide. By similarity with Equation 2.4.2a) the flux transmitted by the system onto a detector is given by:

$$Y(\sigma) = \mathbf{I\,A}\,\epsilon\,\tau_L \int_0^{2\pi} \int_{\Theta_1}^{\Theta_2} \tau_{e_1}(\sigma)\,\tau_{e_2}(\sigma)\,\sin\Theta\;d\Theta\;d\Phi\;. \qquad 4.2.3$$

In the above equation, the τ_{e_i}s are defined by Equation 2.4.3). In the following, the two τ_{e_i}s will be distinguished by the use of a_ls for the coefficients of the first etalon-source combination and R^ks for the second etalon respectively. However, note that for the first etalon-source combination, $a_l = R^l s_l$. The results of the solution for the flux of Equation 4.2.3) are given below:

$$Y(\sigma) = \mathbf{I}\,\mathbf{A}\,\epsilon\,r_L\,4\pi\,f^{\,*}\,n_0^{\,-1}[1-A\,(1-R)^{-1}]^4(1-R)^2(1+R)^{-2}$$

$$\times\left\{1+2\sum_{l=1}^{l=\infty}a_l\;\mathrm{sinc}(2lf^{\,*})\,\cos[2\pi l\,(n_0-f^{\,*})\,]\right.$$

$$+2\sum_{k=1}^{k=\infty}R^k\;\mathrm{sinc}[2kf^{\,*}\,(1-\Delta n\;n_0^{\,-1})\,]\,\cos[2\pi k\,(1-\Delta n\;n_0^{\,-1})(n_0-f^{\,*})\,]$$

$$+2\sum_{k=1}^{k=\infty}\sum_{l=1}^{l=\infty}R^k\,a_l\;\mathrm{sinc}[(l-k\,(1-\Delta n\;n_0^{\,-1}))2f^{\,*}\,]$$

$$\times\cos[2\pi\,(l-k\,(1-\Delta n\;n_0^{\,-1}))(n_0-f^{\,*})\,]$$

$$+2\sum_{k=1}^{k=\infty}\sum_{l=1}^{l=\infty}R^k\,a_l\;\mathrm{sinc}[(l+k\,(1-\Delta n\;n_0^{\,-1}))2f^{\,*}\,]$$

$$\left.\times\cos[2\pi\,(l+k\,(1-\Delta n\;n_0^{\,-1}))(n_0-f^{\,*})\,]\right\}.\qquad 4.2.4$$

In the above equation, the following simplifications have been made. The first etalon is assumed to be operating at order n_0, while the second etalon is set to be at order $n_1 = n_0 - \Delta n$. Also the angles covered by the circular aperture are those encompassed between the central order n_0 and $n_0 - 2f^{\,*}$. Since the solution for the two identical etalons has already been shown in Equation 4.2.1b), it will not be discussed here except to state it can be obtained in the limit of $f^{\,*} \to 0$ from Equation 4.2.4).

For two nearly identical etalons, i.e., $\Delta n << n_0$, close examination of Equation 4.2.4) shows some interesting behavior. First, a reminder of some of the properties of the sinc function (Bracewell, 1965); the sinc function is equal to zero for integer arguments, is equal to unity for zero argument, and is a symmetrical function (sinc $(-x)$ = sinc (x)). Returning to Equation 4.2.4), if the value of $f^{\,*}$ is made equal to 0.5, or an integer multiple M of 0.5, it is found that the first, second and fourth summation terms tend to vanish. This occurs because, for the assumed condition of $\Delta n << n_0$, the following relations apply:

$$\lim_{\Delta n\;n_0^{\,-1}\to 0}k\,(1-\Delta n\;n_0^{\,-1}) = k\,,\qquad 4.2.5a$$

$$\lim_{\Delta n\;n_0^{\,-1}\to 0}[l+k\,(1-\Delta n\;n_0^{\,-1})] = (l+k)\,.\qquad 4.2.5b$$

The values of l and k are integers greater or equal to unity, and thus their product with $2f^{\,*}$ is also an integer since by definition $2f^{\,*}$ has been made an integer. Therefore, the sinc functions are all equal to zero, except for the third summation term, where the sinc functions contain arguments equal to zero, when $l \equiv k$, since:

$$\lim_{\Delta n\;n_0^{\,-1}\to 0}[l-k\,(1-\Delta n\;n_0^{\,-1})] = l-k \equiv 0\,.\qquad 4.2.5c$$

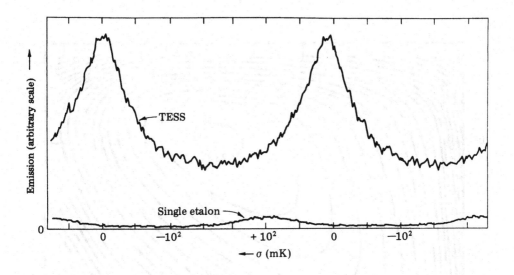

FIGURE 4.5. Experimental flux gain obtained with a TESS device. (Hernandez *et al.*, 1981).

Then, because of the presence of sinc functions with zero argument, the third summation term has a finite value (other than zero). Under the above conditions Equation 4.2.4) becomes:

$$Y(\sigma) = \mathbf{I}\,\mathbf{A}\,\epsilon\,\tau_L\,2\pi\,M\,n_0^{-1}\,[1 - A\,(1-R)^{-1}]^4 (1-R)^2 (1+R)^{-2}$$

$$\times \left\{ 1 + 2 \sum_{k=1}^{k=\infty} R^k\,a_k\,\cos[2\pi k\,(\Delta n - M\,(2n_0)^{-1})\,] \right\}. \qquad 4.2.6$$

The above equation, when compared to Equation 2.4.2c), shows no effects in regard to the presence of a circular aperture other than a substantial increase in the flux received by the detector. Changing Δn will scan the spectrum, yet preserve the information of the line shape of the source being examined.

The modulation of the output flux, as well as the large increase in the flux obtained, is the principle of the TESS (Twin Etalon Scanning Spectrometer) device (Hernandez *et al.*, 1981; Hernandez, 1982a). The TESS device, as described in Equation 4.2.6), allows the user to arbitrarily increase the flux into the detector without loss in the resolving power of the instrument. Equation 4.2.6) can be recast in a more familiar form, remembering that $a_k = R^k\,\mathbf{s}_k$, thus:

$$Y(\sigma) = \mathbf{I}\,\mathbf{A}\,\epsilon\,\tau_L\,2\pi\,M\,n_0^{-1}\,[1 - A\,(1-R)^{-1}]^4 (1-R)^2 (1+R)^{-2}$$

$$\times \left\{ 1 + 2 \sum_{k=1}^{k=\infty} R^{2k}\,\mathbf{s}_k\,\cos[2\pi\,k\,(\Delta n - M\,(2n_0)^{-1})\,] \right\}. \qquad 4.2.7$$

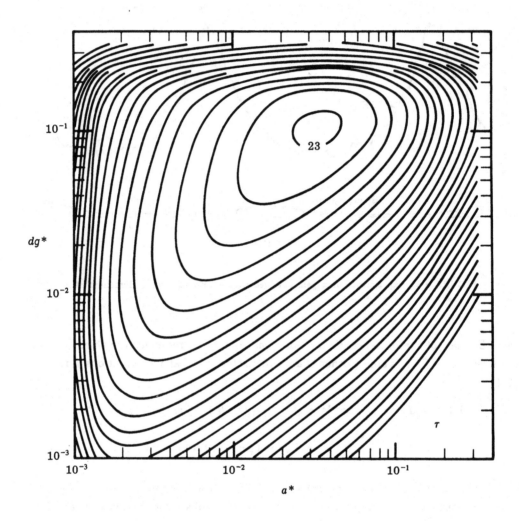

FIGURE 4.6. Uncertainties of determination of temperature from a TESS profile. The uncertainties are expressed as contours of the ratio of the uncertainties of determination over the temperature. Same conditions as Figure 3.3, except for one-order-aperture device. Hernandez (1982a).

Other than showing the effective reflectivity of the TESS device to be lower than either of the etalons from which it is constructed, Equation 4.2.7) indicates the device is otherwise like a single-etalon spectrometer. Figure 4.5 shows the luminosity gain obtained with a TESS device. Therefore, the techniques used in Chapter 3, Section 2, are applicable for determining the source line profile. Because of this similarity, there must exist a best source and instrumental width combination that leads to minimum uncertainties of determination, and indeed, this is the case (Hernandez,

1982a).

Figure 4.6 gives, in the same format as Figure 3.3, the uncertainties of determination of temperature as a function of the normalized source and instrumental line widths. The minimum error of determination is found to be at an instrumental normalized width $a^* = 0.032_4$ and source line width $dg^* = 0.10_6$. Since, in the figure, a^* is defined to be the normalized width of each etalon of the TESS device, the associated values of the reflective finesse and reflection coefficient are about 15 and 0.82, respectively. These results are nearly indistinguishable from those obtained for an optimized single-etalon-interferometric spectrometer, i.e., 14 and 0.80 for the finesse and reflectivity. The results for the TESS instrument (Hernandez, 1982a) also show that the minimum number of samples required to describe unambiguously a measured profile for temperature determinations is equal to 14, which is also the number required for the single etalon device.

The configuration of the TESS spectrometer has rather weak dependence on the line source center wavenumber since the modulating term in Equations 4.2.6) and 4.2.7) is dependent only on the difference in orders Δn between the two etalons, rather than the absolute order n_0 for the single-etalon spectrometer of Equation 2.4.2c). Note that in the latter equation, $n_0 = \sigma \, (\Delta\sigma)^{-1}$. The weak dependence on wavenumber can be appreciated easily because, in the TESS configuration, the spatial resolution of the etalons has been sacrificed by the use of a single large circular aperture. The shift in the angle Θ of the etalons as a function of wavenumber [cf. Equation 2.4.3)] is still present, but is no longer measured. Should a set of annuli(each of width less than one order) be used, then the spatial dependence can be regained. The use of position-sensitive detectors (such as photographic plates, television tubes, and imaging photomultiplier devices) will recover this spatial information, while the total flux determines the line shape. Note that the behavior of two etalons in series towards white light has been used since 1897 by Perot and Fabry to determine the equality and/or ratio between two etalon spacers.

2.3. Unequal etalons

Thus far, the discussion on two-etalon systems has been limited to the identical, or nearly identical, etalon configuration. When the two etalon spacers are considerably different, other useful properties of the general two-etalon system appear. For convenience, the two-unlike-etalon configuration can be divided into two broad categories, namely when the etalon spacings are related in size by either large factors (≥ 2) or small differences. These two categories are usually called the simple and vernier-compound-etalon systems.

The simple compound-etalon system, with a binary ratio between the spacers, provides a good starting point for the investigation of the properties of compound etalons. For convenience, it is assumed that both etalons have the same reflectivity, thus a similar expression to Equation 4.2.1b) is obtained:

$$Y(\delta) = [1 - A (1 - R)^{-1}]^4 (1 - R)^4$$

$$\times \left\{ [1 + R^2 - 2R \cos(\delta)] [1 + R^2 - 2R \cos(2\delta)] \right\}^{-1}. \qquad 4.2.8a$$

This equation can, in turn, be simplified into the following:

$$Y(\delta) = [1 - A (1 - R)^{-1}]^4$$

$$\times \left\{ [1 + \mathbf{F} \sin^2(\delta/2)] [1 + \mathbf{F}/2 (1 - \cos(2\delta))] \right\}. \qquad 4.2.8b$$

The usual meaning of $\mathbf{F} = 4R (1 - R)^{-2}$ has been used in the above expression. Another convenient simplification is given below:

$$Y(\delta) = \left\{ 1 + [\mathbf{F} \sin^2(\delta/2)] \Big[1 + [4\cos^2(\delta/2)] [1 + \mathbf{F} \sin^2(\delta/2)] \Big] \right\}^{-1}. \qquad 4.2.8c$$

The equations presented above allow investigation of the compound interferometer either as a product of two etalons or as a single etalon with a more complex modulation than usual [cf. Equation 2.1.7d)]. The results given in Equation 4.2.8c) show that the extra modulation term has two maxima near $\pm \pi/2$, or more correctly, at $\pm \cos^{-1}(\mathbf{F}^{-1})$. Therefore Equation 4.2.8c) should show two minima at approximately $\pm 0.6 \pi$, and the value at these points will be lower than for the simple etalon. This effect is shown in Figure 4.7, where it should be noted that the 'peak' at π between the two minima has the same transmission as the single etalon minimum. See Figure 2.4 for comparison. The location of the minima for the binary-etalon combination is found at:

$$\delta_{min} = \cos^{-1}\left[\left((1 + R^2) - [(1 + R)^4 - R (1 - R)^2]^{1/2} \right) (6R)^{-1} \right]. \qquad 4.2.9$$

Note that for $0 < R < 1$,

$$0.5804 \pi \leqslant \delta_{min} \leqslant 0.6081 \pi . \qquad 4.2.10$$

As would be expected, when the ratio of the etalon spacers is changed, still in integer steps, it is found that the number of both minima and 'peaks' are increased. This naturally leads into the vernier configuration where it then becomes possible to locate the minima and 'peaks' at will. The expression for the vernier is a simple variant of Equation 4.2.8b), i.e.,

$$Y(\delta) = [1 - A (1 - R)^{-1}]^4$$

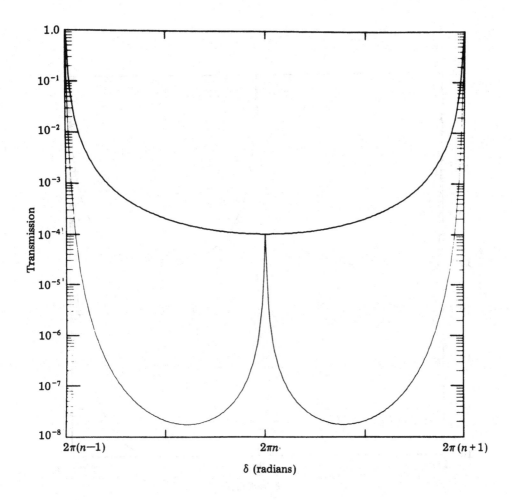

FIGURE 4.7. Transmission through two etalons with gaps differing by a factor of 2. The smaller spacer etalon transmission is shown for comparison. The scale on the abscissa is given for the larger spacer etalon.

$$\times \left\{ \; [1+\mathbf{F}\; \sin^2(\delta/2) \;] \; [1+\mathbf{F}/2\; (1-\cos(a\delta)) \;] \; \right\}^{-1} . \qquad 4.2.8d$$

In the above, a represents the ratio of the two etalon gaps. The resultant of two etalons with a 1.0000 : 0.666666 spacer ratio is shown in Figure 4.8.

As described by Houston (1927) and Gehrcke and Lau (1927), the usefulness of the two-etalon scheme is to increase the effective free spectral range of the larger spacer etalon, while keeping its resolving power. In exchange for this, a set of parasitic peaks has been gained. These parasitic peaks have only nuisance value (Meissner, 1942), since they tend to interfere with the measured spectra, but they are easily

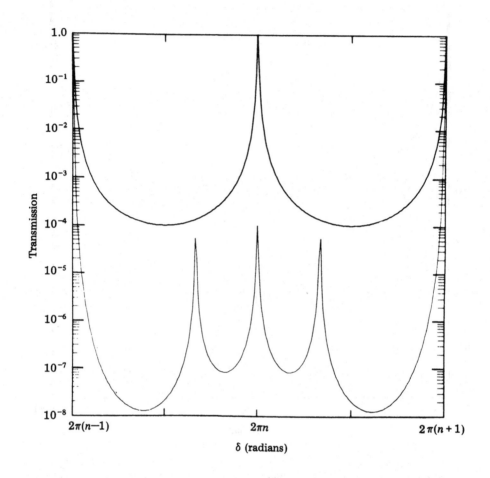

FIGURE 4.8. Transmission through two etalons with a 1:0.6666666 spacer ratio. The upper graph shows the unity spacer profile, and the abscissa also refers to this etalon.

taken into account since their position is well-known. In extreme cases, where these peaks overlap a line being measured, a change in the etalon ratios will then clarify the ambiguity. The parasitic 'peaks' can be further decreased in magnitude by the use of another etalon (Chabbal, 1958a, b, c), and this leads into poly-etalon systems of which the best example is the PEPSIOS (Poly-Etalon Pressure Scanned Interferometric Optical Spectrometer) device (Mack *et al.*, 1963).

A general treatment for two, or more, etalons in series has been discussed by Moghrabi and Gaume (1974, 1977). For two equal-reflectivity etalons in series, the function that describes their transmission is:

$$Y(\delta_1,\delta_2) = [\, 1 - (\, 1 - R\,)^{-1}]^4 \, (\, 1 - R\,)^4$$

$$\times \, (\, 1 + R^2 + 2R\, \cos\delta_1\,)^{-1} \, (\, 1 + R^2 + 2R\, \cos\delta_2\,)^{-1}, \qquad 4.2.11a$$

where the subscripts are employed to distinguish the individual etalons. This equation can be simplified if the absorption coefficient, A, is set to zero and a reflective parameter, b, is defined as:

$$b = (\, 1 + R^2\,) \, (\, 2R\,)^{-1}$$

$$= (\, C + 1\,) \, (\, C - 1\,)^{-1} \simeq 1 + \pi^2 \, (\, 2N_R^2\,)^{-1}, \qquad 4.2.12$$

where C is the single etalon contrast defined in Equation 2.1.14), and N_R is the approximate reflective finesse, given in Equation 2.1.13b). Replacing b into Equation 4.2.11a), the following is obtained:

$$Y(\delta_1,\delta_2) = (\, b - 1\,)^2 \, (\, b - \cos\delta_1\,)^{-1} \, (\, b - \cos\delta_2\,)^{-1}. \qquad 4.2.11b$$

The phases δ_1 and δ_2 can be defined in terms of the number (N_i) of free spectral ranges of each etalon necessary for mutual overlap of orders to occur, i.e., $N_1 \Delta\sigma_1 = N_2 \Delta\sigma_2$, or:

$$\delta_i = N_i \, \delta, \qquad 4.2.13$$

where $0 \leqslant \delta \leqslant 2\pi$. For completeness, N_i is an integer and $N_2 > N_1$. The transmission function is then:

$$Y(\delta) = (\, b - 1\,)^2 \, [\, b - \cos(N_1\delta)\,]^{-1} \, [\, b - \cos(N_2\delta)\,]^{-1}, \qquad 4.2.11c$$

For the appropriate choice of N_1 and N_2, $Y(\delta)$ is a periodic function with unity maxima at $\delta = 2\pi k$, $k = 0, 1, 2, 3,...$, and, as expected, also has a number of small transmission, or parasitic, peaks at values of δ other than $2\pi k$. For an integer n_2, defined as $0 \leqslant n_2 \leqslant N_2$, the phase δ can be replaced with:

$$\delta = 2\,\pi\,n_2\,N_2^{-1}, \qquad 4.2.14$$

and then Equation 4.2.11c) is expressed as:

$$Y(n_2) = (\, b - 1\,) \, [\, b - \cos(2\pi\,n_2 N_1 N_2^{-1}\,)\,]. \qquad 4.2.11d$$

This equation gives the transmission values of the parasitic peaks located at n_2. Note that similar results would have been obtained if the phase was defined in Equation 4.2.14) in terms of n_1, which is associated with N_1 in the same fashion as n_2 is associated with N_2.

The values of $Y(n_2)$ in Equation 4.2.11d) are less than unity as long as $n_2 N_1 N_2^{-1}$ is not an integer. Note that $n_2 N_1 N_2^{-1}$ can be identified with the order n_1, $0 \leqslant n_1 \leqslant N_1$, associated with n_2; however, in this case, n_1 need not be an integer. The lowest possible value of $Y(n_2)$, for n_2 an integer, is given when the argument of the cosine is equal to $(2k+1)\pi$, for k an integer, including zero. For these values of the argument, $Y(n_2) = (b-1)(b+1)^{-1} = \mathbf{C}^{-1}$. Thus, the smallest parasitic peak(s) will have a transmission equal to $\mathbf{C}^{-1}$.

Therefore, for two etalons in series, it is possible to arbitrarily limit the transmission of the parasitic peaks to some specified criterion, l, such that:

$$Y(n_2) \leqslant l , \qquad \mathbf{C}^{-1} \leqslant l \leqslant 1 . \qquad\qquad 4.2.15$$

Replacing $Y(n_2)$ of this inequality with Equation 4.2.11d) and solving for $n_2 N_1 N_2^{-1}$:

$$n_2 N_1 N_2^{-1} \leqslant k + (2\pi)^{-1} \cos^{-1}[b - (b-1) l^{-1}] , \qquad\qquad 4.2.16a$$

where k can be recognized to be equal to n_1 for integer values, and:

$$N_1 N_2^{-1} = n_1 n_2^{-1} \pm (2\pi n_2)^{-1} \cos^{-1}[b - (b-1) l^{-1}] . \qquad\qquad 4.2.16b$$

Since the purpose of crossing two, unequal spacing, etalons is to increase the effective free spectral range of the etalon with the highest resolving power, while retaining this resolving power, the optimum combination of etalon spacers is that combination of N_1 and N_2 which has no integer values of n_1 for any corresponding n_2. Thus, the choice of a ratio $N_1 N_2^{-1}$ which gives the lowest combined parasitic peak transmission with the largest effective free spectral range (N_2 in the present discussion), will give the optimum combination of etalons. As discussed earlier, integer values of $n_2 N_1 N_2^{-1}$ give rise to maximum transmission, which defeats the purpose of having chosen a given ratio $N_1 N_2^{-1}$. The existence of ratios of $n_1 n_2^{-1}$, for both n_1 and n_2 integers, gives rise to 'forbidden zones' (Moghrabi and Gaume, 1974) where these ratios occur. Equation 4.2.16b) provides both the location ($n_1 n_2^{-1}$) and the width $\{ 2 (2\pi n_2)^{-1} \cos^{-1}[b - (b-1) l^{-1}] \}$ of these forbidden zones. Note that the width of these forbidden zones is dependent on both the etalon reflectivity and the parasitic peak criterion l. Therefore, it is possible to construct, graphically or otherwise, a table of values of N_2 as a function of N_1, with the forbidden zones superimposed (Moghrabi and Gaume, 1974, 1977). Then the useful values of N_1 and N_2 are those pairs which do not overlap the forbidden zones.

Moghrabi and Gaume (1977) have extended their method to the combination of an interference filter and two etalons, and in principle, to any number of etalons in series. As these authors show, the choice of values of N_1 and N_2 of the etalons depends on the desired effective free spectral range, the criterion l chosen for the parasitic peaks and the reflectivity of the etalons. A quantitative measure of the optimization achieved is given by the global finesse (Moghrabi and Gaume, 1977), which is defined (approximately) as:

$$N_G \simeq N_2 \, N_R \; . \tag{4.2.17}$$

For a reflectivity of 0.90 ($N_R \simeq 30$) and $l \leqslant 0.015$, a global finesse of near 9000 is possible (Moghrabi and Gaume, 1977) for two etalons and one filter.

The analytic treatment presented thus far is not the only method available to investigate the properties of etalons in series. Neuhaus and Nylén (1970) have discussed two etalons in series in terms of matrix formalism.

3. Poly-etalon devices

Although poly-etalon systems had already been used (Series, 1954), the systematic investigations of Chabbal (1958a, b, c) are the foundation of present day poly-etalon devices. In the previous discussion, the use of a device with more than one etalon was treated in the context of emission line spectra measurements, but for completeness, the response of these devices toward white light must also be included.

The techniques used to describe the behavior and properties of multiple-etalon devices have been developed mostly for the PEPSIOS triple-etalon device (Mack *et al.*, 1963, McNutt, 1965, Stoner, 1966, Roesler and Mack, 1967, Roesler, 1974). Since the approach is of general applicability, it will be followed here. As an introduction to these techniques and definitions, the combination of a single etalon and a filter will be discussed first. For convenience, it is assumed that the filter is described by a Lorentz, or dispersion profile, i.e.,

$$F(\sigma) = [\, 1 + [(\sigma - \sigma_0) \, s^{-1}]^2)^{-1} \; . \tag{4.3.1}$$

As the equation shows, the filter has unity transmission at σ_0 and a HWHH equal to s. This description is, in the limit, that of an interference filter where the periodicity of a low-order Fabry-Perot etalon can be neglected (Hernandez, 1974). The etalon function will be given by the following expression:

$$
\begin{aligned}
E(\sigma) &= [1 - A \, (1 - R)^{-1}]^2 (1 - R)(1 + R)^{-1} \\
&\quad \times \left[1 + 2 \sum_{l=1}^{l=\infty} R^l \, \cos[2\pi \; l \; \Delta\sigma^{-1}(\sigma - \sigma_0) \,] \right] \\
&= \tau_E \left[1 + 2 \sum_{l=1}^{l=\infty} R^l \, \cos[2\pi \; l \; \Delta\sigma^{-1}(\sigma - \sigma_0) \,] \right] . \tag{4.3.2}
\end{aligned}
$$

Further simplifications have been made, namely, that a scanning aperture of negligibly small size is being used, and the usual constants associated with its use [cf. Equation 2.4.2b)] are arbitrarily set to unity.

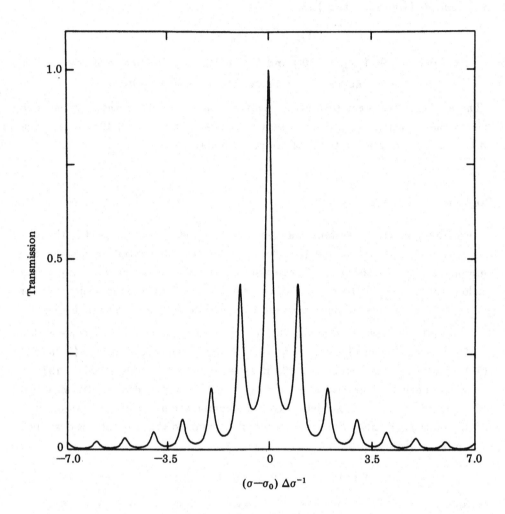

FIGURE 4.9. Transmission of an etalon combined with a Lorentzian filter.

When the filter-etalon combination examines a continuous emission source, the response of the device is given by the product $F \times E$, or:

$$(F \times E)(\sigma) = \tau_E \ [1 + [(\sigma - \sigma_0) \ s^{-1}]^2]^{-1}$$

$$\times \left[1 + 2 \sum_{l=1}^{l=\infty} R^l \ \cos[2\pi \ l \ \Delta\sigma^{-1}(\sigma - \sigma_0) \ ] \right]. \qquad 4.3.3$$

A pictorial representation of the above equation is shown in Figure 4.9. where, as expected, those wavenumbers of the radiation with integral order number have maximum transmission by the etalon, with the total transmission modified by the filter.

The total radiation received by the detector is then the area under the curve of Equation 4.3.3), or Figure 4.9. Since only the radiation passing through the center peak is desired, the extra radiation is called the parasitic light (Mack *et al.*, 1963). Stoner (1966) defines the parasitic light as the difference between the radiation passed by one order of the etalon at the filter peak and the total radiation passed by the filter-etalon combination. This is quantitatively expressed as the ratio of the parasitic light to the radiation of the one order, i.e.,

$$\mathbf{P} = \left\{ \int_{-\infty}^{\infty} (F \times E)(\sigma)\, d\sigma - \int_{-\Delta\sigma/2}^{\Delta\sigma/2} (F \times E)(\sigma)\, d\sigma \right\}$$
$$\times \left\{ \int_{-\Delta\sigma/2}^{\Delta\sigma/2} (F \times E)(\sigma)\, d\sigma \right\}^{-1} . \qquad 4.3.4a$$

For the simple case when $s > \Delta\sigma$, the solution for the above is found to be equal to:

$$\mathbf{P} = s\ \pi\ (1 + R_E)(1 - R_E)^{-1} - 1 . \qquad 4.3.4b$$

In this equation, $R_E = R\ \exp[-2\pi\ s\ \Delta\sigma^{-1}]$, as was done earlier [cf. Equation 2.2.4b)]. The limiting condition of $s \gg \Delta\sigma$ shows the expected result of the etalon behavior towards white light, that is, $R_E \to 0$ and $\mathbf{P} \to s\ \pi$, or all parasitic light. The result given in Equation 4.3.4b) assumes the etalon and filter peak transmission are matched. When this is not the case, the results are:

$$\mathbf{P}(\sigma_1) = s\ \pi\ \left\{ 1 + 2 \sum_{l=1}^{l=\infty} R_E^l\ \cos[2\pi\ l\ \Delta\sigma^{-1}(\sigma_1 - \sigma_0)] \right\} - 1 . \qquad 4.3.4c$$

Another measure for the parasitic light is given by the ratio **Q** of maximum to minimum signal when the etalon is scanned over one order (Stoner, 1966). The value for this ratio is:

$$\mathbf{Q} = (1 + R_E)^2 (1 - R_E)^{-2} . \qquad 4.3.5$$

The ratio **Q** approaches the contrast $\mathbf{C} = (1 + R)^2 (1 - R)^{-2}$ of the single etalon [cf. Equation 2.1.14)] in the absence of parasitic light. For the limiting condition of $s \gg \Delta\sigma$, it is again found that $R_E \to 0$ and consequently $\mathbf{Q} \equiv 1$. This zero modulation is the white light behavior of the etalon discussed in Chapter 2, Section 2, and the basis of Giacomo's (1952) method for the measurement of etalon reflectivity. This approach can be enlarged to encompass any number of etalons in the system, with the thought of minimizing the parasitic light as a function of the filter width, etalon spacers and etalon reflectivities. The minimization procedure has been carried out for the PEPSIOS system with the help of approximations (Stoner, 1966). The approximations used severely limit the sensitivity of the results.

Another approach can be taken if the etalon function, E, of Equation 2.1.7e) is considered, i.e.,

$$E(\sigma) = \sum_{n=-\infty}^{n=\infty} E_n(\sigma) .$$

4.3.6a

$E_n(\sigma)$ is defined by:

$$E_n(\sigma) = (1+R)^{-1}(1-R)\, 2\,[\ln(R^{-1})\,]^{-1}$$

$$\times \left\{ 1 + [\Delta\sigma\,\ln(R^{-1})\,]^{-2}[2\pi\,(\sigma-\sigma_0)-2\pi\,\Delta\sigma\,n]^2 \right\}^{-1} .$$

4.3.6b

Note that an ideal etalon without any absorption or scattering properties is being considered in the above equation. Since Equation 4.3.6a) represents the etalon function as a superposition of Lorentz profiles displaced from each other by integer multiples of the free spectral range $\Delta\sigma$, it is then possible to consider the parasitic light to be associated with those Lorentz profiles when $n \neq 0$, and the useful radiation is that radiation associated with $n \equiv 0$. The total radiation transmitted by the etalon described in Equation 4.3.6a) is, of course, infinite and the ratio **P** of Equation 4.3.4a) is very large and the 'contrast' **Q** of Equation 4.3.5) is equal to unity, that is, the description of the etalon behavior towards white light has been found again. However, when a filter with a finite width is placed in series with the etalon, the amount of parasitic light will be finite and the values of **P** and **Q** will reflect this condition. Quantitatively, assuming a Lorentz-shape filter as done earlier in Equation 4.3.1), the filter-etalon combination is described by:

$$Y(\sigma) = \sum_{n=-\infty}^{n=\infty} Y_n(\sigma) = \sum_{n=-\infty}^{n=\infty} \left\{ E_n(\sigma) \times \left[1 + c^{-2}[2\pi\,(\sigma-\sigma_0)\,]^2\right]^{-1} \right\} .$$

4.3.7

The quantity c is related to s of Equation 4.3.1) by $c = 2\pi\,s$, thus the HWHH of the filter is equal to $c\,(2\pi)^{-1} = s$. Since the filter peak transmission has been arbitrarily set to be at the same wavenumber as the etalon peak with $n = 0$, then the ratio **P** can be defined in the same fashion as in Equation 4.3.4a), namely:

$$\mathbf{P} = \left\{ \int_{-\infty}^{\infty} Y(\sigma)\,d\sigma - \int_{-\infty}^{\infty} Y_{n=0}(\sigma)\,d\sigma \right\} \left\{ \int_{-\infty}^{\infty} Y_{n=0}(\sigma)\,d\sigma \right\}^{-1}$$

$$= \left\{ \left[\sum_{n=-\infty}^{n=\infty} W_n \right] - W_{n=0} \right\} \left[W_{n=0} \right]^{-1} .$$

4.3.8

$$W_n = \int_{-\infty}^{\infty} Y_n(\sigma)\,d\sigma .$$

The solution to this equation can be found by the usual analytical means; however, more physical insight is gained if the elegant treatment of McNutt (1965) is used. Basically, his approach uses the theory of residues to solve the above type of equation, and at the same time allows separation of the useful radiation from the parasitic light. The functions entering into Equation 4.3.8) are, neglecting constants, of the form:

$$(1+c^{-2}z^2)^{-1}\left[1+[\Delta\sigma\ \ln(R^{-1})]^{-2}(z\ -\ 2\pi\ \Delta\sigma\ n\ )^2\right]^{-1}. \qquad 4.3.9$$

The terms in this equation have simple poles at Z_F and Z_{E_n} associated with the filter and the etalon respectively. These poles are found at:

$$Z_F\ =\ \pm\ ic\ , \qquad 4.3.10a$$

$$Z_{E_n}\ =\ \pm\ i\ [\ \Delta\sigma\ \ln(R^{-1})\ ]\ +\ 2\pi\ \Delta\sigma\ n\ ,$$

$$n\ =\ 0,\ \pm\ 1,\ \pm\ 2,\ ....\ . \qquad 4.3.10b$$

The principal value of the integral containing the terms associated with the poles, i.e., W_n of Equation 4.3.8), is given by the theory of residues to be equal to $\pi\ i$ times the sum of the residues associated with the poles. Note that the sense of the integration path must be accounted for in the summation. The value of the residues associated with the poles of Equation 4.3.10) are:

$$K_F\ =\ [\ \Delta\sigma\ \ln(R^{-1})]^2\ (2\pi)^{-1}\left\{\ \pm\ 2ic^{-1}\left[\ (2\pi\ \Delta\sigma\ n)^2\ -\ c^2\right.\right.$$

$$\left.+\ [\ \Delta\sigma\ \ln(R^{-1})]^2\right]\ +\ 8\ \pi\ \Delta\sigma\ n\ \bigg\}^{-1}, \qquad 4.3.11a$$

$$K_{E_n}\ =\ [\ c^2\ \Delta\sigma\ \ln(R^{-1})]\ (2\pi)^{-1}\left\{\ \pm\ 2\ i\ \left[(2\pi\ \Delta\sigma\ n)^2\right.\right.$$

$$\left.-\ [\ \Delta\sigma\ \ln(R^{-1})\ ]^2\ +\ c^2\right]\ -\ 8\ \pi\ \Delta\sigma^2\ n\ \ln(R^{-1})\ \bigg\}^{-1}. \qquad 4.3.11b$$

The value of W_n of Equation 4.3.8) can be explicitly expressed as:

$$W_n\ =\ 2^{-1}\ c\ [\ \Delta\sigma\ \ln(R^{-1})]^2$$

$$\times\left\{\left[\ (2\pi\ \Delta\sigma\ n)^2\ -\ c^2\ +\ [\ \Delta\sigma\ \ln(R^{-1})\ ]^2\right]\right.$$

$$\times\left[\left(\ (2\pi\ \Delta\sigma\ n)^2\ -\ c^2\ +\ [\ \Delta\sigma\ \ln(R^{-1})\ ]^2\right)^2\ +\ (4\pi\ \Delta\sigma\ n\ c\ )^2\right]^{-1}\right\}$$

$$+\ 2^{-1}\ c^2\ [\ \Delta\sigma\ \ln(R^{-1})\ ]$$

$$\times\left\{\left[\ (2\pi\ \Delta\sigma\ n)^2\ +\ c^2\ -\ [\ \Delta\sigma\ \ln(R^{-1})\ ]^2\right]\right.$$

$$\times\left[\left(\ (2\pi\ \Delta\sigma\ n)^2\ +\ c^2\ -\ [\ \Delta\sigma\ \ln(R^{-1})\ ]^2\right)^2\right.$$

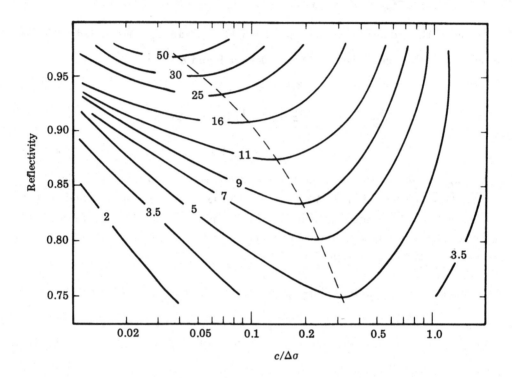

FIGURE 4.10. Contours of the parasitic light for the single-etalon filter combination as a function of the ratio of the filter width to the etalon free spectral range.

$$+ \left[4\pi \; \Delta\sigma^2 \; n \; \ln(R^{-1}) \;]^2 \right]^{-1} \bigg\} . \qquad \qquad 4.3.12$$

In the limit of $R \rightarrow 0$, or absence of the etalon, it can be found that:

$$\lim_{R \rightarrow 0} W_n = c \; 2^{-1} . \qquad \qquad 4.3.13$$

This result is the same as that which would be obtained for the integration of the filter function of Equation 4.3.7), i.e.,

$$(2 \; \pi)^{-1} \int_{-\infty}^{\infty} [\; 1 \; + \; c^{-2} x^2]^{-1} \; dx \; = \; c \; 2^{-1} . \qquad \qquad 4.3.14$$

On the other hand, when the value of c is made very large with respect to $\Delta\sigma$, then Equation 4.3.12) reduces to $2^{-1} \, \Delta\sigma \, \ln(R^{-1})$ which, when the proper constants [cf. Equation 4.3.6)] are used, is equal to $\Delta\sigma \; (1-R)(1+R)^{-1}$. The latter value is the response of the etalon to a continuous source for one order, as described in Chapter 2, Section 2.

The results of Equation 4.3.8) are shown in Figure 4.10 as contours of $c \times W_{n=0} [2 \Delta\sigma (\sum_{n=-\infty}^{n=\infty} W_n - W_{n=0})]^{-1}$ as a function of the reflectivity and the c parameter normalized to the free spectral range, $\Delta\sigma$. The values of the contours are measures of the useful radiation transmitted by the filter. The maximum for each value of the reflectivity is joined by the dashed line. Since the abscissa is, in terms of the filter FWHH $\delta_{1/2}$, equal to $\pi \, \delta_{1/2} \, \Delta\sigma^{-1}$, then the dashed line represents the FWHH of the etalon function normalized to the free spectral range times π. This answer is not surprising, since the filter Lorentzian shape matches the etalon shape. The example of the filter-etalon combination presented above provides the necessary tools to solve for multi-etalon filter devices.

As discussed earlier, after Equation 4.3.6), it is possible to associate the Lorentz profiles with the useful light when $n = 0$, and those when $n \neq 0$ with the parasitic light. In the same manner, the W_n of Equation 4.3.8), which are the areas under the Lorentzian profiles as modified by the filter, provide a quantitative measure of the contribution of each profile to the final result. McNutt (1965) considers each residue to be an equivalent width which provides an intuitive understanding of the behavior of the instrument. When the equivalent width of the useful radiation is compared with that of the parasitic light, a quantitative estimate of the 'parasitic' peaks, or ghosts, is obtained. Each of the equivalent widths allow the determination of the position and relative strength of the individual ghosts which, as discussed in the two-etalon section, can be a nuisance in hyperfine structure measurements.

The residues approach, shown in detail above, can be extended to any number of etalons in tandem with a filter. This method easily allows the addition of any number of extra etalons without much difficulty since each etalon contributes a new set of residues, such as those of Equation 4.3.11b) to the W_n term. The specific case for the PEPSIOS three-etalon device has been solved by McNutt (1965). The solution for minimum parasitic light, as a function of the ratios of the etalon spacers (relative to the first etalon), has been obtained by McNutt (1965) for the implicit assumption of large reflectivities. His results show that the ratios 1.0000 : 0.8831 : 0.7244 exhibit the best instrumental transmittance with high spectral purity. The selectivity thus attained by a PEPSIOS instrument is useful not only in absorption spectra measurements but in emission spectra measurements as well (Roesler and Mack, 1967). Therefore, a three-etalon device like the PEPSIOS with the McNutt spacer configuration preserves the high resolving power of a single-etalon Fabry-Perot spectrometer, while avoiding the overlap problems associated with it, yet also keeping the high luminosity associated with a Fabry-Perot. An example of a PEPSIOS device, with additional filtering of the radiation by gratings, is shown in Figure 4.11, and the transmission bands of the system elements are illustrated in Figure 4.12 (Burnett and Burnett, 1983).

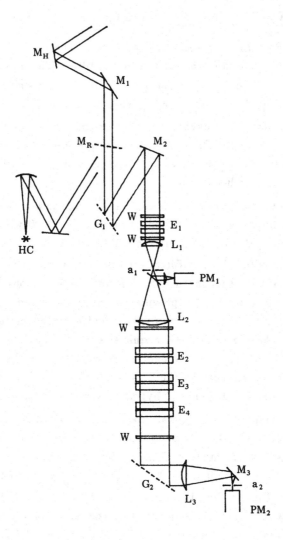

FIGURE 4.11. Modified PEPSIOS arrangement, where additional filtering by gratings (G_1, G_2) is employed. (Burnett and Burnett, 1983).

4. Multi-etalon special considerations

Surface defects in the etalons forming a multi-etalon configuration have the profile-broadening effects already discussed in Chapter 2, Section 3, as well as another effect peculiar to multi-etalon devices (Mack *et al.*, 1963; Roesler, 1969). This latter effect can be visualized when two ideal etalons are placed in series, but one of the etalons is slightly misaligned in a wedge fashion. Thus, restricting the discussion to an arbitrary wavenumber and to the direction of the optical axis of such a system, it is found there is one section of the wedged-etalon where the maximum transmission

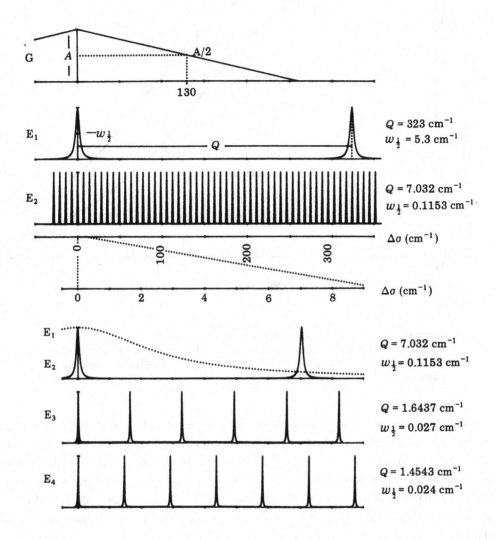

FIGURE 4.12. Transmission passbands of the PEPSIOS spectrometer elements shown in Figure 4.11. (Burnett and Burnett, 1983).

matches the maximum transmission of the perfect etalon. Everywhere else, the two-etalon transmissions will be mismatched to some extent. When the combination of the two etalons is considered to be the sum of pairs of elementary etalons, some of which are mismatched in maximum transmission and thus of lower transmission than the maximum possible, then the sum of these elementary etalons pairs leading to the final etalon system will have a lower transmission than would exist for two perfectly

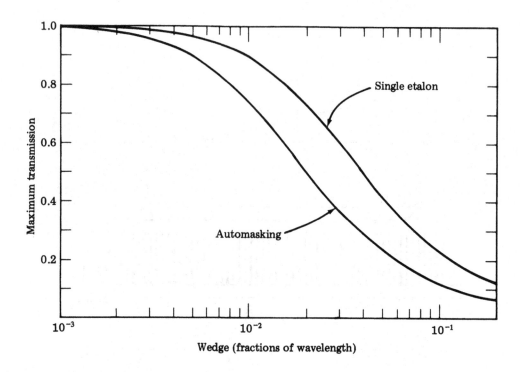

FIGURE 4.13. Automasking effects between two etalons when one of these etalons is misaligned. Otherwise the two etalons are perfect. The abscissa is given in fractions of a wavelength for the height of the wedge formed by this misalignment, and thus is independent of the etalon dimensions.

aligned ideal etalons. For real etalons, with arbitrary surface defects, the total transmission will behave as the wedged-etalon example in the sense that the combination can be considered to consist of the sum of pairs of elementary etalons mismatched in transmission with a distribution as defined by the surface defects. This property of multi-etalon devices is called 'automasking' of the etalons (Mack *et al.*, 1963; Roesler, 1974) and has rather adverse effects on the resultant quality of the final instrument.

   Figure 4.13 shows the automasking effects of the previously described wedged- and perfect ideal-etalon combination, as well as the transmission of the imperfect wedged-etalon by itself. For the purposes of this figure, it has been assumed that the mirror substrates are square, and the wedge is parallel to one of the edges of the etalons. Also a reflectivity of 0.9 has been used for both etalons, which are assumed to be operating at a central order of 47,000. The figure shows that the automasking effect is still important at small wedges, or their equivalent of surface defects. As would be expected, lower-reflectivity etalons will show a less-pronounced automasking effect. Using the example in the figure, it is shown easily that the transmission of the automasked etalon can be made equal to that of a single etalon when the reflective finesse of the automasked etalon is equal to one-half the reflective finesse of the single etalon. This finding is in agreement with the results of Roesler (1969) for the PEP-SIOS device, but with more realistic surface defects. Note that the price of lower reflectivity is lower spectral purity of the multi-etalon combination. Therefore, for a given performance level, the quality of the optical flats used in multi-etalon devices must be higher than that required for single-etalon applications.

As discussed near the end of Section 2.1, the formation of 'extra' etalons by the reflective surfaces can cause some difficulties in multi-etalon devices. In the PEP-SIOS instrument, these effects are minimized by tilting one of the etalons, usually the one with lowest resolving power, with respect to the optical axis, and thus throwing the reflections out of the field of view (Mack *et al.,* 1963).

When two, or more, etalons of unequal spacing are placed in series, they need not have equal areas for maximum flux transmission. As Roesler (1974) has discussed, coupling of unequal area and unequal spacer etalons with afocal systems can be as efficient as two equal area etalons in series, as long as the proper relationship among the etalon spacers, d_i, the etalon diameters, D_i, and the focal lengths, f_i (of a two-lens afocal system) is followed, i.e.:

$$d_1 d_2^{-1} = D_1^2 D_2^{-2} = f_1^2 f_2^{-2} \; . \qquad\qquad 4.4.1$$

Afocal coupling is not very useful with etalons with vernier spacing since $d_1 \simeq d_2$.

As would be expected, multiple etalons are not restricted to the series configuration studied here. The parallel configuration of etalons is discussed in Chapter 8, Section 3.

Chapter 5

Luminosity and resolution considerations

1. Fabry - Perot and other spectrometers

The Fabry-Perot spectrometer has been discussed, thus far, independently of other spectrometric devices. A comparison of the relative usefulness of a Fabry-Perot relative to other spectrometric devices has been made by Jacquinot (1954), on the basis of the flux transmitted at a given resolving power by each instrument.

The arguments used by Jacquinot (1954) to compare the Fabry-Perot with prism and grating spectrometers are given below. Consider both prism and grating spectrometers to have slit-angular widths α_i, angular height β_i, and angular dispersion D_i while they are operated at a resolving power $\mathbf{R}$ [cf. Equation 2.1.15)]. Ignoring diffraction effects, the maximum flux received by a prism spectrometer is:

$$\Phi_p = \tau \, \mathbf{I} \, \mathbf{A}_B \, \beta \, \alpha = \tau \, \mathbf{I} \, \mathbf{A} \, \mathbf{R}^{-1} \, \beta \, \lambda \, d\mu/d\lambda , \qquad 5.1.1$$

where $\mathbf{A}_B$ is the area of the beam. For the grating spectrometer the flux is given by:

$$\Phi_g = \tau \, \mathbf{I} \, \mathbf{A} \, \mathbf{R}^{-1} \, \beta \, 2 \sin\psi . \qquad 5.1.2$$

In the above two equations, $\mathbf{I}$ is the irradiance of the (monochromatic) source, τ includes transmission losses in the instrument, ψ is the blaze angle of the grating and $\mathbf{A}$ is the area of the base of the prism or the area of the grating. For the (single aperture) Fabry-Perot spectrometer the flux is:

$$\Phi_{fp} = \tau\, \mathbf{I}\, \mathbf{A}\, \pi^2\, [\, 2.8\, \mathbf{R}\,]^{-1}\, . \qquad\qquad 5.1.3$$

In order to derive Equation 5.1.3), it has been assumed that the relation $\mathbf{R} = 0.7\, \mathbf{R_0}$ produces the best compromise of operation for a Fabry-Perot spectrometer [cf. Figure 2.11].

The ratio of the flux received by the prism spectrometer to that received by the grating spectrometer is (Jacquinot, 1954):

$$P = \Phi_p\, [\, \Phi_g\,]^{-1} = \lambda\, d\mu/d\lambda\, [\, 2\sin\psi\,]^{-1}\, . \qquad\qquad 5.1.4a$$

For a nominal blaze angle $\psi \approx 30°$, the ratio P becomes:

$$P = \lambda\, d\mu/d\lambda\, . \qquad\qquad 5.1.4b$$

As shown by Jacquinot, the value of the right side of Equation 5.1.4b) is smaller than 0.2, and typically 0.01 to 0.02. Thus, under similar circumstances, the grating instrument is more luminous than the prism instrument by at least a factor of 5, and typically closer to a factor of 50.

On the other hand, the ratio of the fluxes of the Fabry-Perot spectrometer to the grating spectrometer is (for a 30° grating blaze angle):

$$\mathbf{G} = \Phi_{fp}\, [\, \Phi_g\,]^{-1} \simeq 3.4\, \beta^{-1}\, . \qquad\qquad 5.1.5$$

Since, for a very luminous grating spectrometer, β seldom exceeds a value of 0.1, Equation 5.1.5) shows the Fabry-Perot spectrometer to be the most efficient of these two spectrometers by a factor of at least 30. This rather large factor is the reason why the Fabry-Perot is attractive for high-resolution work, because the sources used for this type of investigation are generally weak. Note that the results given in both Equations 5.1.4b) and 5.1.5) are independent of the resolving power chosen to operate the spectrometers.

As discussed by Jacquinot (1954), the above comparisons with the Fabry-Perot are given for a single-etalon device with its inherently poor selectivity. The use of multiple-etalon devices, as described in Chapter 4, can increase the selectivity to the level of the grating spectrometer, yet the losses incurred are small enough so that the (multiple-etalon) Fabry-Perot is still a more efficient spectrometer than the prism or grating devices of equal $\mathbf{A}$. The same argument used for the Fabry-Perot can be made for the simple Michelson (1881, 1882) interferometer, and similar results are obtained.

The result shown in Equation 5.1.3), namely, that the flux is inversely proportional to the resolving power used, leads to the existence of a maximum flux which an etalon of arbitrary area can collect at a given resolving power. Following Jacquinot (1960), the luminosity L can be defined as:

$$\mathbf{L} = \Phi_{fp}\, \mathbf{I}^{-1} = \mathbf{A}\, \Omega\, \tau\, . \qquad\qquad 5.1.6a$$

Luminosity is then the flux delivered to the detector when the instrument is examining a unit irradiance source. Replacing the solid angle Ω by its definition given in Equation 2.4.2b), the luminosity becomes:

$$L = 2 \pi \, A \, \tau \, f \, / \sigma .$$ 5.1.6b

The resolving power, as given in Equation 2.1.15), can be rewritten as follows:

$$R = \sigma \, / \delta\sigma = \sigma \, / \, (\, a \oplus f \,) .$$ 5.1.7

The product of the luminosity and the resolving power is then given by:

$$L \times R = A \, \Omega \, \tau \, R$$

$$= 2 \pi \, A \, \tau \, f \, (\, a \oplus f \,)^{-1} .$$ 5.1.8a

Since τ and $f \, (\, a \oplus f \,)^{-1}$ have limiting values of unity, the limiting value of Equation 5.1.8a) becomes:

$$\lim \, L \times R = 2 \pi \, A .$$ 5.1.8b

For practical cases, where τ and $f \, (\, a \oplus f \,)^{-1}$ have values less than unity, the luminosity-resolving power product (normally referred to as the luminosity-resolution-product or LRP) is:

$$L \times R \leqslant 2 \pi \, A .$$ 5.1.8c

The maximizing of the product $\tau \times f \, (\, a \oplus f \,)^{-1}$, or in more general terms, $\tau \times f \, (\, s \oplus e \oplus f \,)^{-1}$, forms the basis of the LRP criterion, which has been discussed in Chapter 2, Section 4 (cf. Figure 2.11). Note that the normalized widths a^{*} and f^{*} can also be used in the ratios. Once the $L \times R$ product has been maximized, the only free parameter left is the area of the etalon mirrors. In principle, this area can be made as large as desired although, in practice, fabrication difficulties of high-quality optical flats used as mirror substrates limit this area to about 500 cm^{2}, or approximately 25 cm diameter apertures.

As Jacquinot (1954, 1960) has indicated, the limitation of the Fabry-Perot is the finite angular range at which a given state of interference is useful, and which, in turn, limits the size of the aperture that may be used and, consequently, the received flux. The above is a restatement of the LRP criterion. However, as reported earlier by Jacquinot and Dufour (1948), more than one angular aperture may be used, effectively increasing the total flux received by the number of apertures. Jacquinot (1954) later proved, as it is implicitly stated in Equation 5.1.3), that the flux received does not depend on the particular fringe employed, thus placing the use of multiple apertures on a sound theoretical basis. Although, as discussed earlier, the number of annuli, or apertures, that may be used is equal to the operating order of the etalon, in practice, the number of apertures used is limited by fabrication tolerances to about 10 for mechanical devices (Sipler and Biondi, 1978; Meaburn, 1976;

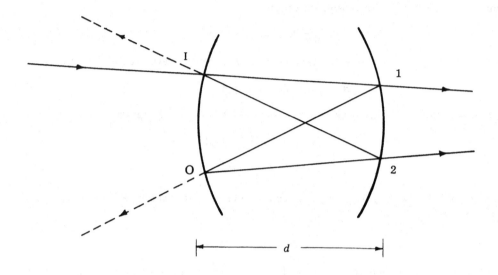

FIGURE 5.1. Schematic illustration of a confocal spherical-etalon.

Okano *et al.*, 1980), and about 50 for photographic masks (Dupoisot and Prat, 1979). As given in the nearly identical etalon section of Chapter 4, the TESS device (Hernandez *et al.*, 1981) luminosity can be increased by large factors without reducing the resolving power, and indeed, the $L \times R$ product increases as the resolving power increases. However, as discussed earlier, this occurs at the cost of the loss of the spatial information. Another approach to increase the flux measured by an etalon was reported by Ascoli-Bartoli *et al.* (1967). These authors recovered the (normally lost) etalon reflected light by means of properly shaped mirrors. Gains of near 5, over a simple Fabry-Perot, were obtained with this method.

2. The Connes spherical Fabry - Perot

In 1956, Connes introduced practical methods to obtain ray-path lengths which are independent of the angle-of-incidence for various interference devices. Among these, a Fabry-Perot device consisting of two (identical) spherical mirrors was included. For confocal operation, that is, when the separation between the mirrors is equal to the common radius of curvature of the mirrors and, using paraxial optics, each mirror can be shown to image the other upon itself. In this idealized picture, shown in schematic form in Figure 5.1, a ray entering at point I in the figure will fall back upon itself after the first four reflections. This cycle will repeat itself, giving rise to an infinite number of coincident outgoing rays at point 1. In comparison, the plane Fabry-Perot has only parallel outgoing rays (cf. Figure 2.1). Looking at Figure 5.1 in more detail, it is found that outgoing rays also exist at points 2, O and I. The latter two are the equivalent of the reflected rays in a plane Fabry-Perot of Figure 2.1.

Considering only the light beams transmitted by a confocal etalon at points 1 and 2, it is found that the latter lag those at point 1 by the equivalent phase given by the two extra reflections. Therefore, following Hercher's (1968) notation, the rays exiting at point 1 after undergoing 4m reflections will be called Type 1, while those rays exiting at point 2 after (4m + 2) reflections will be called Type 2. In both cases m = 0 ,1, 2, 3, 4, ...

Using similar arguments to those in Equation 2.1.3) for the plane Fabry-Perot, the transmitted intensities of Type 1 rays are:

$$Y_{t_1}(\delta_o) = \tau^2 (1 + R^4 - 2 R^2 \cos\delta_o)^{-1} . \qquad 5.2.1a$$

For Type 2 rays, the transmitted intensity is found to be equal to:

$$Y_{t_2}(\delta_o) = (\tau R)^2 (1 + R^4 - 2 R^2 \cos\delta_o)^{-1} . \qquad 5.2.2a$$

In both of the previous equations the phase δ_o is defined as follows:

$$\delta_o = 2 \pi \nu 4 \mu d c^{-1} = 2 \pi \sigma 4 \mu d . \qquad 5.2.3a$$

In the above definition, the symbols have the same meaning as in Chapter 2, Section 1, and d is still the separation between the two (spherical) mirrors. Equation 5.2.3a) shows that, when compared with Equation 2.1.1a), for a given separation between the mirrors for plane and spherical etalons, the optical gap of the spherical etalon is twice as large as that for the plane etalon. Using the usual definition of $\tau = 1 - R - A$, where A is the absorption and/or scattering of the etalon coatings, Equations 5.2.1a) and 5.2.2a) are transformed into:

$$Y_{t_1}(\delta_o) = (1 - R)^2 [1 - A (1 - R)^{-1}]^2$$

$$\times (1+ R^4 - 2 R^2 \cos\delta_o)^{-1}$$

$$= [1 - A (1 - R)^{-1}]^2 (1 + R)^{-2}$$

$$\times [1 + 4R^2 (1-R^2)^{-2} \sin^2(\delta_o 2^{-1})]^{-1} , \qquad 5.2.1b$$

$$Y_{t_2}(\delta_o) = R^2 (1 - R)^2 [1 - A (1 - R)^{-1}]^2$$

$$\times (1 + R^4 - 2 R^2 \cos\delta_o)^{-1}$$

$$= [1 - A (1 - R)^{-1}]^2 (1 + R)^{-2} R^2$$

$$\times \left[\, 1 + 4\, R^{2}(1-R^{2})^{-2}\, \sin^{2}(\delta_{o}\, 2^{-1})\, \right]^{-1} \, . \qquad \text{5.2.2b}$$

The properties of both Type 1 and Type 2 transmitted beams are alike those found earlier for the plane etalon, namely maxima at $\delta_{o} = 2\,\pi\,k$, and minima at $\delta_{o} = 2\,\pi\,(k\,+\,1/2)$, where $k = 1, 2, 3,....$. The transmission at the maxima for $0 \leqslant R \leqslant 1$, is therefore equal to:

$$Y_{t_{1}}(\delta_{o\,\max}) = [1 - A\,(1-R)^{-1}]^{2}(1+R)^{-2} \leqslant 1\,, \qquad \text{5.2.4}$$

$$Y_{t_{2}}(\delta_{o\,\max}) = R^{2}[1 - A\,(1-R)^{-1}]^{2}(1+R)^{-2} \leqslant 0.25\,. \qquad \text{5.2.5}$$

Note that for the limiting case when $A \to 0$ and $R \to 1$, the values of both Equations 5.2.4) and 5.2.5) asymptotically approach a value of 0.25, or equal intensities for both paths. The contrast C is given in the same manner as Equation 2.1.14), namely:

$$C_{1,2} = (1+R^{2})^{2}\,(1-R^{2})^{-2}\,. \qquad \text{5.2.6}$$

A cursory look at Equations 5.2.4) through 5.2.6), shows the maximum transmission and contrast of the spherical Fabry-Perot to be lower than those of a plane Fabry-Perot of the same reflectivity, since this reflectivity enters as the second power in the spherical etalon rather than as a first power as in the plane etalon case [cf. Equation 2.1.14)]. Thus it appears it would be necessary to redefine the finesse N_{R} of Equation 2.1.13a) and the **F** parameter of Equation 2.1.13c) in terms of the reflectivity to the second power.

Equations 5.2.1) and 5.2.2) can be recast into more familiar form when the following identity is applied:

$$(\, 1 + R^{4} - 2\,R^{2}\,\cos\delta_{o}\,)$$

$$= (1-R)^{4}\,[\, 1 + \mathbf{F}\,\sin^{2}(\delta_{o}\,4^{-1})\,]\,[\, 1 + \mathbf{F}\,\cos^{2}(\delta_{o}\,4^{-1})\,]\,. \qquad \text{5.2.7}$$

In the above expression **F** has been defined earlier as $4R\,(1-R)^{2}$ in Equation 2.1.13c), and thus:

$$Y_{t_{1}}(\delta_{o}) = [\, 1 - A\,(\, 1 - R\,)^{-1}\,]^{2}\,(\, 1 - R\,)^{-2}$$

$$\times \left\{\, [1 + \mathbf{F}\,\sin^{2}(\delta_{o}\,4^{-1})\,]\,[1 + \mathbf{F}\,\cos^{2}(\delta_{o}\,4^{-1})\,]\,\right\}^{-1}\,, \qquad \text{5.2.1c}$$

$$Y_{t_{2}}(\delta_{o}) = [\, 1 - A\,(\, 1 - R\,)^{-1}\,]^{2}\,(\, 1 - R\,)^{-2}\,R^{2}$$

$$\times \left\{\, [1 + \mathbf{F}\,\sin^{2}(\delta_{o}\,4^{-1})\,]\,[1 + \mathbf{F}\,\cos^{2}(\delta_{o}\,4^{-1})\,]\,\right\}^{-1}\,. \qquad \text{5.2.2c}$$

When a comparison is made between δ_{o} of Equation 5.2.3a) and δ of Equation

2.1.1b) for the central order of a plane Fabry - Perot etalon, it is found that:

$$\delta = \delta_0 \ 2^{-1} = 2 \ \pi \ \sigma \ 2 \ \mu \ d \ .$$ 5.2.3b

Therefore, replacing the value of δ_0 with its equivalent in δ, Equations 5.2.1c) and 5.2.2c) become:

$$Y'_{t_1}(\delta) = [\ 1 - A \ (\ 1 - R \ )^{-1}\ ]^2 \ (\ 1 - R \ )^{-2}$$

$$\times \left\{ [1 + \mathbf{F} \ \sin^2(\delta \ 2^{-1})\ ] \ [1 + \mathbf{F} \ \cos^2(\delta \ 2^{-1})\ ] \right\}^{-1} \ ,$$ 5.2.1d

$$Y_{t_2}(\delta) = [\ 1 - A \ (\ 1 - R \ )^{-1}\ ]^2 \ (\ 1 - R \ )^{-2} \ R^2$$

$$\times \left\{ [1 + \mathbf{F} \ \sin^2(\delta \ 2^{-1})\ ] \ [1 + \mathbf{F} \ \cos^2(\delta \ 2^{-1})\ ] \right\}^{-1} \ .$$ 5.2.2d

The results in these two equations show, in the same notation used earlier in Chapter 2, the basic properties of the confocal etalon relative to the plane Fabry-Perot etalon in the same form as given in Equation 2.1.7d). The obvious difference is found to be the double response of the confocal interferometer given by the $\sin^2$ and $\cos^2$ terms in Equations 5.2.1d) and 5.2.2d), when compared to the single response found for the plane etalon in Equation 2.1.7d). Therefore, the use of expressions of the confocal etalon in the form given in Equations 5.2.1d) and 5.2.2d) provides internal consistency with the plane Fabry-Perot notation employed in Chapter 2. The main advantage obtained with this consistent notation is the removal of the ambiguity in the definitions of a device-dependent free spectral range, finesse and $\mathbf{F}$ parameter.

For the paraxial approximation discussed thus far, Equations 5.2.4) and 5.2.5) are applicable for all incidence angles of the incoming rays. This constant state of interference over a large solid angle then gives rise to a $\mathbf{L} \times \mathbf{R}$ product that increases with resolving power.

In his original reports, Connes (1956, 1958) removed the Type 2 ray bundle by simply masking, or blocking, the lower half of the confocal cavity, as indicated in Figure 5.2. The reports of Jackson (1961), Hercher (1968), Johnson (1968) and Bradley and Mitchell (1968) treat the general case of the unmasked spherical etalon coated with semi-transparent mirrors, where both Type 1 and Type 2 rays occur. When the spherical etalon is used with these semi-transparent coatings, with the consequent presence of both Type 1 and Type 2 beams, automatically a two-beam interferometer has been formed, since the Type 2 rays are delayed with respect to Type 1 rays by the path given by two internal reflections in the cavity. When these two beams leave the cavity at a small angle from the axis, they will form an additional interference pattern made up of equally spaced straight fringes whose separation is determined by the angle at which the beams are brought to focus (Hercher, 1968).

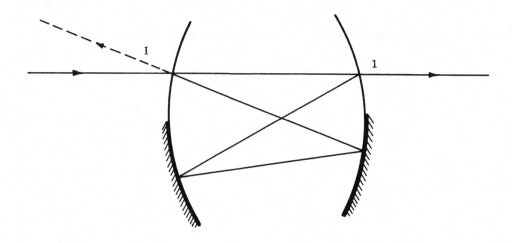

FIGURE 5.2. The masked confocal-etalon used by Connes (1956).

This two-beam interference effect is easily observable when the source is highly coherent, such as a laser.

In the previous discussion, Type 1 and Type 2 beams have been treated separately, which does not allow the treatment of other effects, such as the two-beam interferometer mentioned earlier. Following Johnson (1968), assume an off-axis source, S_1, which radiates only over a narrow forward lobe, illuminating a confocal etalon as shown in Figure 5.3. Lens L_1 focuses the beams at plane P_o, and these beams form an angle ψ with the axis as indicated. As seen in the figure, the two beams have a region of overlap at the back focal plane of L_1, which is then examined by eyepiece L_2. Note that the image at P_o appears to an observer as if there were sources at S_1 and S_2. Returning to the overlapping region, the amplitude of the beams add and the resultant intensity distribution at this plane at a distance x away and perpendicular to the axis is given by (Johnson, 1968):

$$Y_t(\delta,x,\psi,\sigma) = [1-A(1-R)^{-1}]^2 [1+F \sin^2(\delta\,2^{-1}+2\pi\sigma x\,\sin\psi)]$$

$$\times \left\{ [1+F \sin^2(\delta\,2^{-1})]\,[1+F \cos^2(\delta\,2^{-1})] \right\}^{-1}. \qquad 5.2.8a$$

The result given above shows the two-beam interference to appear as a modulation of the output in the form of straight lines in a plane perpendicular to the axis and the figure plane, just as in Fresnel's bi-prism experiment (1826). The position of these fringes with respect to x = 0, or the axis, and their brightness, is determined by

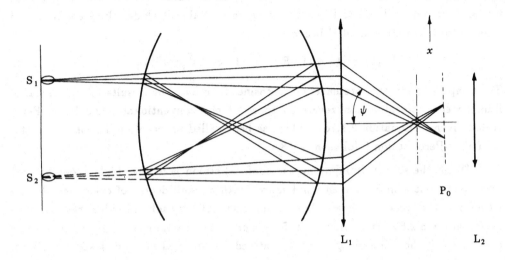

FIGURE 5.3. Confocal-etalon illumination used in the description of two-beam and transmission effects. After Johnson (1968).

$\delta\ 2^{-1}$. Thus, as δ is monotonically changed, the two-beam-pattern fringes shift smoothly and their maximum intensity changes drastically (Johnson, 1968). The maxima are found at $\delta = k\pi$, or the same conditions found in Equations 5.2.1) and 5.2.2), but the value of the maxima are different. For even values of k, or $k = 2n$, $\delta = 2\pi n$ and Equation 5.2.8a) becomes:

$$Y_{even}\ (\ x\ ,\ \psi\ ) = (\ 1 + \mathbf{F}\ )^{-1}\ [\ 1 - A\ (\ 1 - R\ )^{-1}\ ]^2$$

$$\times\ [\ 1 + \mathbf{F}\ \sin^2(2\pi\ x\ \sigma\ \sin\psi\ )\ ]\ . \qquad 5.2.9a$$

For odd values of k, or $\delta = 2\pi\ (n + 1)$, the result is:

$$Y_{odd}\ (\ x\ ,\ \psi\ ) = (\ 1 + \mathbf{F}\ )^{-1}\ [\ 1 - A\ (\ 1 - R\ )^{-1}\ ]^2$$

$$\times\ [\ 1 + \mathbf{F}\ \cos^2(2\pi\ x\ \sigma\ \sin\psi\ )\ ]\ . \qquad 5.2.10a$$

The value of ψ determines the maximum intensities and the apparent spacing between adjacent two-beam fringes. In the limit of $\psi \to 0$, the maxima of Equations 5.2.9a) and 5.2.10a) are:

$$Y_{even}\ (\text{max}) = [\ 1 - A\ (\ 1 - R\ )^{-1}\ ]^2\ (\ 1 + \mathbf{F}\ )^{-1}\ , \qquad 5.2.9b$$

$$Y_{odd}\ (\text{max}) = [\ 1 - A\ (\ 1 - R\ )^{-1}\ ]^2\ . \qquad 5.2.10b$$

The above two equations demonstrate that, for coaxial illumination with a highly directional source, such as a laser, the confocal etalon favors the odd values of k, and for large values of $\mathbf{F}$, tends to eliminate the even values. Under these conditions of illumination, Equation 5.2.8.a) becomes:

$$Y_t(\delta) = [1 - A(1-R)^{-1}]^2 [1 + \mathbf{F} \cos^2(\delta\, 2^{-1})]^{-1} . \qquad 5.2.8b$$

Thus, apart from the presence of a cosine function, the results of a coaxially-illuminated confocal etalon revert to those of the conventional plane Fabry-Perot etalon given in Equation 2.1.7d). This has been called proper illumination of a confocal Fabry-Perot etalon (Johnson, 1968).

   When the source illuminating the confocal etalon is neither highly coherent nor directional, but a more conventional source with a small degree of coherence along a wavefront, the treatment is as follows (Johnson, 1968): each individual ray, as given in Equation 5.2.8a) is treated as before, and then the intensities are subsequently combined. If the radiation pattern is confined to a conical volume of apex ψ_o, symmetric about the axis, and $I(\psi)$ is the intensity per unit solid angle within this cone, then the net intensity at a point x in the back focal plane of L_1 in Figure 5.3 is found to be:

$$Y_t(\delta) = [1 - A(1-R)^{-1}]^2 \left\{ [1 + \mathbf{F}\sin^2(\delta\,2^{-1})]\,[1 + \mathbf{F}\cos^2(\delta\,2^{-1})] \right\}^{-1}$$

$$\times \int_0^{2\pi} d\phi \int_0^{\psi_o} I(\psi)\,[1 + \mathbf{F}\sin^2(\delta\,2^{-1} + 2\pi\,z\,\sigma\,\sin\psi)]\,d\psi . \qquad 5.2.11a$$

The integration over ψ causes the sine term in the integral to oscillate very rapidly, since $2\pi\,z\,\sigma$ is a very large number. Therefore, the value of the integral is given by the average value (Johnson, 1968), or:

$$Y_t(\delta) = [1 - A(1-R)^{-1}]^2 [1 + \mathbf{F}\,2^{-1}]$$

$$\times \left\{ [1 + \mathbf{F}\sin^2(\delta\,2^{-1})]\,[1 + \mathbf{F}\cos^2(\delta\,2^{-1})] \right\}^{-1}$$

$$= [1 - A(1-R)^{-1}]^2 [1 + R^2]\,[1 - R]^2$$

$$\times [1 + R^4 - 2R^2 \cos(2\delta)]^{-1}$$

$$= [1 - A(1-R)^{-1}]^2 [1 - R]\,[1 + R]^{-1}$$

$$\times \left[1 + 2 \sum_{k=1}^{k=\infty} R^{2k} \cos(2k\,\delta) \right] . \qquad 5.2.11b$$

As would be expected, the two-beam-interferometer effects have disappeared in the above expression. Note that the value of $I(\psi)$, integrated over the range 0 to ψ_o, has been set to unity. The peak transmission of the confocal etalon illuminated with

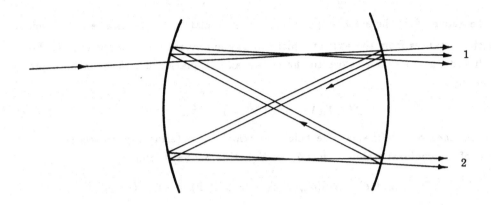

FIGURE 5.4. Spherical etalon showing the effects of spherical aberration on the ray paths. After Hercher (1968).

a nearly incoherent source is given by:

$$Y_t (\text{max}) = [1 - A (1-R)^{-1}]^2 [1 + \mathbf{F}\, 2^{-1}] [1 + \mathbf{F}\,]^{-1}$$

$$= [1 - A (1-R)^{-1}]^2 (1 + R^2) (1 + R)^{-2} \leqslant 1 \ . \qquad 5.2.12$$

Although the result of Equation 5.2.11b) could have been obtained by addition of Equations 5.2.4) and 5.2.5), the more detailed treatment presented thus far has provided a physical explanation for the ongoing process. Illumination of a confocal etalon with an incoherent light source has been called improper illumination (Johnson, 1968). Without the treatment employed above, the apparent ambiguity of a device-dependent free spectral range and finesse would have remained.

The ideal instrument so far described is useful in the sense it provides an understanding of the basic properties of the confocal etalon. The idealized paraxial picture presented earlier is limited by spherical aberration of the mirrors. When spherical aberration is taken into account, a typical ray does not fall back upon itself, but will follow a path such as that shown in Figure 5.4. If the angle that this ray makes to the axis is not too large, the ray will continue to intersect itself in the vicinity of a point, as illustrated in the figure. The position of the points at which such rays continue to intersect themselves determines the position of the fringe pattern (Connes, 1958; Hercher, 1968).

Because of spherical aberration, the four-reflection ray path differs from the paraxial ray path by an amount Δ (Connes, 1956, 1958; Jackson, 1961; Hercher, 1968; Johnson, 1968). For the general case of mirror separation $d + \epsilon$, where d is the mirror's radius of curvature and $\epsilon << d$, this difference in path is (Hercher, 1968):

$$\Delta = \rho_1^2 \, \rho_2^2 \, d^{-3} \, \cos 2\Theta + 2 \, \epsilon \, (\, \rho_1^2 + \rho_2^2 \,) d^{-2} \, . \qquad\qquad 5.2.13a$$

In the above, Θ is the skew angle the ray makes relative to the axis, and ρ_1 and ρ_2 are the distances from the axis at which the ray enters and departs the etalon. For a small and distant source, close to the etalon axis, Equation 5.2.13a) can be approximated by:

$$\Delta(\rho,d,\epsilon) \simeq \rho^4 \, d^{-3} + 4 \, \epsilon \, \rho^2 \, d^{-2} \, . \qquad\qquad 5.2.13b$$

The distance ρ is defined as the height at which an entering ray crosses the central plane of the confocal etalon. Thus δ_o of Equation 5.2.3a) becomes:

$$\delta_o \, (\sigma,\rho,d,\epsilon) = 2 \, \delta(\sigma,\rho,d,\epsilon) = 2\pi \, \sigma \, [\; \Delta(\rho,d,\epsilon) + 4(d+\epsilon) \;]$$

$$= 2\pi \, \sigma \, [\; \rho^4 d^{-3} + 4\epsilon\rho^2 d^{-2} + 4(d+\epsilon) \;] \, . \qquad\qquad 5.2.3c$$

Bright fringes will be formed when δ_o is a multiple integer of 2π, or $\delta_o = 2\pi \, j$. If it is further assumed that $4\sigma \, (d+\epsilon)$ is also an integer, or exact order of interference at the axis, then $\Delta(\rho,d,\epsilon)$ must also satisfy an integer value m in order to obtain bright fringes in the central plane of the etalon:

$$\Delta(\rho,d,\epsilon) = \rho^4 d^{-3} + 4\epsilon\rho^2 d^{-2} = m \, \sigma^{-1} \, . \qquad\qquad 5.2.14$$

Thus, fringes associated with a given value of m will have a radius ρ_m given by:

$$\rho_m = [\; -2\epsilon d \pm (4\epsilon^2 d^2 + m \sigma^{-1} d^3)^{1/2} \;]^{1/2} \, . \qquad\qquad 5.2.15a$$

The solutions for ρ_m are such that for $\epsilon > 0$, ρ_m is single-valued and $m > 0$. For $\epsilon < 0$, ρ_m is double-valued for $m \leqslant 0$, and single-valued for $m > 0$ (Hercher, 1968). For values of $| \; \epsilon \; | > 0$, the system is called a defocused etalon (Bradley and Mitchell, 1968); however, for the present purposes only the near-confocal etalon, i.e., $\epsilon \simeq 0$, will be considered. Then, for the near-confocal etalon, the radius of the fringes is:

$$\rho_m = [\; (m-\varsigma) \, \sigma^{-1} d^3 \;]^{1/4} \, . \qquad\qquad 5.2.15b$$

For the sake of generality, ς in the previous expression is defined as $\varsigma < 1$.

Therefore, the effects of spherical aberration are limited diameter fringes, rather than the constant state of interference of the ideal confocal etalon derived with paraxial methods. As such, a reduced $\mathbf{L} \times \mathbf{R}$ product is obtained, since now there exist fringes of limited width. This is graphically shown in Figure 5.5 as a function of the normalized distance, $\rho d^{-3/4}$, with $\epsilon = 0$. The non-linearity of the graph is noticeable, since, as noted in Equation 5.2.3c) the phase is proportional to ρ^4. As the figure indicates, there must exist an aperture at which the $\mathbf{L} \times \mathbf{R}$ product is at a maximum, and this is given (Connes, 1956, 1958) (by similarity with a plane etalon) by the LRP criterion to be for an aperture, in this case etalon aperture, equal in

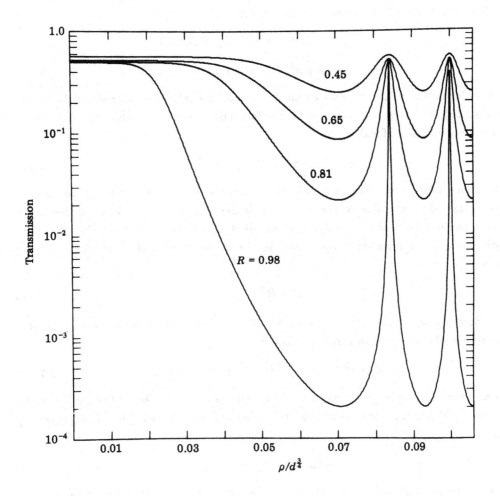

FIGURE 5.5. Transmission of a confocal etalon as a function of the normalized distance and various reflectivity values.

radius to the HWHH of the fringe and usually called ρ_o.

The use of strict confocality, i.e., $\epsilon = 0$, provides a fixed phase difference between marginal and axial rays for a given aperture radius. This difference in phase can be deduced from Equation 5.2.3c) to be equal to:

$$\Delta\delta_o(\sigma,\rho,d,0) = \delta_o(\sigma,\rho,d,0) - \delta_o(\sigma,0,d,0)$$

$$= 2\pi\sigma\rho^4 d^{-3}. \qquad \text{5.2.3d}$$

This constant phase difference is alike the phase difference between the axial and edge rays in the aperture of a Jacquinot and Dufour (1948) plane Fabry-Perot spectrometer. When the spherical etalon is used away from confocality, or $\epsilon \neq 0$, there is phase difference which is dep ndent on ϵ, or:

$$\Delta\delta_0 = \delta_0 (\sigma,\rho,d,\epsilon) - \delta_0 (\sigma,0,d,\epsilon)$$

$$= 2\pi\sigma (\rho^4 d^{-3} + 4\epsilon\rho^2 d^{-2}) . \qquad 5.2.3e$$

This ϵ-dependent phase appears as an undesirable variable-size aperture. But Equation 5.2.3e) indicates that as long as the following inequality holds, the instrument will behave like a confocal etalon:

$$\rho^4 d^{-3} >> | 4\epsilon\rho^2 d^{-2} | . \qquad 5.2.16a$$

The above expression shows that slight departures from confocality are acceptable. The actual physical value of these acceptable departures is dependent on the curvature of the mirrors and the resolving power used. First, as a reference, the size of the optimum aperture for a strict confocal etalon is made equal to the width of the etalon function, or:

$$\rho_0^4 d^{-3} = (1 - R^2) (2\pi\sigma R)^{-1} . \qquad 5.2.17a$$

Introducing this result into Equation 5.2.16a), it becomes possible to estimate the magnitude of the departures from confocality:

$$(1-R^2)^{1/2} (2\pi\sigma R)^{-1/2} 4^{-1} d^{1/2} >> | \epsilon | . \qquad 5.2.16b$$

As an example, assume a reflectivity of the mirrors of $R = 0.95$, in the visible range, i.e., $\sigma = 20\,000\,K$, from which the following can be derived (for d expressed in cm.):

$$2.2 \times 10^{-4} d^{1/2} >> | \epsilon | . \qquad 5.2.16c$$

A more useful approach is to express the left side of Equation 5.2.16b) in terms of wavelengths (λ):

$$4.5 \lambda d^{1/2} >> | \epsilon | . \qquad 5.2.16d$$

For a lower value of the reflectivity, namely $R = 0.90$, the result in terms of wavelengths is:

$$.26 \lambda d^{1/2} >> | \epsilon | . \qquad 5.2.16e$$

The above expressions show the possible tolerances of ϵ required to obtain near confocality. Note in particular, that, as the radius of curvature d is increased, the value of the tolerance increases. The existence of this tolerance indicates that there are limits on mechanical scanning of a near-confocal etalon. Assume a true, or properly illuminated, free spectral range is to be scanned, or $| \epsilon | = \lambda 2^{-1}$. If it is arbitrarily presumed that the $>>$ symbol means at least a factor of 50, then Equations

5.2.16d) and 5.2.16e) can be transformed into:

$$(4.5 \, \lambda \, d^{1/2}) = 25 \, \lambda \,, \qquad\qquad 5.2.16f$$

$$(26 \, \lambda \, d^{1/2}) = 25 \, \lambda \,. \qquad\qquad 5.2.16g$$

Solving for d, it is found that the mirror radii must be at least 31 cm and 1 cm for the $R = 0.95$ and the $R = 0.90$ spherical etalons respectively. The meaning of these tolerance limits can be understood in terms of the effects scanning the etalon will produce. If mechanical scanning is to be performed at etalon separations smaller than those found in the examples above, the etalon is no longer confocal; that is, the size of the aperture is now variable and dependent on the value of ϵ. Therefore, if confocality is to be preserved, another means of scanning must be adopted. Index of refraction scanning would be a suitable choice since, as has been mentioned previously, with this method the wavelength of the radiation is changed. In the following text it is assumed that near confocality of an etalon has been attained, or

$$\left| \, 4\epsilon\rho^2 d^{-2} \, \right| \, <<< \, \rho^4 \, d^{-3}.$$

The general expression for a confocal ($\epsilon = 0$) etalon with a finite aperture is found by the same methods used in Chapter 2, Section 4, for the plane Fabry-Perot etalon. For convenience, the Fourier representation of Equation 5.2.11b) will be used here. For some arbitrary aperture radius ρ_0, the resultant normalized behavior is:

$$Y (\sigma ,d ,\rho_0) = (\pi \, \rho_0^2 \, d^{-2})^{-1} \int_0^{2\pi} d\alpha \int_0^{\rho_0/d} Y_t [\, \delta_0 \, (\sigma ,\rho ,d ,0)] \, \beta \, d\beta$$

$$= (1 - R) (1 + R)^{-1} [1 - A (1 - R)^{-1}]^2 \, 2 \, (\rho_0^2 \, d^{-2})^{-1}$$

$$\times \int_0^{\rho_0/d} \left\{ 1 + 2 \sum_{k=1}^{k=\infty} R^{2k} \, \cos[k \, \delta_0 \, (\sigma ,\rho ,d ,0)] \right\} \beta \, d\beta \,, \qquad 5.2.18a$$

where $\beta = \rho \, d^{-1}$, and thus, the final result is:

$$Y (\sigma ,d ,\rho_0) = (1 - R) (1 + R)^{-1} [1 - A (1 - R)^{-1}]^2$$

$$\times \left\{ 1 + 2 \sum_{k=1}^{k=\infty} R^{2k} \, \cos(k \, 2\pi \, \sigma \, 4d) \, \mathrm{C}(z_k) \, z_k^{-1} \right.$$

$$\left. -2 \sum_{k=1}^{k=\infty} R^{2k} \, \sin(k \, 2\pi \, \sigma \, 4d) \, \mathrm{S}(z_k) \, z_k^{-1} \right\}$$

$$= (1 - R) (1 + R)^{-1} [1 - A (1 - R)^{-1}]^2$$

$$\times \left\{ 1 + 2 \sum_{k=1}^{k=\infty} R^{2k} \, z_k^{-1} \, [\, \mathrm{C}^2(z_k) + \mathrm{S}^2(z_k) \,]^{1/2} \right.$$

$$\left. \times \cos(k \, 2 \, \pi \, \sigma \, 4 \, d - \Phi_k) \right\}. \qquad 5.2.18b$$

For convenience in the last result, the substitution of z_k for the following quantity has been made:

$$z_k = (\, 4 \, \sigma \, d^{-3} \, \rho_0^{\,4} \, k \,)^{1/2} \, . \qquad\qquad 5.2.19$$

In Equation 5.2.18b), $C(z)$ and $S(z)$ are the Fresnel integrals defined by (Gautschi, 1964):

$$\mathbf{C}(z) = \int_0^z \cos(\, 2^{-1}\pi \, t^{\,2}) \, dt \qquad < 0.78 \, , \qquad\qquad 5.2.20a$$

$$\mathbf{S}(z) = \int_0^z \sin(\, 2^{-1}\pi \, t^{\,2}) \, dt \qquad < 0.72 \, , \qquad\qquad 5.2.20b$$

where the limiting values in the two equations are for real values of z.

As can be seen from Equation 5.2.18b), the peak response of the etalon with a finite aperture has been shifted for each term by:

$$\Phi_k = - \, \tan^{-1} [\, \mathbf{S}(z_k) \, \mathbf{C}^{-1}(z_k) \,] \, . \qquad\qquad 5.2.21$$

Because of the properties of Fresnel integrals, Φ_k does not monotonically change as a function of k, leading to asymmetry of the profiles.

Figure 5.6 displays the absolute value of Φ_k as a function of z_k, showing the oscillatory behavior of the phase angle. Figure 5.7 shows the effects the aperture function has on the R^{2k} coefficients of Equation 5.2.18b). For values of $R \rightarrow 1$, the curve in the figure becomes the coefficients of the Fourier series representation given in Equation 5.2.18b), that is, the coefficients of the aperture function. The decreasing value of the aperture function as a function of z_k leads to the expected broadening of the resultant profile, relative to the infinitesimally small aperture etalon. This effect, coupled with the oscillating phase of Figure 5.6, gives rise to the observed asymmetrical broadened profiles obtained with spherical etalons. As can be deduced from these two figures, the use of either low-reflectivity coatings or small apertures will diminish the asymmetrical response of spherical etalons. For the low-reflectivity example, this is attained because of the negligible contribution of the higher harmonic coefficients, and their oscillatory behavior, to the final profile. In the small aperture example, or small values of z_k, the aperture function is described by the left side of Figures 5.6 and 5.7 where the coefficients are near unity and the phase is smoothly changing. These aperture effects are graphically shown in Figures 5.8 and 5.9. Another version of the compensated etalon is shown in Figure 5.10. As seen in the figure, the behavior of this configuration is very much alike the two-spherical-mirror etalon.

Usage of a finite-aperture spherical etalon gives results like those obtained for the plane Fabry-Perot in Chapter 2, Section 4, except the confocal aperture function does not have the same mathematical description. The inclusion of sources with

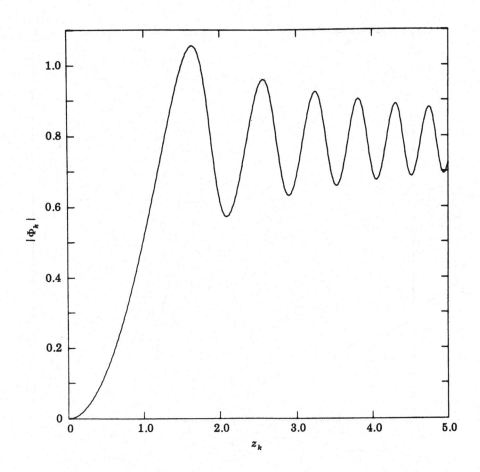

FIGURE 5.6. Absolute value of the phase angle Φ_k, as a function of the parameter z_k, as defined in the text.

arbitrary shapes and finite widths, surface imperfections, etc., follow the method already given in Chapter 2 and will not be repeated here.

Since the Connes (1956) spherical Fabry-Perot has been introduced as a more luminous version of the Fabry-Perot interferometer, it is now necessary to show, in a quantitative manner, the extent and limitations of this device. Referring to the defin-ition of the $\mathbf{L} \times \mathbf{R}$ product in Equation 5.1.8a), the terms in the right side can be replaced by their physical equivalents. For the plane Fabry-Perot the result is:

$$(\mathbf{L} \times \mathbf{R})_p = (4^{-1}\pi \, D^2) \, (2\pi \, f \, \sigma^{-1}) \, [\sigma \, (a \oplus f \,)^{-1}] \tau_p$$

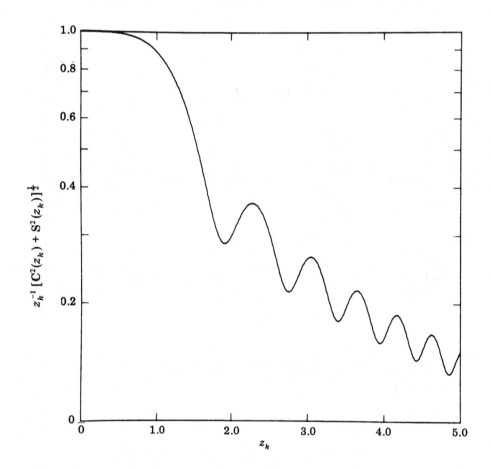

FIGURE 5.7. Aperture function coefficients as a function of the parameter z_k.

$$= 2^{-1}\, \pi^2\, D^2\, f\, (\, a \oplus f\,)^{-1}\, \tau_p\, .   \qquad\qquad 5.2.22a$$

In the above expression, D is the diameter of the mirrors, τ_p is the transmission at the peak of the fringe and σ is the wavenumber of the light. For a spherical confocal etalon it can be shown (Connes, 1958; Hercher, 1968) that:

$$(\mathbf{L} \times \mathbf{R})_s = (\pi\, \rho^2)\, (\pi\, \rho^2 d^{-2})\, \sigma\, (a \oplus f\,)^{-1}\, \tau_s\, .   \qquad 5.2.23a$$

In terms of the phase relations of Equation 5.2.3c), the following equality is obtained:

$$\rho^2\, d^{-2} = \Delta n_s\, d\, \sigma^{-1}\, .   \qquad\qquad 5.2.24$$

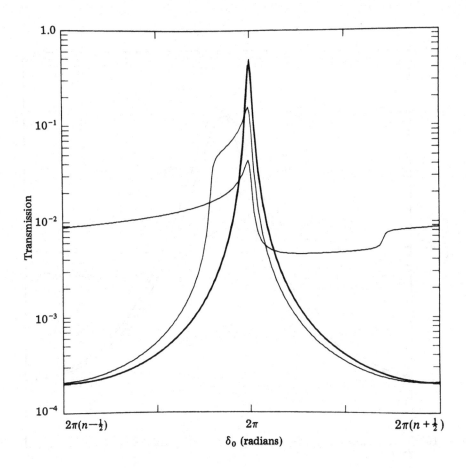

FIGURE 5.8. Aperture size effects on the transmitted profile of a confocal etalon. A reflectivity of 0.95 and apertures equal to 1/2, 1, 2 and 4 times the optimum aperture have been used. Note the asymmetrical behavior at large apertures.

In the previous two equations, ρ is the spherical-etalon-aperture radius and Δn_s is the fraction of an order the aperture covers. Note in particular that Δn_s, in Equation 5.2.24), is defined in terms of the degenerate, or improperly illuminated etalon, free spectral range, i.e., $\Delta\sigma = (4\,d)^{-1}$. In terms of the usual free spectral range, namely $\Delta\sigma = (2\,d)^{-1}$, and after replacement of the equality in Equation 5.2.24), the following is obtained:

$$(\mathbf{L} \times \mathbf{R})_s = 2^{-1}\,\pi^2\,d\,(\Delta\sigma)^{-1}\,f\,(a \oplus f)^{-1}\,\tau_s\;. \qquad 5.2.23b$$

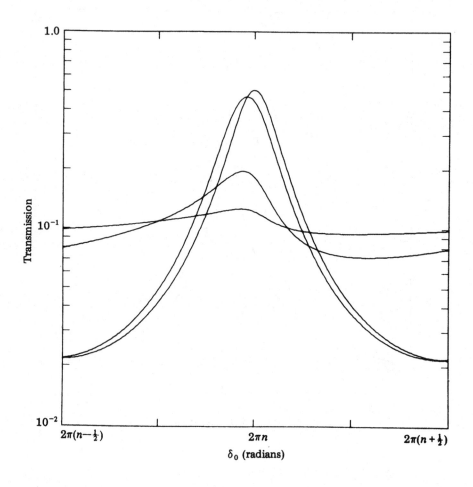

FIGURE 5.9. Same as Figure 5.8, except a reflectivity of 0.85 has been used in the calculations.

The peak transmission, τ_p, for an ideal plane etalon is equal to unity [cf. Equation 2.1.7)], while the peak transmission of an ideal spherical etalon is $\approx 2^{-1}$, as given by Equation 5.2.12) for reasonably high values of R. Since the effects of the aperture on the plane etalon will reduce this transmission by some fraction, by similarity, it can be said that a finite-aperture spherical etalon with the same wavenumber extent will affect the transmission in the same fashion. Using this argument, it is then possible to say that $\tau_p \approx 2\tau_s$. If, in addition, the value for the free spectral range is replaced by its equivalent, $(2\,d\,)^{-1}$, then Equation 5.2.23b) becomes:

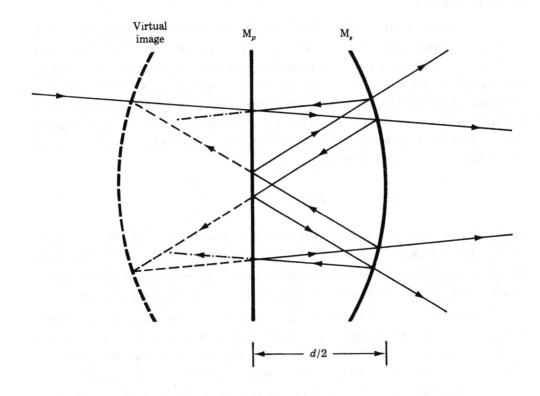

FIGURE 5.10. Hemispherical version of the confocal etalon.

$$(\mathbf{L} \times \mathbf{R})_s = 2^{-1} \, \pi^2 \, d^2 \, f \, (a \oplus f)^{-1} \, \tau_p \; . \qquad\qquad 5.2.23c$$

Using Equations 5.2.22a) and 5.2.23c) directly leads to the comparison between the plane and spherical etalons, expressed as the gain ($\mathbf{G}$), namely:

$$\mathbf{G} = (\mathbf{L} \times \mathbf{R})_s \, (\mathbf{L} \times \mathbf{R})_p^{-1} = d^2 \, D^{-2} \; . \qquad\qquad 5.2.25a$$

Had the degenerate free spectral range been used, the customary answer would have been obtained:

$$\mathbf{G}_d = 2 \, d^2 \, D^{-2} \; . \qquad\qquad 5.2.25b$$

As derived earlier in Equations 5.1.8) and 5.2.22a), the $\mathbf{L} \times \mathbf{R}$ product of a plane etalon can be increased if the area, or diameter, of the etalon is increased. For the spherical etalon, this $\mathbf{L} \times \mathbf{R}$ product can be increased by using larger radius-of-curvature mirrors, in other words, for the spherical etalon the $\mathbf{L} \times \mathbf{R}$ product increases with increasing resolving power $\mathbf{R}$. This result is implicit in Equation 5.2.24) where it is shown that the ratio $\rho^2 \, d^{-2}$ is constant for a given $\Delta n_s \, \sigma^{-1}$, that is, the value of ρ can increase when d increases and thus the area and solid angle of

Equation 5.2.23a) can increase.

According to Equation 5.2.25), the $\mathbf{L} \times \mathbf{R}$ product for the spherical and plane Fabry-Perot etalons becomes comparable when $d = D$ or $d = 2^{-1/2} D$. Since there exists a reasonable limit for the maximum diameter of high-quality plane etalons, typically 15 cm, a plane etalon of this aperture is comparable to a spherical etalon in the range 10 cm to 15 cm radius of curvature. Thus, the spherical etalon is more luminous at higher resolving powers than a practical plane etalon. As has been stated previously in Equation 5.2.17a), the optimum aperture of a spherical etalon is a function of the reflectivity and mirror radius, or:

$$\rho_0 = \left\{ (1-R^2)\,(2\,\pi\,\sigma\,R)^{-1}\,d^3 \right\}^{1/4}. \qquad 5.2.17b$$

For a nominal radius d of 12 cm, reflectivity $R = 0.95$, and $\sigma = 20\,000\,K$, it can be easily calculated that ρ_0 is equal to a 0.194 cm radius. This shows that similar results can be obtained with a spherical etalon of area roughly 1500 times smaller than its plane etalon counterpart, which makes for a more compact instrument.

The comparison made earlier presumes ideal etalons, perfectly aligned. In practice, real plane etalons of large aperture suffer from finite surface defects and are not usually aligned to perfect parallelism, thus lowering the effective resolving power of this device. On the other hand, the useful area of the spherical etalon is sufficiently small such that the extent of the surface defects, or departure from sphericity, is quite small. More importantly, a spherical etalon only requires near-confocality for alignment, since the equivalent motion that misaligns a plane etalon by forming a wedge does not affect the spherical etalon; it merely redefines the etalon axis. The outcome of these limitations is that the practical gain is larger than the derived gain given in Equation 5.2.25) (Jackson, 1961; Hercher, 1968).

In the same vein as Chapter 3, Section 2.1, it would be expected that the spherical confocal etalon has an optimum point of operation. This has been done (Hernandez, 1985d); however, using a change in the notation. This change has been to define the normalized HWHH relative to the improperly illuminated, or degenerate, free spectral range of the spherical etalon and is denoted by the superscript $+$. For example, the normalized HWHH of a Doppler profile, of absolute width dg, is given by:

$$dg^+ = dg\,(\Delta\sigma_s)^{-1} = dg\,4\,d_s, \qquad 5.2.26$$

where the subscript s will be used to denote the spherical etalon, as opposed to the plane etalon, which will be labeled by the subscript p. The normalized HWHH, dg^+, is related to the temperature and the plane etalon normalized HWHH, dg^*, of Equation 3.2.26) by:

$$\gamma\,\tau = \pi^2\,(dg^*)^2\,[\ln(2)]^{-1} = \pi^2\,(2^{-1}\,dg^+)^2\,[\ln(2)]^{-1}. \qquad 5.2.27$$

The etalon normalized width is given in terms of the mean reflectivity of the mirrors, i.e., $R = (R_1 R_2)^{1/2}$, or:

$$a^+ = \pi^{-1} \sin^{-1}[\ (1-R^2)\ (2R)^{-1}\ ]\ . \qquad 5.2.28a$$

which, in terms of the approximate finesse of Equation 2.1.13b), can also be expressed as:

$$a^+ = (1+R)\ R^{-1/2}\ (\ 2\ N_R\ )^{-1}\ . \qquad 5.2.28b$$

The spherical etalon finite aperture can be identified with the physical aperture used with the plane Fabry-Perot etalon. The normalized aperture HWHH will be denoted by f^+.

Since the mathematical derivations are alike those of the plane etalon, given in detail in Chapter 3, Section 2.1, it is only necessary to give the final results. Here, as in Chapter 3, it is assumed that a Doppler broadened line is being examined by an arbitrary aperture spherical etalon and the latter has negligible surface defects. The flux collected by the spherical etalon spectrometer just described is:

$$P(\sigma) = \mathbf{I}\ \mathbf{A}\ \epsilon\ \tau_L \int_0^{2\pi} d\alpha \int_0^{\rho_0/d} Y_t(\sigma)\ \beta\ d\beta$$

$$= \mathbf{I}\ \pi^2\ \rho_0^4\ d^{-2}\ \epsilon\ \tau_L\ [1-A\ (1-R)^{-1}]^2\ (1-R)\ (1+R)^{-1}$$

$$\times \left[\ 1+2\ \sum_{k=1}^{k=\infty} R^{2k}\ \exp\{-k^2\pi^2\ (dg^+)^2\ [\ln(2)]^{-1}\}\ z_k^{-1} \right.$$

$$\left. \times [\ \mathbf{C}^2(z_k)+\mathbf{S}^2(z_k)\ ]^{1/2}\ \cos(\ 2\pi\ k\ \sigma\ 4d - \Phi_k)\ \right]\ . \qquad 5.2.29$$

Note the similarity between this equation and Equations 2.4.2a and 5.2.18b). The etalon aperture radius can be easily transformed into the equivalent normalized aperture, since:

$$2\ f^+\ \sigma^{-1} = \rho_0^4\ d^{-3}\ , \qquad 5.2.30$$

which can be derived from Equation 5.2.3d). Therefore, z_k of Equation 5.2.19) becomes:

$$z_k = (\ 8\ f^+\ k\ )^{1/2}\ . \qquad 5.2.31$$

Further simplification is obtained when the Fourier coefficients are redefined and the retardation is described in terms of orders (thus, easily converted into arbitrary units x and $\mathbf{T}$):

$$a_k = R^{2k}\ z_k^{-1}\ [\ \mathbf{C}^2(z_k)+\mathbf{S}^2(z_k)\ ]^{1/2}\ \exp(-k^2\ 4\ \gamma\ \tau)\ , \qquad 5.2.32$$

$$P(x) = \mathbf{I}\ \pi^2\ f^+\ \epsilon\ \tau_L\ (2\ \Delta\sigma_s^2\ n_{os}\ )^{-1}\ [1-A\ (1-R)^{-1}]^2\ (1-R)\ (1+R)^{-1}$$

$$\times \left\{ 1+2 \sum_{k=1}^{k=\infty} a_k \, \cos[2\pi k \, (x-x_0)\mathbf{T}^{-1}] \right\}. \qquad\qquad 5.2.33$$

In the above equation n_{os} is defined as $\sigma \, (\Delta\sigma_s)^{-1}$. Using the same derivations of Chapter 3, Section 2.1, the uncertainties in Doppler shift (expressed as winds and denoted by the subscript v) and in Doppler width (given as temperature and marked with the subscript τ) are (Hernandez, 1985d):

$$\sigma_v^2 \, \mathbf{I} \, n_{os} \, \mathbf{T} \, \epsilon \, \tau_L \, (\Delta\sigma_s \, dg^+)^{-2} =$$

$$\left\{ 4 \, \pi^4 \, c^{-2} \, f^+ (dg^+)^2 \, [1-A \, (1-R)^{-1}]^2 \, (1-R)(1+R)^{-1} \, \mathbf{E}_k \right\}^{-1}, \qquad 5.2.34a$$

$$\sigma_\tau^2 \, (\tau^2 \, \Delta\sigma_s^2 \, n_{os})^{-1} \, \mathbf{I} \, \epsilon \, \tau_L \, \mathbf{T} =$$

$$\left\{ \pi^6 f^+ (dg^+)^4 \, (\ln 2)^{-2} \, [1-A \, (1-R)^{-1}]^2 \, (1-R)(1+R)^{-1} \, \mathbf{W}_k \right\}^{-1}, \qquad 5.2.35a$$

$$\mathbf{E}_k = \sum_{k=1}^{k=N} (a_k \, k)^2 \, [1 - a_{2k}]^{-1}, \qquad\qquad 5.2.36$$

$$\mathbf{W}_k = \sum_{k=1}^{k=N} k^4 \, w_k - \left(\sum_{k=1}^{k=N} k^2 \, w_k \right)^2 \left(\sum_{k=1}^{k=N} w_k \right)^{-1}, \qquad\qquad 5.2.37$$

$$w_k = a_k^2 \, (1 + a_{2k})^{-1}. \qquad\qquad 5.2.38$$

The uncertainties given in Equations 5.2.34a) and 5.2.35a) are the maximum uncertainties, since the usual $\cos(4\pi x_0 \mathbf{T}^{-1})$ term has been set to unity [cf. Equation 3.2.34)]. N is again the minimum number of coefficients necessary to describe unambiguously the profile and related to the critical number of samples by the Nyquist theorem (Bracewell, 1965), i.e., $\mathbf{T} \geqslant 2 \mathbf{N}$. Equations 5.2.34a) and 5.2.35a) have been numerically solved for a number of combinations of the etalon width, a^+, line width, dg^+, and etalon aperture, f^+ (Hernandez, 1985d) and the results are shown in Figures 5.11 and 5.12, in the same format as Figures 3.2 and 3.3, for the winds (Doppler shift) and temperature (Doppler width), respectively. Figures 5.11 and 5.12 show, just as the figures for the plane etalon, minima of uncertainty which vary as a function of f^+. This is illustrated more clearly in Figures 5.13 and 5.14, while Figures 5.15 and 5.16 show the values of a^+ and dg^+ associated with the appropriate minima of uncertainty, also as a function of f^+. The values of $\mathbf{N}$ to describe unambiguously the profile at the minima of uncertainty are illustrated in Figure 5.17.

The least uncertainty points for the winds (temperature) are found for $f^+ =$ 0.140 (0.065), $dg^+ = 0.16$ (0.11) and $a^+ = 0.072$ (0.035). The value of a^+ corresponds to a reflectivity of $R = 0.80$ (0.90) and finesse $N_R = 14.$ (28.5).

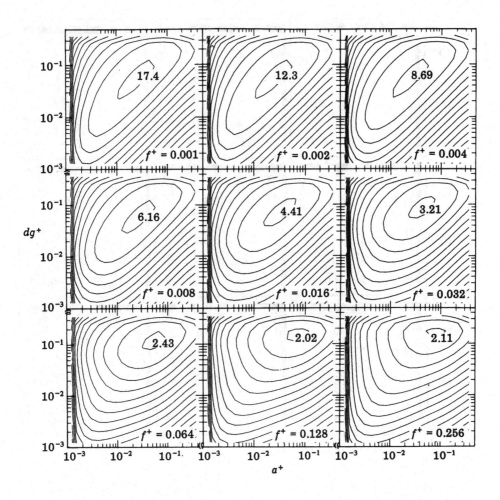

FIGURE 5.11. Uncertainties of determination of the location of the peak of a Fabry-Perot profile measured with a spherical etalon device. The uncertainties are expressed as contours of the standard deviation of the equivalent wind (10^9 m/s units), as a function of the normalized source and etalon widths with the normalized etalon aperture as a parameter. The last is indicated in the lower right corner of each panel. The increasing value contours are separated by a factor of $2^{1/2}$, beginning with the marked value. The uncertainties are given for a lossless etalon examining a unit irradiance source in the first order at unity free spectral range with a unity quantum efficiency detector. Hernandez (1985d).

Since there are four reflections per ray for the spherical etalon, against two in the plane etalon, a comparison between the reflectivity of these devices must take this into account. The 'equivalent' reflectivity for the spherical etalon, i.e., for two reflections, is then $R = 0.64$ (0.80), which is in excellent agreement with the optimum

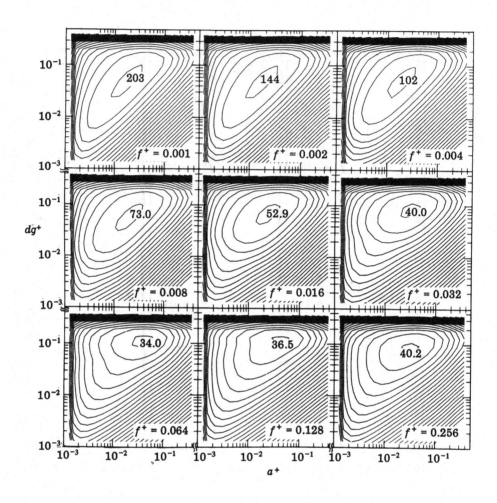

FIGURE 5.12. Uncertainties of determination of the temperature of a Fabry-Perot profile measured with a spherical etalon device. The uncertainties are expressed as the ratio of the temperature uncertainty to the temperature. Same conditions as Figure 5.11. Hernandez (1985d).

reflectivity of the plane device in Chapter 3, Section 2.1. The results of the normalized HWHH of the spherical spectrometer parameters can be directly compared with those obtained for the plane spectrometer in Chapter 3, Section 2.1 and the agreement is excellent, except that $dg^+ \simeq 1.14 \ dg^*$ for the winds. In fact, the ability to directly compare the results of the two devices has been the reason why the normalized HWHHs were chosen with respect to the degenerate free spectral range of the spherical etalon. The values of the minimum uncertainties of determination for the spherical etalon are larger than those for the plane etalon, and require further

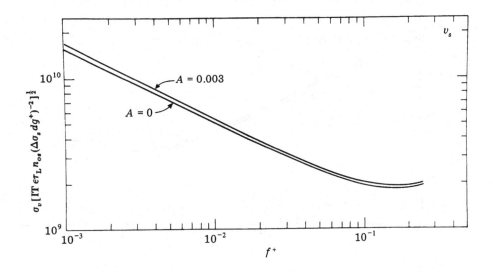

FIGURE 5.13. Minimum uncertainties of determination of the position of a Fabry-Perot profile measured with a spherical etalon device as a function of the normalized etalon aperture. Same units as Figure 5.11. Results for mirrors with and without absorption are indicated. Hernandez (1985d).

examination. Equations 3.2.24), 3.2.33), 5.2.34a) and 5.2.35a) can be abbreviated as follows:

$$\sigma^2_{vp} \; n_{op} \; \mathbf{I} \, \mathbf{A} \, \mathbf{T} \, \epsilon \, \tau_L \; (\, dg^* \,)_p^{-2} = \sigma^2_{vpc} \; , \qquad 5.2.39$$

$$\sigma^2_{\tau p} \; (\tau^2 \, n_{op} \,)^{-1} \mathbf{I} \, \mathbf{A} \, \mathbf{T} \, \epsilon \, \tau_L = \sigma^2_{\tau pc} \; , \qquad 5.2.40$$

$$\sigma^2_{vs} \; \mathbf{I} \, n_{os} \; \Delta\sigma_s^{-2} \; \mathbf{T} \, (dg^+)_s^{-2} \, \epsilon \, \tau_L = \sigma^2_{vsc} \; , \qquad 5.2.34b$$

$$\sigma^2_{\tau s} \; (\tau^2 \, \Delta\sigma_s^2 \, n_{os} \,)^{-1} \mathbf{I} \, \epsilon \, \tau_L \, \mathbf{T} = \sigma^2_{\tau sc} \; , \qquad 5.2.35b$$

where the subscript c indicates calculated. From the above equations it can be shown that:

$$\sigma^2_{vs} \; \sigma^{-2}_{vp} = \sigma^2_{vsc} \; \sigma^{-2}_{vpc} \; \mathbf{A} \, \Delta\sigma_s \, \Delta\sigma_p \; , \qquad 5.2.41a$$

$$\sigma^2_{\tau s} \; \sigma^{-2}_{\tau p} = \sigma^2_{\tau sc} \; \sigma^2_{\tau pc} \; \mathbf{A} \, \Delta\sigma_s \, \Delta\sigma_p \; . \qquad 5.2.42a$$

Taking the appropriate values of the (calculated) uncertainties from Figures 3.4, 3.5, 5.13 and 5.14 and describing the area of the plane etalon in terms of its associated diameter:

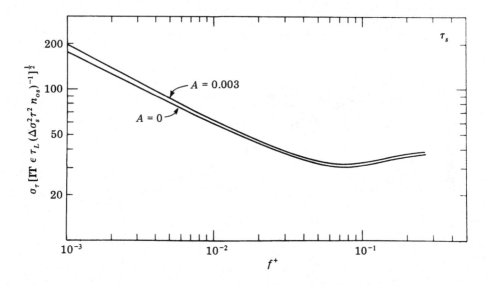

FIGURE 5.14. Minimum uncertainties of determination of the temperature of a Fabry-Perot profile measured with a spherical etalon device as a function of the etalon normalized aperture. Results for mirrors with and without absorption are indicated. Hernandez (1985d).

$$\sigma_{vs}^{2}\,\sigma_{vp}^{-2} = 3.43\;D^{2}\,\Delta\sigma_{s}\,\Delta\sigma_{p}\;,\qquad\qquad 5.2.41b$$

$$\sigma_{\tau s}^{2}\,\sigma_{\tau p}^{-2} = 3.35\;D^{2}\,\Delta\sigma_{s}\,\Delta\sigma_{p}\;.\qquad\qquad 5.2.42b$$

The values of $\Delta\sigma_{s}$ and $\Delta\sigma_{p}$ are fixed by the optimum values of dg^{+} and dg^{*} for both etalons. From the definition of normalized widths, given in Equations 2.1.12a) and 5.2.26), the following is obtained:

$$\Delta\sigma_{s} = dg\;(\,dg^{+}\,)^{-1} = 4^{-1}\,d_{s}^{-1}\;,\qquad\qquad 5.2.43$$

$$\Delta\sigma_{p} = dg\;(\,dg^{*}\,)^{-1} = 2^{-1}\,d_{p}^{-1}\;.\qquad\qquad 5.2.44$$

However, since the two devices are measuring the same line of HWHH dg, the following relationship applies:

$$d_{p} = 2\;d_{s}\;dg^{*}\;(\,dg^{+}\,)^{-1}\;.\qquad\qquad 5.2.45a$$

Introducing the appropriate values of dg^{+} and dg^{*} for both wind and temperature determinations, from Figures 3.6, 3.7, 5.15 and 5.16, it is then found that:

$$\left[\,dg^{+}\,(\,dg^{*}\,)^{-1}\,\right]_{v} \simeq 1.14\;,\qquad\qquad 5.2.46$$

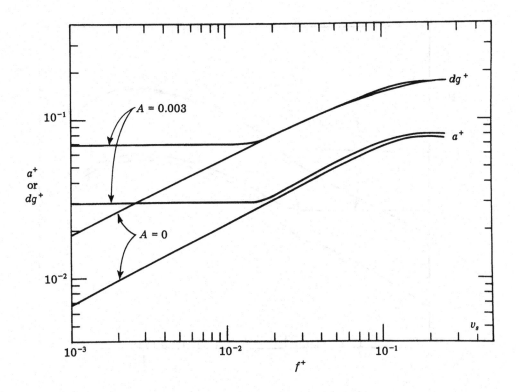

FIGURE 5.15. Optimum normalized etalon and source widths as a function of the etalon normalized aperture. (Winds). Hernandez (1985d).

$$[dg^{+} (dg^{*})^{-1}]_{r} \simeq 1.0 .$$ 5.2.47

Replacing the values of the $\Delta\sigma$'s with their equivalent in terms of d_{s} and the ratio of the normalized line widths, of Equations 5.2.46) and 5.2.47), into Equations 5.2.41b) and 5.2.42b), we have:

$$\sigma^{2}_{vs} \, \sigma^{-2}_{vp} \simeq 0.279 \, D^{2} \, d_{s}^{-2} ,$$ 5.2.41c

$$\sigma^{2}_{\tau s} \, \sigma^{-2}_{\tau p} \simeq 0.209 \, D^{2} \, d_{s}^{-2} .$$ 5.2.42c

Therefore, to achieve equal uncertainty of determination with both spectrometers, with the largest commercially available high-quality plane mirrors (15.0 cm D), it is found the two spectrometers behave alike when the spherical etalon spacing and the associated free spectral range are:

$$d_{s} \simeq 7.94 \text{ cm } (6.87 \text{ cm }) ,$$ 5.2.48

$$\Delta\sigma_{s} = 0.0314 \text{ K } (0.0363 \text{ K }) ,$$ 5.2.49

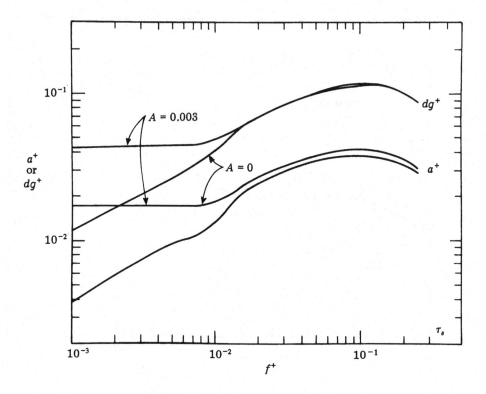

FIGURE 5.16. Optimum normalized etalon and source widths as a function of the etalon normalized aperture. (Temperature). Hernandez (1985d).

for wind (temperature) determinations. It must be pointed out that the plane etalon separation d_p needs not be twice the spherical separation d_s. For the wind example, Equations 5.2.45a) and 5.2.46), indicate that for equal uncertainty:

$$d_{p_v} = 1.754 \, d_{s_v} \, . \qquad\qquad 5.2.45\text{b}$$

From Equations 5.2.41c) and 5.2.42c), and in the sense of Equation 5.2.25), it is possible to define a gain in terms of the ratio of the uncertainties of determination (Hernandez, 1985d):

$$\mathbf{G}_v = \sigma_{vp}^{\,2} \, \sigma_{vs}^{\,-2} \simeq 3.56 \, d_s^{\,2} \, D^{-2} \, , \qquad\qquad 5.2.50$$

$$\mathbf{G}_\tau = \sigma_{\tau p}^{\,2} \, \sigma_{\tau s}^{\,-2} \simeq 4.75 \, d_s^{\,2} \, D^{-2} \, , \qquad\qquad 5.2.51$$

which are directly comparable with Equation 5.2.25b). Note that the gains of Equations 5.2.50) and 5.2.51) are the best gains possible, since both etalons are operating at their optimum conditions. The same comments, made earlier, with respect to the

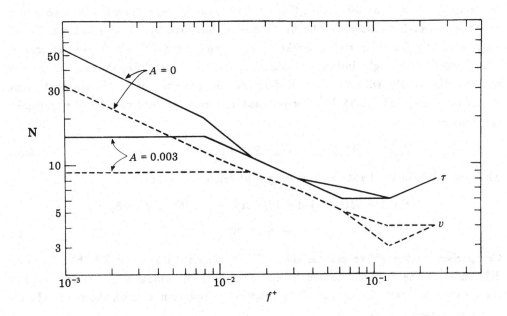

FIGURE 5.17. Critical number of coefficients required to determine unambiguously
the properties of a Fabry-Perot profile (measured with a spherical etalon device) for
the minima of uncertainty given in Figures. 5.13 and 5.14. Hernandez (1985d).

surface defects and parallelism alignment of the plane Fabry-Perot are also applicable
here. The optimum radius of the spherical Etalon, ρ_0, for the conditions of Equations
5.2.48) and σ = 20 000 K, is $\simeq$ 0.29 cm (0.21 cm) or an area $\simeq$ 0.25 cm^2 (0.14
cm^2) for wind (temperature) measurements. That is, similar uncertainties of determination can be obtained with a spherical etalon whose area is 675 (1200) times
smaller than the largest available diameter plane etalon. Note that this comparison is
at the optimum point of operation for both types of etalons, not at an arbitrary
reflectivity as given in the earlier example.

As the reader has noticed, diffraction effects have been neglected in the discussion, since, for most practical applications, they are negligible. The method presented
in this section is only one of the possible means to describe a spherical etalon. Boyd
and Gordon (1961) have treated this topic in terms of confocal resonators and a
general review of these has been given by Koppelmann (1969) (cf. Chapter 6, Section 4). The optical coupling necessary for the efficient use of more than one spherical etalon in series has been reported by Connes (1958).

3. Practical plane Fabry-Perot etalons

In the derivation of the transmitted-light response of a plane Fabry-Perot in Chapter 2, Section 1, a source of considerable light loss was not discussed. Referring to Figure 2.1 and Equation 2.1.3), this loss is found to occur upon first entrance of a ray into the etalon cavity, or the first transmission loss, t_1. This effect was recognized early by Lummer and Gehrcke (1902) and Barnes (1904), who suggested introduction of the light into the etalon as shown in Figure 5.18, that is, for the ray to enter the cavity without loss. Rederiving the response of the instrument using Equations 2.1.2) and 2.1.3), less the relevant t_1 term, the behavior for the transmitted light is:

$$Y_t(\delta) = (1-R)\,[1-A(1-R)^{-1}]\,(1+R^2-2R\,\cos\delta\,)^{-1}\,, \qquad 5.3.1$$

while for the reflected light the following expression is obtained:

$$Y_r(\delta) = R(1-R)\,[1-A(1-R)^{-1}]\,(1+R^2-2R\,\cos\delta)^{-1}$$

$$= R\ Y_t(\delta)\,. \qquad 5.3.2$$

Comparison of the above expressions with Equations 2.1.6a) and 2.1.7b) shows the differences in the response caused by the change in illumination of the cavity. For the transmitted path, Equation 5.3.1), relative to Equation 2.1.7b), shows the change which expressed as a ratio is:

$$Y_t(\delta)_{t_1=1}\,[\ Y_t(\delta)_{t_1}]^{-1} = (1-R)^{-1}[1-A(1-R)^{-1}]^{-1}\,. \qquad 5.3.3$$

Since both of the right-side terms of Equation 5.3.3) inside the parenthesis are ≤ 1, the numerical value of this equation is ≥ 1. In the days of silvered mirrors, the gain afforded by the last term of Equation 5.3.3) was important, since the absorption, A, by silver coatings is large and dependent on the thickness of the deposited silver (Dufour, 1945). Other than this gain in the transmission, no other changes in the resultant fringes are observed. Note, for instance, that the contrast, C [cf. Equation 2.1.14)], has not changed relative to that of the normal Fabry-Perot. Because of the character of the illumination of the etalon, only off-axis fringes are obtained, but this was not considered to be a drawback in photographic recording. The changes observed in the reflected ray path are large; in particular, note the reflected pattern now resembles the transmitted fringe pattern, as indicated in Equation 5.3.2), rather than being its complement (cf. Figure 2.3). This behavior is easily understood after examination of Figure 5.18. Again, off-axis fringes are obtained in the reflected pattern.

In the side-illuminated Fabry-Perot, the intensity of the reflected and transmitted fringes tend to asymptotically become equal as $R \to 1$. When the etalon cavity has an index of refraction larger than the surrounding medium, high reflectivities

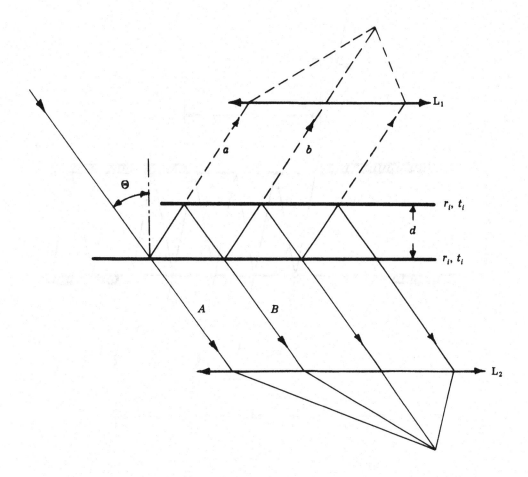

FIGURE 5.18. Side illumination of a Fabry-Perot etalon.

can be obtained by internal reflection, as the critical angle is approached. Under these conditions, absorption losses are negligible, when compared to those associated with silver coatings. As the reader has discerned, the above describes the Lummer - Gehrcke plate (Lummer and Gehrcke, 1903). This device, as described earlier, proved to be quite useful; however, it was limited by the inhomogeneities of the glass slabs from which it was fabricated, as well as being inflexible when compared with an air-spaced Fabry-Perot etalon, since the gap spacing was fixed during manufacture. Since large angles to the normal to the mirrors are used in side-illuminated Fabry-Perots (and Lummer - Gehrcke plates), the physical size of these devices needs to be large, in order to allow sufficient number of internal reflections to occur.

Light can also be introduced into the etalon by means of a beam splitter within the cavity. This is the principle of the Fox-Smith interferometer (Fox, 1970; Smith,

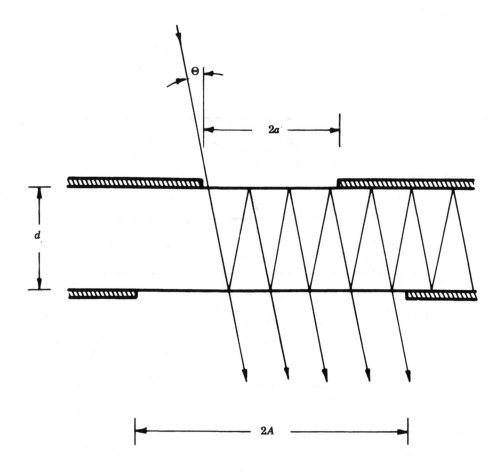

FIGURE 5.19. Schematic diagram of a finite-size etalon. After Vander Sluis and
McNally (1956).

1965), which is illustrated in Figure 6.17.

3.1. Finite-size etalons

Implicit in the derivations of Chapter 2, as well as those earlier in this section,
is the existence of mirrors large enough in physical size, so a large (i.e., infinite)
number of reflections can occur for a given light ray. Under this assumption, the
results given by Equations 2.1.4) and 2.1.5) hold true. Intuitively, the size of a given
etalon needs not be as large when only the rays near the axis are used, as would be
necessary for the examples of a side-illuminated Fabry-Perot and Lummer-Gehrcke
plates, where off-axis rays are used. Then, depending on the intended use of an

etalon, there is a minimum finite size for the mirrors which will give near-theoretical behavior. For instance, a Jacquinot and Dufour (1948) spectrometer needs smaller mirrors than those necessary in imaging applications, where many off-axis fringes are utilized, since in the former only the fringe near the axis is used. The side-illuminated Fabry-Perot and the Lummer-Gehrcke plate fall under the same category as the imaging applications, since in all these instances the off-axis fringes are used.

The effects of the finite size of etalons on the resultant profiles of Fabry-Perot measurements have been treated in detail by Vander Sluis and McNally (1956). Their approach employs a general plane etalon with an entrance aperture of size $2\,a$, and an exit aperture $2\,A$ wide, as shown in Figure 5.19. In the figure, a ray entering the etalon at an angle Θ to the etalon normal undergoes a number of internal reflections. These internal reflections are, for convenience, divided into two categories (Vander Sluis and McNally, 1956): those reflections that occur within the entrance aperture, and those which occur within the exit aperture, but outside the entrance aperture. As the figure illustrates, there is a limited number of internal reflections that can occur, and this can be quantitatively expressed. For an etalon with spacer d, and receiving a ray at an angle Θ to the normal, the stepping distance per reflection is defined as:

$$l = 2\,d\,\tan\Theta\,. \qquad\qquad 5.3.4$$

Therefore, the number of reflections occurring within the entrance aperture (to the closest integer value k), are given by:

$$k = 2\,a\,/\,l = 2\,a\,(\,2\,d\,\tan\Theta\,)^{-1} \simeq a\,(\,d\,\Theta\,)^{-1}\,. \qquad\qquad 5.3.5$$

Similarly, the number of reflections within the exit aperture, but outside the entrance aperture, is defined as:

$$n \simeq (\,A\,-\,a\,)\,(\,2\,d\,\Theta\,)^{-1} \qquad\qquad 5.3.6$$

Therefore, consider an entrance aperture, now divided into k sections, where a ray entering the jth section produces a series of rays transmitted by the exit aperture, with amplitudes and phase retardation given by:

$$t_1 t_2 + t_1 t_2 r_2^2\,e^{i\,\delta} + \ldots\ldots + t_1 t_2\,(\,r_2^2\,)^{(j+n-1)} e^{i\,\delta\,(j+n-1)}\,. \qquad 5.3.7a$$

This expression can be recognized as Equation 2.1.3), given earlier, except it now has a limited number of terms. The value of this limited-terms sum is:

$$A_{t,j,n}(\delta) = \tau\,[\,1\,-\,(\,R\,e^{i\,\delta}\,)^{j+n}\,]\,(\,1 - R\,e^{i\,\delta}\,)^{-1}\,. \qquad 5.3.7b$$

In the above, the same assumptions used to obtain Equation 2.1.5) have been made. The resultant transmitted intensities are obtained, as before, by the product of the amplitude and its complex conjugate, except now this is done for all the j rays averaged over the possible k beams. The result of this summation is:

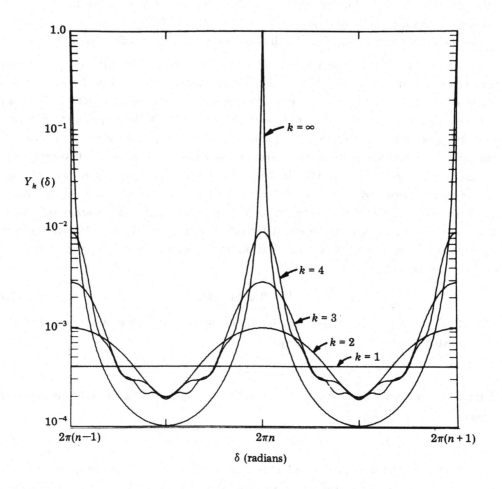

FIGURE 5.20. Instrumental transmission for a finite-size matched-aperture etalon. A reflectivity of 0.98 has been used in the calculations, and the absorption, A, has been set equal to zero. The effective-etalon size k used is indicated in the profiles. The ideal etalon ($k = \infty$) is shown for comparison.

$$
Y_{n,k}(\delta) = Y_t(\delta) \left\{ 1 + R^{2n+2}(1 - R^{2k}) \left[k (1 - R^2) \right]^{-1} \right.
$$

$$
- 2 Y_{t_{A=0}}(\delta) R^{n+1} \left[k (1 - R)^2 \right]^{-1} \times \left[\cos[(n+1)\delta] - R \cos(n\delta) \right.
$$

$$
\left. \left. - R^{k} \cos[(n+k+1)\delta] + R^{k+1} \cos[(n+k)\delta] \right] \right\} . \qquad 5.3.8
$$

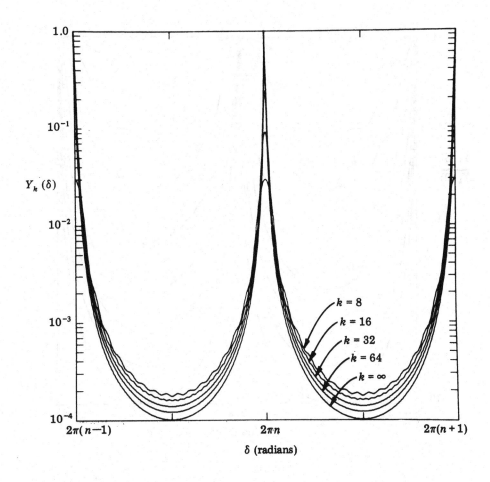

FIGURE 5.21. Instrumental transmission for a finite-sized matched-aperture etalon. Same conditions as Figure 5.20, except for the values of k.

In the above expression, $Y_t(\delta)$ and $Y_{t_{A}=0}(\delta)$ are the ideal-etalon functions, given by Equation 2.1.7b), with and without the presence of absorption/scattering. Equation 5.3.8) is applicable for the conditions when both apertures are matched in size, ($n = 0$), and when the exit aperture is larger than the entrance aperture ($n > 0$).

The matched-aperture condition is the more commonly found situation, as etalons are usually available with equal-size mirror substrates, and will be discussed first. This matched aperture description follows from Equation 5.3.8), with the condition that $n = 0$, i.e.:

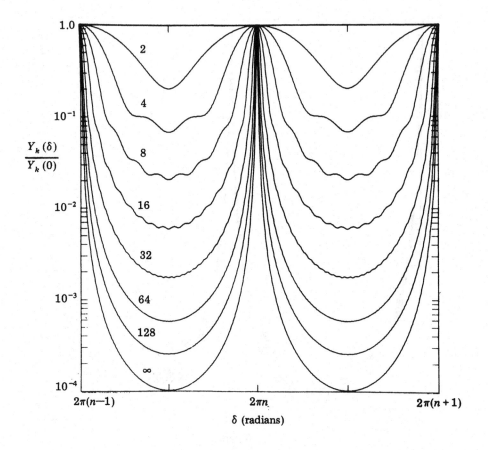

FIGURE 5.22. Normalized transmission for a finite-size matched-aperture etalon. Note both the broadening of the profiles and decrease of the contrast with decreasing effective-etalon size parameter k. Same conditions as Figure 5.20.

$$Y_k(\delta) = Y_t(\delta) \left\{ 1 + R^2(1 - R^{2k}) \left[k(1 - R^2) \right]^{-1} \right.$$

$$- 2 R\, Y_{t_{A=0}}(\delta) \left[k(1-R)^2 \right]^{-1}$$

$$\times \left. \left[\cos\delta - R - R^k \cos[(k+1)\delta] + R^{k+1} \cos(k\delta) \right] \right\}. \qquad 5.3.9$$

This expression shows the basic properties of the finite-size Fabry-Perot etalon. Note, for instance, the absence of phase shifts relative to the ideal etalon. Other properties are better discussed by referring to Figures 5.20 and 5.21, which graphically show specific solutions of Equation 5.3.9) for a reflectivity of 0.98. The two figures show the rather substantial loss in transmission associated with a finite-size etalon, as well as oscillations in the resultant profile. As seen in Equation 5.3.8), the oscillations are associated with the presence of cosine terms with k in their argument. As would be expected, these oscillations are more noticeable for the smaller values of k, or smaller effective-size etalons, and they persist into large values of the etalon effective-size parameter size k. Figure 5.22 shows the same profiles as those of the earlier figures, but normalized to unity transmission at $\delta = 0$. This form of presentation shows both the appreciable loss in the contrast, as defined in Equation 2.1.14), as well as an increase in the width of the profiles, relative to the width of the ideal etalon, with decreasing etalon size. This broadening translates into a loss of resolving power [cf. Equation 2.1.15)]. Other properties of the limited-size etalon can be investigated when the absorption, A, is arbitrarily set to zero, i.e., $Y_t(\delta) \equiv Y_{t_{A=0}}$, and, when k is assumed to be large enough, such that the terms containing R^k become negligibly small. Under these assumptions, Equation 5.3.9) can be approximated by:

$$ Y_k(\delta) \cong Y_t(\delta) \left\{ 1 + [k \ (1 - R^2)]^{-1} \right. $$

$$ \left. \times \left[R^2 - 2 R \ Y_t(\delta) \ (1 + R) \ (1 - R)^{-1} \ [\cos\delta - R \ ] \right] \right\} . \qquad 5.3.10\text{a} $$

With further manipulation, the above expression becomes:

$$ Y_k(\delta) \cong Y_t(\delta) \left[1 + [k \ (1 - R^2)]^{-1} \right] $$

$$ - (1 + R \ ) [k \ (1 - R \ )]^{-1} \ Y_t^{2}(\delta) . \qquad 5.3.10\text{b} $$

The above (approximate) expression shows, that a practical-limited-size etalon behaves like the ideal etalon of Equation 2.1.7); however, as modified by two terms inversely proportional to k, one of which is proportional to the second power of the ideal-etalon function. Therefore, for large values of k, the practical etalon asymptotically approaches the ideal-etalon behavior, as expected. The transmission properties of the finite-size etalon can be easily deduced from Equation 5.3.10a), where the term in braces can be shown to be smaller than unity for values of δ near $2 \pi m$, $m = 0, \pm 1, \pm 2 \ \ldots\ldots$, and larger than unity for values of δ near $2 \pi \ (m + 1/2)$. Thus, as shown in Figures 5.20 and 5.21, the maximum transmission of a finite-size etalon will have lower value than an ideal etalon, while the minimum transmission will have a larger value for the finite-size etalon. Quantitatively, for the approximations of Equation 5.3.10), this is expressed by:

$$ Y_k(\delta)_{\max} \cong \left\{ 1 - R \ (2 + R \ ) [k \ (1 - R^2)]^{-1} \right\} , \qquad 5.3.11 $$

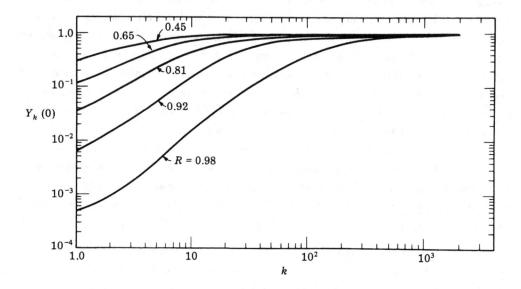

FIGURE 5.23. Maximum transmission of a finite-size matched-aperture etalon, as a function of the etalon effective-size k, and various reflectivities.

$$Y_k(\delta)_{min} \cong Y_t(\delta)_{min} \left\{ 1 + R(2-R)[k(1-R^2)]^{-1} \right\}$$

$$\cong (1-R)^2(1+R)^{-2} \left\{ 1 + R(2-R)[k(1-R^2)]^{-1} \right\} . \qquad 5.3.12$$

The contrast, C_k, for a finite-size etalon is defined to be:

$$C_k = Y_k(\delta)_{max} [Y_k(\delta)_{min}]^{-1}$$

$$\cong C \times \left\{ 1 - 4R[k(1-R^2) - R(2-R)]^{-1} \right\} , \qquad 5.3.13$$

where C has been defined earlier in Equation 2.1.14) for the ideal etalon. Note that C_k is smaller than C for all finite values of k.

The effects of the finite-etalon size on the maximum transmission, $Y_k(0)$, the contrast, C_k, and profile width are shown in Figures 5.23, 5.24, and 5.25 as a function of the effective-etalon size k. For convenience, the widths are shown as the ratio of the normalized ideal-etalon width, a^*, to the normalized width of the finite-size etalon, a_k^*. The normalized width a_k^* is defined as the ratio of the HWHH of the finite-etalon profile to the free spectral range, or Fabry-Perot period [cf.

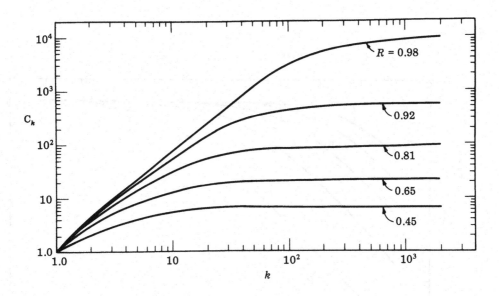

FIGURE 5.24. Contrast of a finite-size matched-aperture etalon as a function of the etalon effective-size k and various reflectivities.

Equation 2.1.12a)]. The values shown in these figures have been calculated using Equation 5.3.9), rather than the approximations of Equations 5.3.11) through 5.3.13). Note that 99% of the ideal-etalon behavior is not reached, even for the lowest reflectivities, for values of k less than about 100, and this value increases as the reflectivity increases.

For a given wavenumber, σ, the relationship between k and the aperture size is (Vander Sluis and McNally, 1956):

$$k = 2 a \left(\sigma \, \Delta\sigma \, 2^{-1} \, \Delta n^{-1} \right)^{1/2} . \qquad\qquad 5.3.14$$

In the above, Δn is defined as the number of orders, away from the central order, for which k is desired. For example, for $\sigma = 20\,000$ K, $\Delta\sigma = 0.5$ K , and $2\,a = 1$ cm, the values of k for the first, third and fifth fringes away from the axis are 71, 41, and 32 respectively. Changing $\Delta\sigma$ to 0.05 K reduces k to 22, 13 and 10 for the same fringes previously mentioned. As can be determined from Figures 5.20 through 5.22, the previous examples show that the effects of a limited-size etalon are more deleterious for imaging experiments, where the fringes away from the central order are used, and for higher resolving powers, where the effective-etalon aperture, k, becomes small because of the large etalon gaps used.

Employing the relations derived earlier, it is now possible to relate k with a given physical aperture size, which will give near-ideal etalon behavior. For a

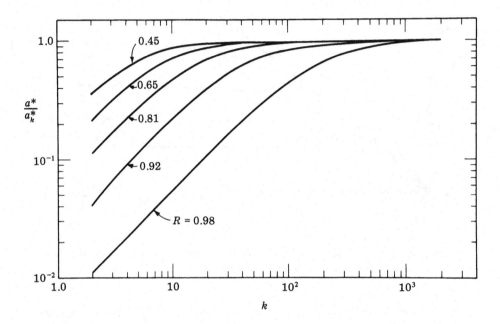

FIGURE 5.25. Width of the finite-size matched-aperture etalon, expressed as the ratio of the ideal-etalon function width to the finite-size etalon function width, as a function of the etalon effective-size k, at various reflectivities.

Jacquinot and Dufour (1948) spectrometer, with a single aperture and a given reflectivity, the aperture size in orders can be described in terms of the reflectivity using the LRP criterion, as discussed in Chapter 2, Section 4. Under these circumstances, the aperture function is equal to $2\, a^{*}$, or N_R^{-1} [cf. Equations 2.1.13a) and 2.1.13b)], and Equation 5.3.14) is equal to:

$$k = 2\, a\, (\, \sigma\, \Delta\sigma\, 2^{-1}\, N_R\,)^{1/2}$$

$$= 2\, a\, [\sigma\, \Delta\sigma\, \pi\, 2^{-1}\, R^{1/2}\, (\, 1 - R\,)^{-1}\,]^{1/2} . \qquad 5.3.15$$

A near-ideal etalon behavior is arbitrarily defined as an etalon which provides 0.99 of the transmission of an ideal etalon (Vander Sluis and McNally, 1956). Using this criterion, it is then found from Equation 5.3.11), after some re-ordering, that the necessary value of k is:

$$k = 100\, R\, (\, 2 + R\,)\, (\, 1 - R^2\,)^{-1} . \qquad 5.3.16$$

Inserting the expression obtained above into Equation 5.3.14), the etalon size expression is then:

$$2\,a = 100\,R\,(\,2 + R\,)\,(\,1 - R^2)^{-1}\,(\,\sigma\,\Delta\sigma\,2^{-1}\,N_R\,)^{-1/2}\,. \qquad 5.3.17$$

For a single-aperture Jacquinot and Dufour spectrometer, with $N_R \simeq 20$ ($R = 0.85$), $\sigma = 20\,000$ K, and $\Delta\sigma = 0.5$ K, the value for the near-ideal etalon aperture is found to be 2.76 cm. When the same spectrometer is used with a spatial detector (Sivjee *et al.*, 1980; Abreu *et al.*, 1981; Killeen *et al.*, 1983), and a spectral element is observed one order away from the axis, the size of the etalon aperture is now found to be 12.35 cm, which is an etalon of appreciable aperture. If the finesse is increased to $N_R \simeq 30$, or $R = 0.90$, keeping the same conditions as the previous example, the etalon apertures can be calculated to be 3.55 cm and 19.43 cm respectively. As these examples have shown, lowering the etalon reflectivity reduces the need for rather large etalon apertures, but the necessity of using sizeable-aperture etalons cannot be avoided when using off-axis fringes. Implicit in the previous discussion and examples, is the existence of an instrumental function which changes as operation away from the central order is made. This effect is particularly severe when undersized-aperture etalons are used, as in multipassing (Sandercock, 1975), where an etalon is effectively masked-down into a few small etalons. Also, note that the finite-size etalon results are, in particular, applicable to series etalons (Vander Sluis and McNally, 1956). In this instance, the entrance aperture of the second and higher etalons is predetermined by the exit aperture of the previous etalon and the divergence of the rays between etalons. This effective entrance aperture will reduce the efficiency of the subsequent etalon(s), and consequently, the efficiency of the multiple etalon device.

Another solution is available for the minimum-aperture problem (Vander Sluis and McNally, 1956), when the exit aperture is made larger than the entrance aperture of the etalon. Using 0.99 of the theoretical peak transmission as their criterion, these authors were able to find that for any value of k of the entrance aperture, there exists a value of n for the exit aperture that satisfies their criterion. Introducing this criterion into Equation 5.3.8), and solving for n, it is found that:

$$n_k + 1 = (\,\ln R\,)^{-1} \ln \left\{ [\,(\,1 + R\,)(\,1 + R^k\,)^{-1}\,] \right.$$

$$\left. \times \left[1 - \left(1 - 0.01\,(1-R)(1+R^k)\,k\,[\,(1+R)\,(1-R^k)\,]^{-1} \right)^{1/2} \right] \right\}, \qquad 5.3.18$$

where n_k is the desired value. It follows from Equations 5.3.5), 5.3.6) and 5.3.14), that the exit aperture is:

$$2\,A = (2\,n_k + k)\,(\sigma\,\Delta\sigma\,2^{-1}\,\Delta n^{-1})^{-1/2}\,, \qquad 5.3.19$$

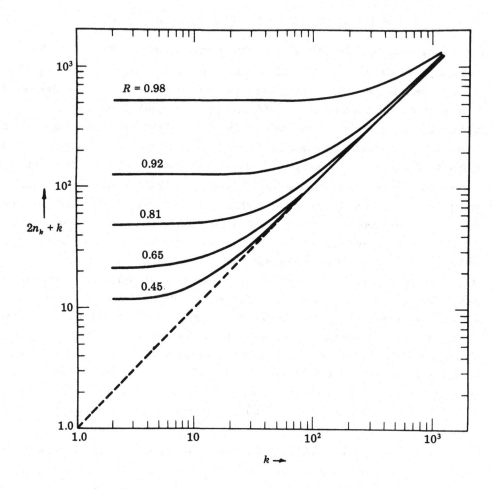

FIGURE 5.26. Effective size for the exit aperture, $2n_k + k$, of a finite-size etalon as a function of the entrance aperture effective-size k, for various reflectivities.

from which similar relations to those given in Equations 5.3.16) and 5.3.17) can be derived. The relationship between ($2\,n_k + k$) and k, obtained from Equation 5.3.18), is given in graphical form in Figure 5.26, for the indicated reflectivity values. This figure shows that, for large k, the value of $2n_k + k$ asymptotically approaches k, or the equal aperture solution given earlier. As seen in the figure, there is no advantage in using an exit aperture larger than the entrance aperture for values of k greater than about 100, except for the higher reflectivities. As Vander Sluis and McNally (1956) have noted, the use of an entrance aperture larger than the exit aperture does not improve the quality of the resultant fringes to a significant extent, and will not be discussed here.

As discussed in Chapter 3, Section 2, optimum operation of a Fabry-Perot spectrometer for the determination of Doppler shifts and widths requires a low value ($\cong 0.80$) for the reflectivity. When this result is coupled with the findings in this section, i.e., for near-theoretical operation a smaller etalon size is needed at lower reflectivities, then the most efficient etalon for a given physical size is that etalon operated at the lower reflectivities.

3.2. Temporal behavior

Implicit in the description of the Fabry-Perot in the previous sections of this Chapter, as well as Chapter 2, Section 1, is the presence of a static wavefront. This was necessary because the etalon gives rise to fringes by interference between a number of beams with a fixed phase difference between consecutive beams. This phase difference is attained by a delay time between these consecutive beams. Therefore, the interference fringes are formed after a finite time, or response time, after incidence of a wavefront has elapsed, because there is a finite time for the interfering beams to build up. Also the beams must be simultaneously present in time for their interference to occur (Bradley *et al.*, 1964). Kastler (1974) has addressed this topic, and Roychoudhuri (1975) has expanded it in detail. Their approach will be followed here. Referring to Equation 2.1.1b), the delay, or transit time between consecutive beams has been defined for normal incidence to be (Steel, 1967):

$$t_0 = 2 \, \mu \, d \, c^{-1} = \mu \, \Delta \nu^{-1} = \mu \, c^{-1} \, \Delta \sigma^{-1} \; , \qquad 5.3.20a$$

where, as usual, μ is the index of refraction of the medium between the etalon mirrors, d is the distance between them, c is the speed of light, and $\Delta \nu$ ($\Delta \sigma$) is the free spectral range, as defined in Equation 2.1.10), in frequency (wavenumber) units. For other than normal incidence, the delay time is defined as:

$$t = t_0 \cos\Theta = 2 \, \mu \, d \, c^{-1} \cos\Theta \; . \qquad 5.3.20b$$

The characteristic time for the etalon response is then given by $m \, t$, where m is the number of beams necessary to give rise to near-ideal static wavefront fringes. By similarity with Equation 5.3.7b), for a finite number of beams, the transmission amplitude for m interfering beams is:

$$A_{t,m}(\delta) = \tau \, [\, 1 - (Re^{i\delta})^m] \, (\, 1 - Re^{i\delta})^{-1} \; , \qquad 5.3.21$$

while the reflection amplitude is:

$$A_{r,m}(\delta) = R^{1/2}[\, 1 - (R+\tau)e^{i\delta} + \tau R^m \, e^{i\delta(m+1)}]$$

$$\times \ (1 - R e^{i \delta})^{-1} \ , \qquad\qquad\qquad 5.3.22$$

where $\delta = 2 \pi \nu t$. The intensity, for both transmission and reflection, is then obtained:

$$Y_{t,m}(\delta) = Y_t(\delta)\,[\,1 + R^{2m} - 2R^m \cos(m\,\delta)\,]\,, \qquad\qquad 5.3.23$$

$$Y_{r,m}(\delta) = Y_r(\delta) + [\,1 - 2R\cos(\delta) + R^2\,]^{-1}$$

$$\times \left\{ 2\tau R^{m+1} \left[\cos[\delta(m+1)] - (R+\tau)\cos(m\,\delta) + \tau R^m 2^{-1} \right] \right\}. \qquad 5.3.24$$

In the above expressions, $Y_r(\delta)$ and $Y_t(\delta)$ have been defined in Equations 2.1.6a) and 2.1.7a). Equation 5.3.23) can also be cast in terms of Equation 2.1.7d) (Roychoudhuri, 1975) as:

$$Y_{t,m}(\delta) = \tau_m\,[\,1 + \mathbf{F}_m \sin^2(2^{-1}m\,\delta)\,]\,[\,1 + \mathbf{F}\sin^2(2^{-1}\delta)\,]^{-1}\,. \qquad 5.3.25$$

$\mathbf{F}$ has already been defined in Equation 2.1.13c), and the two other quantities in Equation 5.3.25) are:

$$\tau_m = [\,\tau\,(1 - R^m)\,(1 - R)^{-1}\,]^2 < [\,\tau\,(1 - R)^{-1}\,]^2\,, \qquad\qquad 5.3.26$$

$$\mathbf{F}_m = 4R^m\,(1 - R^m)^{-2} < \mathbf{F}. \qquad\qquad 5.3.27$$

Like the finite-size etalon of the earlier section, Equations 5.3.23) and 5.3.25) show the existence of transmission maxima of value smaller than those of the ideal etalon, as well as higher transmission minima (with the associated lower contrast) and wider profile widths. Note, in particular, the oscillatory behavior of the minima as a function of m. Equation 5.3.24) indicates higher reflection than the ideal etalon for $\delta = 0$, while for $\delta = \pi$, the reflection again shows the oscillatory behavior associated with m.

The number of beams, $\mathbf{M}$, necessary to give a near-ideal transmission etalon behavior, can be found after application of an arbitrary criterion defining the meaning of near-ideal behavior. Using the transmission maxima of the fringes as an example, i.e., $\delta = 2\pi k$, it is then possible to define a quantity $f < 1$, to be the general criterion for this near-ideal etalon behavior. Then, from Equation 5.3.23):

$$Y_{t,m}(\delta_{\max}) = f\ Y_t(\delta_{\max}) = Y_t(\delta_{\max})\,(1 - R^{\mathbf{M}})^2\,. \qquad 5.3.28a$$

The solution for $\mathbf{M}$ is:

$$\mathbf{M} = \ln(\,1 - f^{1/2})\,[\,\ln(R)\,]^{-1}\,. \qquad\qquad 5.3.28b$$

Since the reflectivity can be related to the width of the fringe profile using the finesse N_R of Equation 2.1.13), the number of beams [using the approximate finesse of Equation 2.1.13b)] is:

$$\mathbf{M} = \ln(1 - f^{1/2}) \left\{ \ln \left[1 + \pi^2 (2 \, N_R)^{-2} \right. \right.$$

$$\left. \left. \times \ [1 - (1 + 4 \, N_R^2 \, \pi^{-2})^{1/2}] \right] \right\}^{-1} . \qquad \text{5.3.28c}$$

For near-ideal etalon behavior, given by a value of $f = 0.99$, the relation between $\mathbf{M}$ and the finesse is:

$$\mathbf{M} \cong 1.7 \, N_R \, , \qquad \text{5.3.28d}$$

while for $f = 0.997$, the number of beams is:

$$\mathbf{M} \cong 2.0 \, N_R \, . \qquad \text{5.3.28e}$$

Since this last value is in agreement with the finding of Equation 3.2.36), i.e., $2 \, N_R$ coefficients are necessary to unambiguously determine a Fabry-Perot profile, as well as to be consistent with Roychoudhuri (1975), the value of $\mathbf{M}$ of Equation 5.3.28e) will be adopted here.

The response time of a Fabry-Perot is then given as the product of the individual beam time-delay and the number of beams necessary to give near-ideal etalon fringes, or:

$$t_r = \mathbf{M} \, t = \mathbf{M} \, t_0 \, \cos\Theta = \mathbf{M} \, 2 \, \mu \, d \, c^{-1} \cos\Theta$$

$$\cong 2 \, N_R \, \Delta\nu^{-1} \, \cos\Theta \, . \qquad \text{5.3.29}$$

The outermost fringes ($\Theta > 0$) will be completely formed first, since $\mathbf{M} \, t_0 \, \cos\Theta < \mathbf{M} \, t_0$, thus, observations with a fast detector will give the appearance of the fringes walking toward the central fringe. Similarly, the central fringe will tend to disappear last upon removal of the beam (Roychoudhuri, 1975). The specific behavior at an arbitrary angle of incidence can be easily derived from Equations 5.3.23) or 5.3.24), for either constructive (maxima) or destructive (minima) interference. For the central order, this is given by:

$$Y_{t,m} (0) = [\tau \, (1 - R^m) \, (1 - R)^{-1}]^2$$

$$= [\tau \, (1 - R)^{-1}]^2 \, [1 - \exp(m \, \ln R)]^2 \, , \qquad \text{5.3.30a}$$

$$Y_{t,m} (\pi) = \left[\tau \, [1 - (-R)^m] \, (1 + R)^{-1} \right]^2$$

$$= [\tau \, (1 + R)^{-1}]^2 \left[1 - \exp[m \, (\ln R + i\pi)] \right]^2 \, , \qquad \text{5.3.31a}$$

$$Y_{r,m,A=0}(0) = R^{2m+1} = \exp[(2m+1) \ln R] \, , \qquad \text{5.3.32}$$

$$Y_{r,m,A=0}(\pi) = R \, (1 + R)^{-2} \left\{ 4 + (1 - R) \, [\, (1 - R) R^{2m} - 4(-R)^m] \right\} . \quad \text{5.3.33}$$

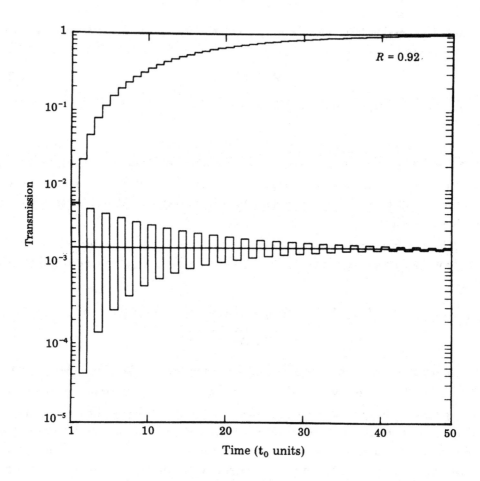

FIGURE 5.27. Response of an etalon to a pulse of length δt greater than $\mathbf{M} t$. The upper curve is $Y_{t,m}(0)$ and the oscillating curve is $Y_{t,m}(\pi)$.

For large values of m, expansion of the exponential term in both Equations 5.3.30a) and 5.3.31a), coupled with the neglect of second order terms, provides the following approximation (Kastler, 1974):

$$Y_{t,m}(0) \simeq [\,\tau\,(1-R\,)^{-1}]^2\,[1-2\,\exp(\,m\,\ln R\,)\,]\,, \qquad 5.3.30b$$

$$Y_{t,m}(\pi) \simeq [\,\tau\,(1+R\,)^{-1}\,]^2\,\Big[1-2\,\exp[\,m\,(\,\ln R\,+i\,\pi\,)\,]\,\Big]\,. \qquad 5.3.31b$$

These approximations show the transmission intensities, at $\delta = 0$, to asymptotically approach steady state with a characteristic time equal to $[\ln(R^{-1})]^{-1}$, while the transmission intensities at $\delta = \pi$ show the oscillatory behavior described earlier. These oscillations can be separated into their even and odd envelopes, namely:

$$Y_{t,k,even}(\pi) = [\tau(1+R)^{-1}(1-R^{2k})]^2 , \qquad 5.3.31c$$

$$Y_{t,k,odd}(\pi) = [\tau (1+R)^{-1}(1-R^{2k+1})]^2 . \qquad 5.3.31d$$

A graphical solution of the behavior of the transmission intensity as a function of $m t_0$ is given in Figure 5.27, for a value of $R = 0.92$.

The effects on the resolving power of a given Fabry - Perot by a limited-time pulse can be ascertained as follows: the resolving power, **R**, as defined in Equation 2.1.15), can be written as:

$$\mathbf{R} = \sigma \, a^{-1} = \nu \, (\delta\nu)^{-1} = n_0 \, N_R$$

$$= N_R \, 2 \, \mu \, d \, \lambda^{-1} \cos\Theta = N_R \, \nu \, t_0 \, \cos\Theta$$

$$\simeq \lambda^{-1} \mathbf{M} \, 2 \, \mu \, d \cos\Theta = \nu \mathbf{M} \, t_0 \, \cos\Theta , \qquad 5.3.34$$

where n is the order number. The last two expressions indicate that the resolving power is the number of wavelengths between the first and last interfering wavefronts, and is proportional to the time difference between these two wavefronts. Thus, in order to provide near-ideal etalon fringes, a light pulse must be δt long such that:

$$\delta t \geq \mathbf{M} \, t_0 \, \cos\Theta = (\, \delta\nu \,)^{-1} . \qquad 5.3.35a$$

From this expression it is then found:

$$\mathbf{M} \leq \delta t \, (\, t_0 \, \cos\Theta \,)^{-1} , \qquad 5.3.36$$

or, the longer the duration of the pulse, the greater the resolving power, provided the Fabry - Perot has the potential for such a resolving power (Roychoudhuri, 1975). Note, in particular, when $\delta t < t_0$, no interferometric response is possible.

Equation 5.3.35a) can be rearranged to give the following inequality:

$$\delta t \, \delta\nu \geq 1 . \qquad 5.3.35b$$

Also, for a pulse to give rise to any interference, its length must be:

$$\delta t \, > \, t_0 . \qquad 5.3.37$$

This last expression, when combined with Equation 5.3.35b), leads to:

$$(\, t_0 \,)^{-1} \, > \, \delta\nu . \qquad 5.3.38$$

Replacing t_0 by its definition, given in Equation 5.3.20), Equation 5.3.38) becomes:

$$\Delta\nu \, > \, \delta\nu ,$$

$$\Delta\sigma \; > \; \delta\sigma \; . \hspace{6cm} 5.3.39$$

This last equation gives the (expected) result that, in order to have an interferometric response, it is necessary to have the instrumental free spectral range larger than the width of the profile to be analyzed.

The explicit introduction of time in this section allows examination of the Fabry-Perot etalon as a timing device. As defined in Equation 2.1.1b) and 5.3.20), there is a finite transit time for radiation associated with an etalon gap of spacing d and direction of incidence Θ. For instance, at normal incidence and unity index of refraction, a 1.0 cm gap has a transit time of 66.6 picoseconds. For incidence other than normal, this transit time changes as described in Equation 5.3.20b).

The Kastler (1974) treatment given earlier in Equations 5.3.30) through 5.3.33) provides a good example for the timing properties of a Fabry-Perot etalon. Suppose a finite-length light pulse, of square shape in this instance, of duration $q\,t_0$ being examined by an etalon at normal incidence and with characteristic transit time t_0. At first the transmission intensity will rise as given in Figure 5.27 and, after the pulse has elapsed, this intensity will decrease as given by the interference of the remaining rays in the etalon. The expression describing this decay can be ascertained by the use of Equation 5.3.21), and the result is:

$$Y_{t,k,q}(0) \; = \; R^{2k}(\,1 - R^{q}\,)^2 \; , \hspace{4cm} 5.3.40$$

where k is the number of reflections elapsed since the termination of the pulse. Figure 5.28 gives the graphical solution for a pulse 10t units long. Note that, in principle, the use of a short pulse will directly provide the reflectivity of the etalon mirrors, since the slope of the decay curve, as shown in Figure 5.28, is equal to $2\,\ln R$.

On the other hand, for a series of equally spaced coherent pulses, it is possible to find an etalon spacer that matches the pulse repetition time. Although each pulse produces a chain of non-overlapping pulses, the overlapping pulses from different chains will interfere, since by definition all of the original pulses are coherent. Therefore, a slowly scanning Fabry-Perot etalon at the proper spacing will reproduce the spectrum of the original coherent source (Roychoudhuri, 1975). When the etalon gap is set at a spacing other than t_0, there exists an angle of incidence Θ at which the transit time of the etalon is equal to the interpulse time. At this angle a near-ideal fringe will be formed, thus the time between pulses can be obtained by application of 5.3.20b) (Roychoudhuri, 1975):

$$\Delta t \; = \; t_0 \, \cos\Theta \; = \; n \; \nu^{-1} \; , \hspace{4cm} 5.3.41$$

where n is the order at which the near-ideal fringe is formed. The width of the pulses can be estimated by measuring the properties of fringes away from this near-ideal fringe (Roychoudhuri, 1975).

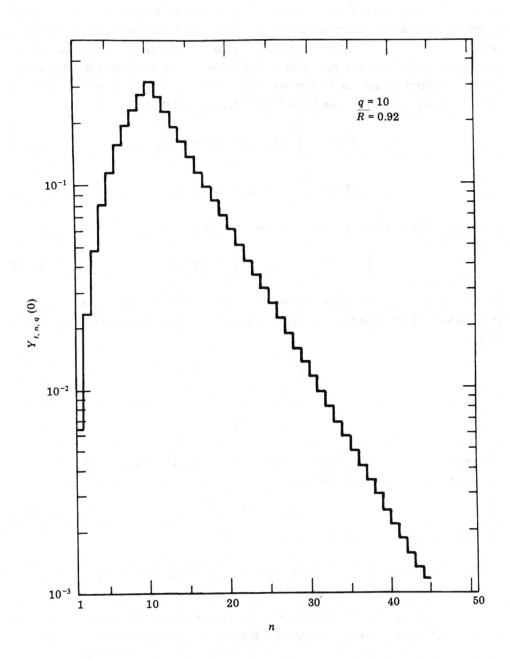

FIGURE 5.28. Response of an etalon to a pulse of length equal to 10t, as a function of the number of reflections.

This temporal behavior of the Fabry-Perot can be carried to its logical conclusion, namely the investigation of the etalon response to infinitesimally short light pulses. Although no interference effects will occur, the response of the etalon to these pulses will provide the instrumental function (Connes, 1961; Stoner, 1966) just as is done in electrical filter theory. This treatment relates the transient response of the instrument to its transmittance as described below. First, a recollection of the properties of a function and its Fourier transform is necessary for this method. A function $\mathbf{f}$ is associated with its transform F by (Bracewell, 1965):

$$F(s) = \int_{-\infty}^{\infty} \mathbf{f}(t) \, \exp(-i\, 2\pi\, ts\,) \, dt \; , \qquad\qquad 5.3.42$$

$$\mathbf{f}(t) = \int_{-\infty}^{\infty} F(s) \, \exp(i\, 2\pi\, ts\,) \, ds \; . \qquad\qquad 5.3.43$$

Further, Rayleigh's theorem states (Bracewell, 1965):

$$\int_{-\infty}^{\infty} \big| \; F(s) \; \big|^{2} \, ds \; = \int_{-\infty}^{\infty} \big| \; \mathbf{f}(t) \big|^{2} \, dt \; . \qquad\qquad 5.3.44$$

In order to examine the transient response of a Fabry-Perot etalon, an extremely short pulse of light must be employed. The mathematical properties of this pulse, $\mathbf{f}(t)$, are:

$$\mathbf{f}(t) = \begin{cases} 0 & t \neq 0 \\ \neq 0 & t = 0 \end{cases} , \qquad\qquad 5.3.45$$

$$\int_{-\infty}^{\infty} \big| \; \mathbf{f}(t) \big|^{2} \, dt \; =1 \; . \qquad\qquad 5.3.46$$

The amplitude distribution of this light pulse is $F(s)$ of Equation 5.3.42) and the frequency spectrum $I(s)$ is defined as:

$$I(s) = \big| \; F(s) \big|^{2} \; . \qquad\qquad 5.3.47$$

The total energy of the pulse is then expressed by:

$$\int_{-\infty}^{\infty} I(s) \, ds \; = \int_{-\infty}^{\infty} \big| \; F(s) \big|^{2} \, ds \; = \int_{-\infty}^{\infty} \big| \; \mathbf{f}(t) \big|^{2} \, dt \; = 1 \; . \qquad\qquad 5.3.48$$

When an etalon is illuminated, at normal incidence, by this pulse of light, the response $a(t)$ of the device is a series of delayed pulses given by (Stoner, 1966):

$$a(t) = t_1 t_2 \sum_{k=0}^{\infty} r_2^{2k} \; \mathbf{f}(t - 2kd\, c^{-1}) \; , \qquad\qquad 5.3.49$$

where r_2 and t_i are the complex reflection and transmission coefficients of the coatings, d is the spacing between the mirrors, and c is the speed of light. The frequency response of the etalon, per pulse, is:

$$\hat{A}(s) = \int_{-\infty}^{\infty} a(t)\,\exp(-i\,2\pi\,ts)\,dt\;\left[\int_{-\infty}^{\infty} f(t)\,dt\right]^{-1}, \qquad 5.3.50a$$

while the transmittance is defined (Stoner, 1966) as:

$$T(s) = \left|\,\hat{A}(s)\,\right|^{2}. \qquad 5.3.51a$$

The specific solutions for the above are:

$$\hat{A}(s) = t_1\,t_2\,\sum_{k=0}^{\infty} r_2^{\,2k}\,\exp(i\,4\pi\,skd\,c^{-1}), \qquad 5.3.50b$$

$$T(s) = (t_1\,t_2)^2\,\left|\,\sum_{k=0}^{\infty} r_2^{\,2k}\,\exp(i\,4\pi\,skd\,c^{-1})\,\right|^{2}, \qquad 5.3.51b$$

$$T(\sigma) = \tau^2\,[\,1 + R^2 - 2R\,\cos(2\pi\,2d\,\sigma)\,]^{-1}. \qquad 5.3.51c$$

In the last equation the relations between r_i, t_i, τ and R of Chapter 2, Section 1, and the equality $\sigma = s\,c^{-1}$ have been used. Equation 5.3.51c) can be recognized to be the same as Equation 2.1.7a), obtained by classical methods. Other properties of the etalon can be found with this method. For instance, the average transmittance, or transmitted energy, is:

$$<\,T(s)\,>_{ave} = \int_{-\infty}^{\infty} \left|\,a(t)\,\right|^{2}\,dt = (1-R)(1+R)^{-1}. \qquad 5.3.52$$

In the above result, the absorption coefficient, A, has been set, for convenience, equal to zero. Equation 5.3.52) can be identified with Equation 2.2.6) derived earlier. The same result given in Equation 5.3.52), could have been obtained by integration of Equation 5.3.51c) over one order.

Although in the previous discussion the short light pulses have been described by their mathematical properties, little has been said about the specific functions that would have this behavior. The null function $\delta^{1/2}(x)$ satisfies the requirements set in Equations 5.3.45) and 5.3.46) (Hernandez, 1985b), and it is defined by a limiting sequence (Bracewell, 1965):

$$\delta^{1/2}(x) = \lim_{b\,\to\,0}\,b^{-1/2}\,\Pi(xb^{-1}), \qquad 5.3.53$$

where $\Pi(x)$ has been defined in Equation 2.5.5). The properties of $\delta^{1/2}(x)$ then are found to be:

$$\delta^{1/2}(x) = \begin{cases} 0 & x \neq 0 \\ \infty & x = 0 \end{cases}, \qquad 5.3.54$$

$$\int_{-\infty}^{\infty} \delta^{1/2}(x)\,dx = 0\,; \qquad \int_{-\infty}^{\infty} \left|\,\delta^{1/2}(x)\,\right|^{2}\,dx = 1. \qquad 5.3.55$$

The property expressed in Equation 5.3.50b) can be deduced easily from the defining sequence of Equation 5.3.53), i.e.:

$$\hat{A}(s) = \lim_{b \to 0} \left\{ \left[b^{-1/2} t^2 \sum_{k=0}^{k=\infty} r_2^{2k} \int_{-\infty}^{\infty} \Pi[(t - 2kd\,c^{-1})b^{-1}] \exp(-i\,2\pi\,ts)\, dt \right] \right.$$

$$\left. \times \left[b^{-1/2} \int_{-\infty}^{\infty} \Pi(tb^{-1})\, dt \right]^{-1} \right\}$$

$$= t_1 t_2 \sum_{k=0}^{k=\infty} r_2^{2k} \exp(-i\,4\pi\,skd\,c^{-1}) . \qquad\qquad 5.3.50b$$

The behavior of an etalon to rapid changes in optical length (Gerardo *et al.*, 1965; Dangor and Fielding, 1970) can be investigated using the techniques employed in this section.

4. Conclusions

In this chapter some of the fundamental advantages, and limitations, of the Fabry-Perot interferometer have been investigated in some detail. Section 1 presents the Fabry-Perot as the spectroscopic device with the highest luminosity at a given resolving power, for those instruments which depend on the interference of light for their operation and directly provide a spectrum. In a nutshell, this is the reason for the continued usage of the Fabry-Perot in all its variants. Section 2 shows how the luminosity of the Fabry-Perot can be maximized, however, at the expense of not being able freely to change the instrumental configuration. The last two sections review the behavior of the instrument with a limited number of interfering beams. The findings in these two sections indicate, in a quantitative way, the need for a minimum finite-size etalon for near-theoretical behavior, as well as the changes in response of the device when illuminated by other than a static wavefront.

Chapter 6

Active media in the etalon spacer

1. Introduction

In the previous chapters, the Fabry-Perot has been treated as a purely spectroscopic tool employed to examine a given spectral region; the sources of the radiation (and its alteration) were external to the etalon. In the present chapter a more active role of the etalon in shaping the radiation field will be discussed.

The properties of the radiation field within the etalon gap will be discussed first, as they will provide the background for further development. Consider the radiation reaching a point P, located at a distance z from the upper substrate of an etalon as illustrated in Figure 6.1, where this radiation can be considered to be proceeding from two counter-propagating beams given by the solid and dashed lines (Kastler, 1962). For convenience, the forward propagating beam(s) from the upper substrate associated with the solid lines will be denoted by the subscript 1, while the backward propagating beam(s) from the lower substrate, i.e., the dashed lines, will be associated with the subscript 2. Referring to Figure 6.1, the phase lag of the solid line beams, with respect to ray 0 at P, is found to be:

$$\phi_k = k \ \phi = k \ 2\pi \ 2\mu d \sigma \ \cos \Theta + 2k \chi \ ; \qquad k = 0, 1, 3.... \ , \qquad 6.1.1$$

while the lag of the backward propagating beam(s) is:

$$\psi_k = k\phi - \gamma + (2 k - 1) \chi \ ; \qquad k = 1, 2, 3.....$$

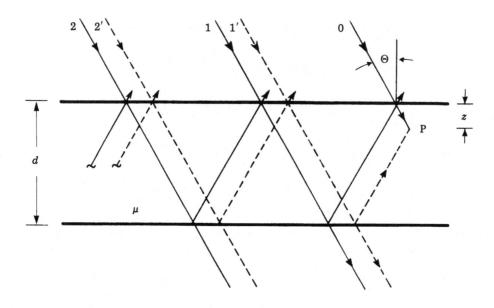

FIGURE 6.1. Geometry used to study the radiation field inside an etalon. After Kastler (1962).

$$= k \ 2\pi \, 2\mu d\sigma \ \cos \Theta \ - \ 2\pi \, 2\mu z\sigma \ \cos \Theta \ + \ (2k-1)\chi \ . \qquad \qquad 6.1.2$$

In the two previous equations, the nomenclature of Chapters 2 and 5 is being used and needs no repetition. Therefore, the amplitude of the two counter-propagating beams is:

$$A_1(\phi) = t_1 \sum_{k=0}^{k=\infty} [\, r_2^2 \ e^{i(\phi+2\chi)} \,]^k = t_1 [\, 1 - r_2^2 \ e^{i(\phi+2\chi)} \,]^{-1} \ , \qquad 6.1.3$$

$$A_2(\phi,\gamma) = t_1 r_2 \ e^{i(\phi-\gamma+\chi)} \sum_{k=0}^{k=\infty} [\, r_2^2 \ e^{i(\phi+2\chi)} \,]^k$$

$$= t_1 r_2 \ e^{i(\phi-\gamma+\chi)} [\, 1 - r_2^2 \ e^{i(\phi+2\chi)} \,]^{-1} \ . \qquad 6.1.4$$

The amplitude and intensity at point P is found to be:

$$A_p(\phi,\gamma) = t_1 [\, 1 + r_2 \ e^{i(\phi-\gamma+\chi)} \,][\, 1 - r_2^2 \ e^{i(\phi+2\chi)} \,]^{-1} \ , \qquad 6.1.5$$

$$Y_p(\phi,\gamma) = A_p(\phi,\gamma) \times A_p^*(\phi,\gamma) = \tau \, [\, 1 + R + 2 \, R^{1/2} \cos(\phi-\gamma+\chi) \,]$$

$$\times \, [\, 1 + R^2 - 2 \, R \ \cos(\, \phi + 2\,\chi \,) \,]^{-1} \ . \qquad 6.1.6$$

The last equation can be simplified in terms of the intensities of the separate counter-propagating beams, i.e.:

$$Y_1(\phi) = \tau \, [\, 1 + R^2 - 2 \, R \ \cos (\, \phi + 2\,\chi \,) \,]^{-1} \ , \qquad 6.1.7$$

$$Y_2(\phi) = \tau \, R \, [\, 1 + R^2 - 2 \, R \, \cos (\, \phi + 2 \, \chi \,) \,]^{-1} = R \; Y_1(\phi) \, , \qquad 6.1.8$$

$$Y_p(\phi, \gamma) = Y_1(\phi) + Y_2(\phi) + 2[Y_1(\phi) \; Y_2(\phi)]^{1/2} \cos (\, \phi - \gamma + \chi \,)$$

$$= \{ \, [Y_1(\phi)]^{1/2} - [Y_2(\phi)]^{1/2} \, \}^2$$

$$+ \, 4 \, [Y_1(\phi) \; Y_2(\phi)]^{1/2} \cos^2 [(\phi - \gamma + \chi) \, 2^{-1}] \, . \qquad 6.19a$$

Note that, for values of $R \to 1$, $Y_1(\phi) \approx Y_2(\phi)$ and the intensity is approximately given by:

$$Y_p(\phi, \gamma) \approx 4 \; Y_1(\phi) \; R^{1/2} \cos^2 [\, (\phi - \gamma + \chi) \, 2^{-1} \,] \, , \qquad 6.1.9b$$

and which for the maximum value of $Y_1(\phi)$, or $Y_1(\phi_{mz}) = (\, 1 - R \,)^{-1}$, then:

$$Y_p(\phi_{mz}, \gamma) \simeq 4 \, R^{1/2} \, (1 - R)^{-1} \cos^2 [\, (\phi_{mz} - \gamma + \chi) \, 2^{-1} \,]$$

$$\simeq 4 \, \pi^{-1} \, N_R \, \cos^2 [\, (\phi_{mz} - \gamma - \chi) \, 2^{-1} \,] \, , \qquad 6.1.9c$$

where N_R is the approximate finesse of Equation 2.1.13b). The approximation given in Equation 6.1.9c) shows near-zero intensity at the nodes and large values at the antinodes. Using a value of $R = 0.98$, the maximum intensity at an antinode at point P will be given by:

$$Y_p(\phi_{mz}, \gamma)_{mz} \simeq 1.27 \, N_R \simeq 198 \, , \qquad 6.1.10$$

which is considerably larger in intensity than the incoming beam. The intensity at the nodes can be easily found from Equation 6.1.9a), using the previous value of $R = 0.98$, to be:

$$Y_p(\phi_{mn}, \gamma)_{mn} = \{ \, [Y_1(\phi_{mn})]^{1/2} - [Y_2(\phi_{mn})]^{1/2} \, \}^2$$

$$= (1 - R)(1 - R^{1/2})^2 \, (1 + R)^{-2} \simeq 5.2 \times 10^{-7} \, . \qquad 6.1.11$$

This value is lower than either of the minima of $Y_1(\phi)$ and $Y_2(\phi)$ or $\sim 5 \times 10^{-3}$.

2. Emission within the cavity

The behavior of an etalon cavity filled with atoms capable of emission (this emission is presumed incoherent and isotropic), can be described in terms of the expressions derived in the previous section. The arrangement is as that already shown in Figure 6.1, namely an elemental volume of atoms at location P is allowed to emit radiation, and the resultant field is calculated (Kastler, 1962). A sketch of the system studied is given in Figure 6.2. As would be expected, the amplitudes of the beams given by the solid and the dashed lines in the figure are those already found in Equations 6.1.3) and 6.1.4) and the resultant intensity by Equation 6.1.9). Using the approximation of Equation 6.1.9b), the intensity of radiation from the elemental

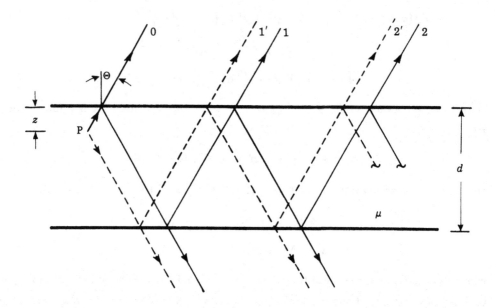

FIGURE 6.2. Schematic of the geometry used to investigate a cavity containing an emitting species. After Kastler (1962).

volume at P, containing N emitting atoms, is:

$$d\,[\,Y_p\,(\phi,\gamma)\,] = 4\ Y_1(\phi)\ R^{1/2}\ N\ \cos^2[(\phi-\gamma+\chi)\ 2^{-1}\,]\ dv\ .\qquad 6.2.1$$

For isotropic radiation, when N is independent of direction, the total intensity is:

$$\int d\,[\,Y_p\,(\phi,\gamma)\,] = \mathbf{I}\,\mathbf{A}$$

$$= 4\ Y_1(\phi)\ R^{1/2}\ N\ \int_v \cos^2\,[(\phi-\gamma+\chi)\ 2^{-1}]\ dv\ .\qquad 6.2.2$$

In this expression $\mathbf{A}$ is an area, $\mathbf{I}$ is the irradiance and the integration is carried over the volume of the emitting atoms in the cavity. When $dv = dx\ dy\ dz$ and $\int_{x,y} dx\ dy$ defines the area $\mathbf{A}$, then:

$$\mathbf{I} = 4\ Y_1(\phi)\ R^{1/2}\ N\ \int_z \cos^2\,[(\phi-\gamma+\chi)\ 2^{-1}]\ dz\ .\qquad 6.2.3a$$

Since γ [cf. Equation 6.1.2)], in the argument of the cosine of the integral, is the only term dependent upon z, integration over a length which is an integer number (m) of half-wavelengths in the direction Θ gives:

$$\mathbf{I} = 2 \ Y_1(\phi) \ R^{1/2} Nm \ . \qquad \text{6.2.3b}$$

Thus, the observed radiation leaving the etalon is found as a set of fringes obeying the usual behavior of the Fabry-Perot etalon. Should the (etalon) mirrors be absent, the radiation emitted will be given by:

$$\mathbf{I}_o \ \mathbf{A} = \int_v N \ dv = N \mathbf{A} m \ . \qquad \text{6.2.4}$$

Therefore, the ratio of $\mathbf{I}$ over $\mathbf{I}_0$, for $Y_1(\phi)_{mz}$, is:

$$\mathbf{I}_{mz} \ \mathbf{I}_0^{-1} = 2 \ R^{1/2} (1 - R)^{-1} \simeq 0.64 \ N_R \ . \qquad \text{6.2.5a}$$

For a mirror reflectivity of $R = 0.98$, used in the earlier example,

$$\mathbf{I}_{mz} \ \mathbf{I}_0^{-1} \simeq 99 \ . \qquad \text{6.2.5b}$$

Therefore, at a given angle, the etalon has redistributed the radiation into fringes, instead of the original isotropic radiation. The mean irradiance per free spectral range can be calculated to be:

$$< \mathbf{I} > = (\Delta\Omega)^{-1} \int_\Omega \mathbf{I} \, d\Omega \ , \qquad \text{6.2.6a}$$

where $d\Omega$ is the elemental solid angle and $\Delta\Omega$ is the solid angle associated with one order. The solution of Equation 6.2.6a), in terms of its ratio over $\mathbf{I}_0$, is:

$$< \mathbf{I} > \mathbf{I}_0^{-1} = 2 \ R^{1/2} (1 + R)^{-1} \ , \qquad \text{6.2.6b}$$

$$\lim_{R \to 1} \ < \mathbf{I} > \mathbf{I}_0^{-1} = 1 \ . \qquad \text{6.2.6c}$$

This result shows that the energy of the system is conserved (Kastler, 1962).

When the emitting gas gives rise to a single emission line, of some arbitrary HWHH, the spectral distribution of the fringes is given by the free spectral range of the etalon and the intensity by the envelope of the emission line profile. This behavior is like the combination of an etalon and an interference filter, discussed in Chapter 4, Section 3. The properties of $Y_1(\phi)$ of Equation 6.1.7) show that, for a given angle Θ, the spectrum of the radiation is as illustrated in Figure 6.3 (cf. Figure 4.9). Thus, depending on the relative widths of the emission line profile and the etalon free spectral range, for an arbitrary angle Θ, the spectral distribution is similar to that given in the figure, except for the specific location of the etalon transmission maxima on the line profile. Note that the spectral width of the fringes is given by $Y_1(\phi)$ of Equation 6.1.7), that is, the resultant width of the profiles is defined by the etalon reflectivity. As would be expected, for an emission line narrower than the etalon profile, the appearance of visual bright fringes will be the resultant effect. Therefore, an etalon cavity filled with an emitting species can be utilized to generate line (emission) profiles whose width is controlled by the etalon parameters.

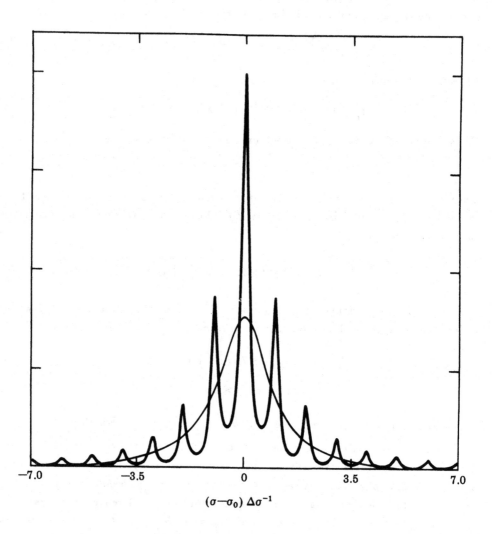

FIGURE 6.3. Resultant spectrum from an etalon containing a line emitting species.
The original emission is also shown.

All the previous derivations presume the emitting gas is dilute and no stimulation of the emission occurs. Also note that the derivations are for an plane ideal
Fabry-Perot of unlimited extent. The limitations on the results given by the use of a
finite size etalon will be investigated in the next section, where an absorbing medium
in the etalon gap will be discussed.

3. Absorption in the cavity

The introduction of an absorbing species in the etalon cavity has been reported by Jackson (1961) and later discussed in detail (Kastler, 1962; Hernandez, 1985c). To model the behavior of an etalon cavity containing an absorbing medium, it will be assumed that the light reaching the etalon is nearly monochromatic, as would be obtained from a tunable laser or a high-resolution predispersing device and, further, its width is negligible with that of the etalon. The absorbing medium in the etalon cavity is considered to have a constant index of refraction and has associated with it a complex transmission, b, which is allowed to be wavenumber-dependent, when necessary. The transmittance of the absorbing species is also defined as:

$$k = b \times b^{*}, \qquad\qquad 6.3.1$$

where k is real and no phase lag is associated with the absorption species. Thus, the transmitted amplitude of a monochromatic ray, incident at an angle Θ to a plane etalon normal, is, by similarity to Equation 2.1.5):

$$A_p(b,\phi_p) = b \; t_1 \, t_2 \sum_{k=0}^{k=\infty} (\, b^2 \, r_2^2 \, e^{i\phi_p} \,)^k$$

$$= b \; t_1 \, t_2 \, (\, 1 - b^2 \, r_2^2 \, e^{i\phi_p} \,)^{-1} . \qquad\qquad 6.3.2$$

The notation employed is that used earlier in Chapters 2 and 5 and in this Chapter and needs no further elaboration. The subscript p is used here to indicate a plane etalon, as opposed to a spherical etalon for which the subscript s will be used later. The resultant intensity is then found to be:

$$Y_p(b,\delta_p) = A(b,\delta_p) \times A^{*}(b,\delta_p)$$

$$= k \; r^2 \, [\, 1 + k^2 \, R^2 - 2 \, k \, R \, \cos(\delta_p) \,]^{-1} , \qquad\qquad 6.3.3$$

$$\delta_p = \phi_p + 2\,\chi = 2\,\pi\,2\,\mu\,d\,\sigma\,\cos\Theta + 2\,\chi . \qquad\qquad 6.3.4$$

For completeness, the equivalent derivation for the Connes' (1956) spherical etalon will be given without further detail, except to note the similarity with the results given in Equation 5.2.11b):

$$Y_s(k,\delta_s) = k \; r^2 \, (\, 1 + k^2 \, R^2 \,) \, [\, 1 + k^4 \, R^4 - 2 \, k^2 \, R^2 \, \cos(\delta_s) \,]^{-1} . \qquad 6.3.5$$

In the above equation $\delta_s = 2^{-1} \delta_p$. Note that in the absence of the absorber, Equations 6.3.3) and 6.3.5) revert to the expressions of the ideal etalon given in Equations 2.1.7a) and 5.2.11b), respectively. For an etalon with mirror coatings with absorption (or scattering) described by A, the definition of r^2 is:

$$r^2 = (1 - R - A)^2 = (1 - R)^2 \, [1 - A \, (1 - R)^{-1}]^2 = (1 - R)^2 \, r_A . \qquad\qquad 6.3.6$$

When the relative transmission of the etalon is defined as the ratio of the intensity transmitted by the etalon cavity in the presence of the absorbing species to the intensity transmitted at the peak by the same cavity in the absence of the absorber, we have:

$$\tau_p (k, \delta_p) = Y_p (k, \delta_p) [Y_p (1,0)]^{-1} , \qquad 6.3.7$$

$$\tau_s (k, \delta_s) = Y_s (k, \delta_s) [Y_s (1,0)]^{-1} . \qquad 6.3.8$$

Insight into the behavior of an etalon containing an absorption species in the cavity can be obtained by investigating the limiting cases when the transmission of the sample, k, approaches both zero and unity. For the purposes of this study, an absorption (ϵ) is associated with the transmission (k), and is defined as:

$$\epsilon = 1 - k . \qquad 6.3.9$$

In the same fashion as in the above expression, it is possible to define the absorption of measurement (at the fringe maximum) by:

$$a (k,0) = 1 - \tau (k,0) . \qquad 6.3.10$$

The relative absorption of measurement is then given as the ratio of Equation 6.3.10) over Equation 6.3.9), or:

$$A (k,0) = \epsilon^{-1} a (k,0) . \qquad 6.3.11a$$

From Equations 6.3.3), 6.3.5), 6.3.6), 6.3.7), and 6.3.8), the relative absorption of measurement for the plane and spherical etalons is given by:

$$A_p (k,0) = [1 + R (1 - R)^{-1}] [1 + R (1 - R)^{-1} \epsilon]^{-2}$$

$$+ [R (1 - R)^{-1}] [1 + R (1 - R)^{-1} \epsilon]^{-1} , \qquad 6.3.12$$

$$A_s (k,0) = \left\{ [1 + R^2 (2 - \epsilon) (1 - R^2)^{-1}] \right.$$

$$\times [1 + R^2 \epsilon (2 - \epsilon) (1 - R^2)^{-1}]^{-2} \Big\}$$

$$+ \left\{ [R^2 (2 - \epsilon) (1 - R^2)^{-1}] \right.$$

$$\times [1 + R^2 \epsilon (2 - \epsilon) (1 - R^2)^{-1}]^{-1} \Big\}$$

$$+ \left\{ [R^2 (2 - \epsilon) (1 - \epsilon) (1 + R^2)^{-1}] \right.$$

$$\times [1 + R^2 \epsilon (2 - \epsilon) (1 - R^2)^{-1}]^{-2} \Big\} . \qquad 6.3.13$$

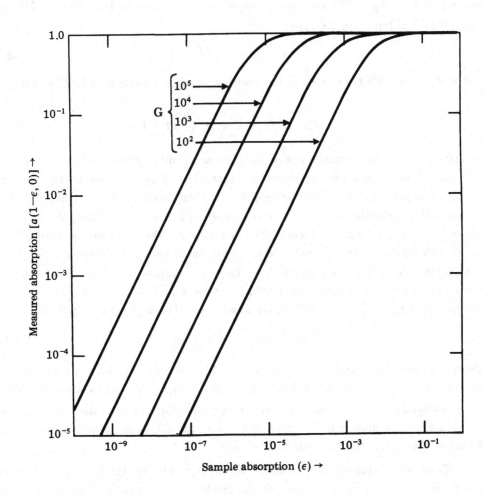

FIGURE 6.4. Etalon-measured absorption as a function of the sample absorption, with the gain, **G**, as a parameter. After Hernandez (1985c).

For a weak absorber, Equations 6.3.12) and 6.3.13) become:

$$\lim_{\epsilon \approx 0} A_p (k,0) = 1 + 2 R (1 - R)^{-1} = 1 + 2 G ,$$ 6.3.14

$$\lim_{\epsilon \approx 0} A_s (k,0) = 1 + 4 R^2 (1 - R^2)^{-1} + 2 R^2 (1 + R^2)^{-1}$$

$$= 1 + 4 G R (1 + R)^{-1} + 2 R^2 (1 + R^2)^{-1}$$

$$\simeq 2 + 2 G ; \qquad R \to 1 ,$$ 6.3.15

$$G = R (1 - R)^{-1} .$$ 6.3.16

The factor **G** has very large values as the reflectivity approaches unity and will be called the gain of the cavity. Note that when $R \to 1$ the value of **G** is given approximately as $\pi^{-1} N_R$. The gain is also associated with the etalon contrast (Hernandez, 1985c) of Equation 2.1.14) by:

$$1 + 2\,\mathbf{G} = \mathbf{C}^{1/2} \,.$$

$$6.3.17$$

For a strong absorber, i.e, $\epsilon \approx 1$, the limiting value of Equations 6.3.12) and 6.3.13) is:

$$\lim_{\epsilon \approx 1} A_p(k,0) = \lim_{\epsilon \approx 1} A_s(k,0) = 1 \,,$$

$$6.3.18$$

or saturation of the system. The weak absorber results given in Equations 6.3.14) and 6.3.15) show that the measured apparent absorption is enhanced relative to a simple absorption cell of the same length. For instance, for a 1 cm cavity spacing etalon with a reflectivity $R = 0.999$ and filled with a weak absorber, the measured absorption would appear to be nearly 2000 times larger than that of the sample for an equal path length cavity. In other words, the etalon cavity behaves as a 20 m path absorption cell. To avoid ambiguity in the interpretation of Equation 6.3.11a), the absorption, ϵ, of the sample should be defined for a unit path length, say 1 cm, and will be called ϵ_0. Therefore the relative absorption of measurement is given by:

$$A_0(k,0) = \epsilon_0^{-1}\, a(k,0) \,.$$

$$6.3.11b$$

Using the previous example of a cavity with $R = 0.999$, but now with a 5 cm long gap spacing, it is found that $A_0(k,0) \simeq 10\ 000$, which is the equivalent of a 100 m path absorption cell. The properties of Equation 6.3.10), and thus also those of Equation 6.3.11a), are illustrated in Figure 6.4. Note the saturation effects deduced from Equation 6.3.18) in the upper part of the figure.

There are examples, other than the absorption cell thus far discussed, where the effects of absorption in the cavity are noticeable. For instance, interference filters (cf. Chapter 2, Section 5.2) are low order etalons with a solid material spacer. When this spacer is a weak absorber (or scatterer), the peak transmission of such a filter relative to an ideal etalon filter, is given by $\tau(k,0)$ of Equation 6.3.7), or:

$$\tau(k,0) = (1 - \epsilon)\, [\, 1 + R\,\epsilon\, (1 - R)^{-1}\,]^{-2}$$

$$= (1 - \epsilon)\, (1 + \mathbf{G}\epsilon)^{-2} \,.$$

$$6.3.19a$$

The width of this filter, again relative to an ideal filter, is found from Equation 6.3.3) equal to:

$$\omega_p(k)\, [\, \omega_p(1)\,]^{-1} = \sin^{-1}\{ (1 - kR)\, [\, 2\, (kR)^{1/2}\,]^{-1} \}$$

$$\times \{ \sin^{-1} [\, (1 - R)\, (\, 2\, R^{1/2}\,)\,]\, \}^{-1}$$

$$\simeq (1 - \epsilon)^{-1/2}\, [1 + R\,\epsilon(1 - R)^{-1}] = (1 - \epsilon)^{-1/2}\, (1 + \mathbf{G}\epsilon)$$

$$\simeq [\ \tau(k,0)\ ]^{-1/2} . \qquad \text{6.3.20a}$$

Thus, for a weak absorber, when $\epsilon \approx 0$ and $G \epsilon < 1$, the previous equations can be simplified to:

$$\lim_{\epsilon \approx 0} \tau_p(k,0) = (1 + G\epsilon)^{-2} \simeq 1 - 2 G \epsilon , \qquad \text{6.3.19b}$$

$$\lim_{\epsilon \approx 0} \omega_p(k) [\ \omega_p(1)\ ]^{-1} \approx 1 + G \epsilon . \qquad \text{6.3.20b}$$

Consider an ideal first order filter in the visible spectrum ($\sigma = 20\ 000$ K) with a FWHH of 5 K. The finesse of this filter is near 4 000, or a value of $G \approx 1\ 300$. If this filter spacer has an absorption $\epsilon \approx 10^{-4}$, the filter transmission $\tau_p(k,0) \simeq 0.78$ and the ratio of the widths is 1.13, which translates to a filter width of 5.65 K. This effect is in addition to others present in interference filters caused by (surface) flatness imperfections, discussed in Chapter 2. Weak absorption by the spacer material may possibly be the limiting factor in achieving high transmission filters. As Equations 6.3.19) and 6.3.20) indicate, higher finesse filters (smaller FWHH) are more sensitive to the effects of spacer absorption.

In the presentation given thus far, it has been implicitly assumed that the etalon under consideration is ideal, namely, it is perfectly aligned, has no surface imperfections and (for the plane etalon) its physical size is as large as is needed. Thus, it is useful to investigate the limitations imposed by not meeting the criteria of such an ideal etalon in a practical situation. As given in Chapter 5, Section 3.1, Vander Sluis and McNally (1956) have shown that for a practical plane etalon to approach the behavior of an ideal etalon, it needs a minimum size which is fixed by the reflectivity and an arbitrary criterion defining a near-ideal etalon. A near-ideal etalon is defined as an etalon having 0.99 of the peak transmission of an ideal etalon (Vander Sluis and McNally, 1956) and the minimum diameter expression has been given in Equation 5.3.17). This equation will be restated here, albeit independent of the etalon spacer:

$$D_{pn} = D\ d^{-1/2} = 200\ R\ (2+R)\ (1-R^2)^{-1}\ (\sigma\ N_R)^{-1/2} . \qquad \text{6.3.21a}$$

The equivalent expression for the spherical etalon can be obtained from Equation 5.2.17b), or:

$$D_{sn} = D\ d^{-3/4} = 2[\ (1-R^2)\ (2\ \pi\ \sigma\ R)^{-1}\ ]^{1/4} . \qquad \text{6.3.22a}$$

For the earlier examples of 1 cm and 5 cm spacing etalons used as absorbing cells, it is found that the plane etalon diameters needed to satisfy the Vander Sluis and McNally (1956) criterion are 37.8 cm and 84.6 cm, respectively. These diameters are larger than presently available high-quality optical flats, i.e., 15 cm diameter. These high-quality substrates have a typical flatness figure of $\lambda/200$ (at $\lambda = 632.8$ nm.), which translates to a limiting finesse of about 100 [cf. Equation 2.3.5)].

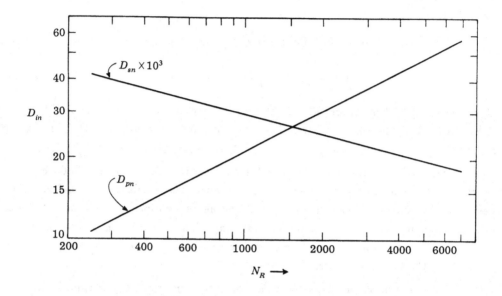

FIGURE 6.5. Useful etalon diameters, normalized to unit spacing, as a function of reflective finesse, for both plane (p) and spherical (s) Fabry-Perot interferometers. The calculations assume σ = 20 000 K. From Hernandez (1985c).

This limiting finesse is quite low, relative to the required finesse of $\sim$ 3 100, which makes impractical the use of plane etalon devices for high gain absorption cells. However, note that interference filters seem to be the exception because the coatings conform to the substrate shape and the useful diameter of these low-order devices is quite small. For the interference filter used in the earlier example, the diameter can be calculated with the help of Equation 6.3.21a) to be no larger than $\sim$ 2 mm diameter. Thus, a practical interference filter is a large collection of elemental filters and, as such, makes the filter relatively independent of the overall quality of the substrate, as attested by the availability of commercial large-diameter (> 10 cm) narrow-band filters.

The Connes' (1956) confocal spherical etalon of Chapter 5, Section 2 has the advantage over a plane etalon, in that the reflected rays are essentially superimposed rather than (spatially) separated as in the plane etalon. Because of this property, the diameter of a spherical etalon, necessary to accomplish a given task, will be smaller than a plane Fabry-Perot of equal spectroscopic properties, as can be found from Equation 6.3.22a) and illustrated in Figure 6.5 for large values of the finesse (N_R) suitable to the absorption cell example. Note in the figure that the normalized diameter decreases with increasing finesse. This result can be obtained from Equation 6.3.22a):

$$\lim_{R \approx 1} D_{sn} \simeq 2 \left(N_R \; \sigma \; \right)^{-1/4} . \qquad\qquad 6.3.22b$$

The increase noted for the plane etalon in the figure can be obtained in the same manner from Equation 6.3.21a), i.e.:

$$\lim_{R \approx 1} D_{pn} \simeq 95 \left(N_R \; \sigma^{-1} \right)^{1/2} . \qquad\qquad 6.3.21b$$

Returning to the useful diameter of the spherical etalon, when used as an absorption cell with 1 cm and 5 cm spacers, it is found that this diameter need be no larger than 0.2 mm and 0.6 mm respectively. The rather small area associated with these diameters makes it possible to find a near-ideal surface section, compatible with the required high-finesse, in a practical spherical etalon. It is important to repeat here that the spherical etalon is less transparent than a plane etalon. As given in Equation 6.3.5) [and in Equations 5.2.4) and 5.2.5)]:

$$\lim_{R \to 1} Y_s \left(1,0 \right) = 2^{-1} \tau_A \; , \qquad\qquad 6.3.23$$

$$\lim_{R \to 1} Y_p \left(1,0 \right) = \tau_A \; . \qquad\qquad 6.3.24$$

The present treatment has assumed that the only interaction of the absorber (emitter) and the radiation within the cavity is linear. As we have seen in Equation 6.1.10), the antinodes of the wave pattern within the cavity of the etalon have considerable power, given approximately by 1.3 N_R ; thus, non-linear effects between the radiation and the medium should be considered (Kastler, 1962). For instance, when an interference filter is used to isolate a feature from a very bright radiation beam, the high intensity at the antinodes can lead to non-linear effects (heating), which can damage the spacer material and possibly destroy the filter.

4. Lasers

The prime contemporary application of the Fabry-Perot is the laser. In this section, the role of the Fabry-Perot in the generation of coherent radiation will be discussed; thus, only the rudiments of the principles associated with laser theory and operation will be treated here and the reader is referred to the literature on these devices (Heavens, 1962; Bennett, 1962; Lamb, 1964; Pátek, 1967; Lengyel, 1971; Svelto, 1982; Stenholm, 1984) for details. The suggestion to use a Fabry-Perot as the cavity of an optical maser was given by Dicke (1958), Prokhorov (1958) and Schawlow and Townes (1958) and experimentally confirmed by Maiman (1960).

The basic background of emission within a Fabry-Perot cavity has been illustrated in Section 2. In particular, there it was found that a Fabry-Perot cavity containing an emitting species redistributes the radiation into fringes, rather than the original isotropic radiation emitted by the species. In the laser, the interaction of the emitting species with the radiation field and the etalon cavity acquire special significance, not apparent from the discussion in Section 2. First, the phenomena which occur when a material interacts with an electromagnetic wave will be discussed. For simplicity of presentation, the example of an atomic species will be employed; extension to molecular species and crystal lattice situations is straightforward.

From the knowledge of atomic spectra (Condon and Shortley, 1957), the discrete energy levels of a set of atoms in thermal equilibrium are populated as given by Boltzmann statistics. For two levels of energies E_1 and E_2, where $E_1 < E_2$, the ratio of the population distributions is:

$$N_2 \, N_1^{-1} = \exp[-(E_2 - E_1) \, (\mathbf{k} \, \tau \,)^{-1}] \, , \qquad \qquad 6.4.1$$

where $\mathbf{k}$ is Boltzmann's constant and τ is the absolute temperature. When a perturbing influence appears, such as an electromagnetic wave of wavenumber σ, defined as $h \, c \, \sigma = (E_2 - E_1)$, there is a finite probability that atoms in level 1 will be raised to level 2. The rate at which these transitions, or absorption, will occur, is dependent on the number of atoms (per unit volume) in level 1, the strength of the electromagnetic wave, F, and the absorption transition cross section, $\sigma_{1\,2}$, for the particular atom species. Note that $\sigma_{1\,2}$ is associated with the coefficient of transition probability of absorption $B_{1\,2}$. The rate at which level 2 is being filled by this process is:

$$dN_2 \, / \, dt = \sigma_{1\,2} \, F \, N_1 \, . \qquad \qquad 6.4.2$$

Once the atoms are raised to level 2, they can return to level 1 by spontaneously emitting radiation at wavenumber σ, and the depopulation rate of level 2 is:

$$dN_2 \, / \, dt = -\, A_{2\,1} \, N_2 \, . \qquad \qquad 6.4.3a$$

If an electromagnetic wave of wavenumber σ is present, which could be due to the spontaneously emitted photons, the atoms can be induced, or stimulated, to radiate such that:

$$dN_2 \, / \, dt = -A_{2\,1} \, N_2 - \sigma_{2\,1} \, F \, N_2 \, . \qquad \qquad 6.4.3b$$

$A_{2\,1}$ is the Einstein spontaneous emission coefficient, $\sigma_{2\,1}$ is the cross-section for stimulated radiation and can be shown to be equal to the cross-section for absorption $\sigma_{1\,2}$ (Pauling and Wilson, 1935; Condon and Shortley, 1957; Svelto, 1982). Therefore, neglecting spontaneous emission, when a photon flux, F, of wavenumber σ, traveling in a direction z passes through a volume of atoms, the change in the flux is (Svelto, 1982):

$$dF = \sigma_B \, F \, (N_2 - N_1) \, dz \, , \qquad \qquad 6.4.4$$

where σ_B is the cross section for absorption and stimulation of radiation. When

$N_1 > N_2$ the atoms behave as absorbers, while when $N_2 > N_1$, the flux increases and the material behaves as an amplifier. For the assumption of thermal equilibrium used in Equation 6.4.1), i.e., $E_2 > E_1$ and $N_1 > N_2$, the material will be considered an absorber. If somehow a non-thermal equilibrium situation can be arranged such that $N_2 > N_1$, then a population inversion exists [or, in terms of Equation 6.4.1), a 'negative' temperature has been achieved], which leads to the increase of the photon flux and the material becomes an amplifier. It must be pointed out that the atoms can be raised to upper energy levels, or lost from these levels, by non-radiative processes. These processes can be physical, such as inelastic collisions, or chemical. In general, the process of raising atoms to an upper state is called pumping.

As has been illustrated in Equation 6.4.4), when a system's population has been inverted, an electromagnetic wave (of the proper wavenumber) will be amplified as long as the inversion is maintained. As given in Equation 6.4.3), spontaneous emission will occur simultaneously and, when the volume of inverted population material is sufficiently large, these spontaneously emitted photons can stimulate further radiation and the process will then build up. When the gain in stimulated radiation exceeds the losses in the system, then an avalanche of radiation will be emitted as given in Equation 6.4.4). When the population inversion is somehow maintained, a steady outpouring of radiation is obtained; otherwise, a pulse is achieved which in turn eliminates the population inversion. A most important property of stimulated radiation is that the induced photon not only has the same wavenumber as the stimulating photon, but also has the same direction and polarization (spatial coherence) as well as the same phase and speed (temporal coherence) (see, for instance: Hecht and Zajac, 1974; Jenkins and White, 1976; Svelto, 1982; Stenholm, 1984). Therefore, the original photon and the stimulated photon can, in turn, stimulate further emission and eventually a coherent wave is obtained. This is true as long as the stimulation process exceeds whatever losses exist in the system. As the reader is aware, the ability to produce a coherent source of optical radiation is the essential characteristic of the laser.

The Fabry-Perot, with its two mirrors, will then allow the reuse of a finite volume medium by effectively introducing a feedback mechanism, which, under the proper circumstances will give rise to coherent oscillations (lasing). Following Svelto (1982), the minimum, or critical, population inversion necessary to obtain coherent emission with a Fabry-Perot cavity can be estimated by defining the gain per pass in the cavity (i.e., the ratio between output and input photon flux), which is obtained by integration of Equation 6.4.4) over the cavity length, l :

$$F_{i+1} \, F_i^{-1} = \exp[\, \sigma_B \, (\, N_2 - N_1 \,) \, l \,] , \qquad 6.4.5$$

When the only losses in the cavity are those associated with the transmission of the mirrors, a threshold for oscillation is reached when:

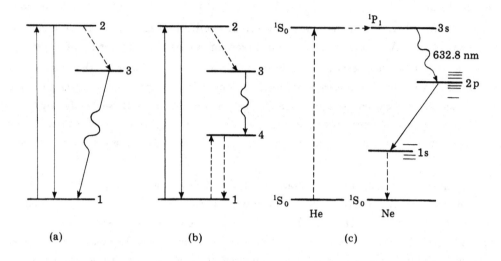

FIGURE 6.6. Schematic of a three (a) and four (b) energy level arrangement for stimulated emission. The solid lines indicate allowed transitions, the dashed lines indicate non-radiative processes and the wavy lines indicate stimulated radiation (lasing) transitions. The *He −Ne* laser level configuration, for the 632.8 nm line, is shown in (c).

$$R^2 \exp \left[2 \sigma_B (N_2 - N_1) l \right] = 1 .$$ 6.4.6

In this expression, R is the power reflectivity of the mirrors, which are presumed to be identical. Note the factor of 2 gain in the argument of the exponential due to the (double) pass. After this threshold is reached, the emission of coherent radiation, or oscillation, will build up from the spontaneous emission and the critical population inversion is (Svelto, 1982):

$$(N_2 - N_1)_{crit} = - (\sigma_B l)^{-1} \ln (R) .$$ 6.4.7

Practical considerations make the two (energy) level laser, thus far discussed, not the best possible arrangement to obtain population inversion. To begin with, at least one-half of the lower state must be pumped to the upper level before a population inversion is achieved. Also, when the lower level is not completely depopulated (or bleached), absorption of emitted photons will occur which, at best, raises the critical population inversion (or threshold) and lowers the efficiency of the laser. Presume that level 2 can not only decay back to level 1, as in the two-level example, but it can also decay down to a third energy level rather quickly by, say, a non-radiative process. These non-radiative processes are independent of the radiation density. As the pumping continues, level 3 will eventually reach the critical population inversion and will be able to emit stimulated coherent radiation while, at the same, time dumping a large part of the atoms at this level down to level 1. Because of the loss of population inversion, the laser output will tend to be a series of pulses occurring as level 3 periodically reaches critical population inversion (Nelson and Boyle, 1962). The periodicity of these pulses can then be seen to be dependent on the strength of the pumping process and the non-radiative losses of level 3. Note that the three-level arrangement still has the limitation that the lower level (level 1) must be at least one-half empty to achieve population inversion. This three-level arrangement was used by Maiman (1960) to obtain the first visible spectrum maser, or laser, and is illustrated in Figure 6.6(a).

When the transition leading to stimulated radiation (and laser action) terminates on a fourth level, which is higher in energy than level 1, the population inversion is now between two normally sparsely populated levels. This is illustrated in Figure 6.6(b). When level 3 is populated in the same fashion as in the three-level arrangement and the fourth level decays fast, preferably by a non-radiative process, a population inversion can be easily maintained. The helium-neon ($He - Ne$) laser of Javan *et al.* (1961) is a variant of the four-level optically pumped laser. In their laser, the Ne energy levels are filled by inelastic collisions with metastable He atoms, due to a resonance between the $^1S_0$ state of He and the $^1P_1$ state of Ne (for 632.8 nm radiation). This is shown in Figure 6.6(c), where the $^1P_1$ level of Ne behaves like level 3 of the optically pumped laser of Figure 6.6b. The metastable He atoms are obtained by inelastic collisions of ground state He with electrons in a gas discharge. These collisions raise the He to many upper levels which, in turn, eventually decay to the long-lived metastable $^1S_0$ and $^3S_1$ states. For other types of lasers, the reader should consult a reference such as Svelto (1982).

The actual physical arrangements for three-level and four-level lasers are given in Figure 6.7. The device at the left of the figure is a schematic of the three-level ruby laser of Maiman (1960), where optical pumping was achieved by a flash from the discharge lamp, while the device on the right is a variation of the Javan *et al.* (1961) $He - Ne$ laser (Rigrod *et al.,* 1962). Note the direct application of the mirrors to the ruby, to form the Fabry-Perot etalon, as well as the physical separation of the mirrors from the gas discharge. The discharge tube has windows at Brewster (1814, 1815) angle to achieve high transmission for light whose electric vector is parallel to the plane of incidence. Since the normal component of the light is partially reflected at the window and removed from the cavity, it is not able to contribute to the stimulation of radiation; therefore, the coherent output of the laser will be linearly polarized. Brewster angle faces can be used with the ruby laser example, although the mirror will no longer be directly applied to the crystal and needs to be placed away from the active ruby medium.

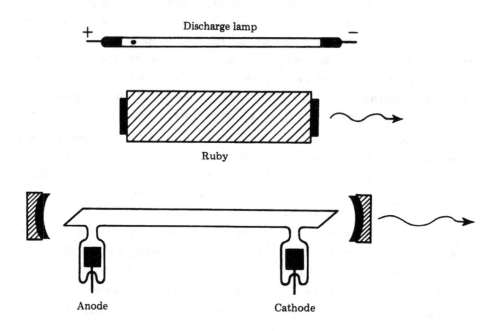

FIGURE 6.7. Schematic diagrams for crystal and gas lasers.

As has been mentioned earlier, the Fabry-Perot, with its two mirrors, allows the reuse of a finite volume of active medium to provide coherent oscillations. The simple theory presented in Section 2 for a classical etalon has shown the basic behavior and properties of an etalon containing an emitting species. The behavior and properties can be summarized by saying the etalon selects to transmit certain wavenumbers at some preferred angles, as dictated by the etalon spacer, while the spectral profiles of this emitted radiation are defined by the etalon reflectivity (for an ideal etalon).

The study of the properties of a cavity, formed by a Fabry-Perot etalon, allows investigation of the differences of the etalon behavior when used as a classical Fabry-Perot or as a laser. The usual illustration of the properties of a cavity begins with the progression of the stretched (bound) string to the two-dimensional membrane and then to a cavity. For the purposes of the present discussion, it is not necessary to repeat the details, which can be found in the literature (Kauzmann, 1957; Koppelmann, 1969; Svelto, 1982; Stenholm, 1984), but only to state the results. For a three-dimensional cavity, the standing electromagnetic field distribution is described in terms of the number of nodes a standing wave has in orthogonal directions x, y

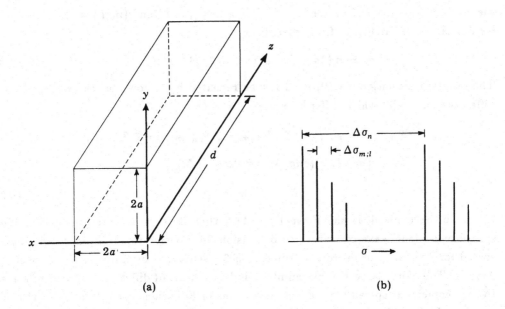

FIGURE 6.8. a) Rectangular cavity used to discuss resonance modes and b) the relative location of the modes in the wavenumber spectrum.

and z. The frequency of the mode, associated with the nodes, is then dictated by the physical configuration of the cavity. As an example, the rectangular cavity shown in Figure 6.8a will be used to illustrate the results from the analysis. The frequencies associated with the quantized resonant modes of the E-field components, expressed as wavenumbers, are (Schawlow and Townes, 1958; Svelto, 1982):

$$\sigma_{n,m,l} = n \ (2d)^{-1} \left\{ 1 + [d \ (2an)^{-1}]^2 (m^2 + l^2) \right\}^{1/2}, \qquad 6.4.8a$$

where n, m, and l are integers which describe the modes in the z, x and y directions. Note that m and l are degenerate because of the symmetry adopted here. For the assumption that $d > 2a$, i.e., $n >> (m,l)$, the modes associated with n can be considered to be decoupled with the modes associated with m and l since:

$$d^2 \ (m^2 + l^2) \ (2an)^{-2} \approx d^2 \ (m^2 + l^2) \ [2a \ (n+1)]^{-2}. \qquad 6.4.9$$

Thus, the modes associated with the direction of the length d, i.e., n, are called longitudinal modes and those associated with $2a$ are called transverse electromagnetic modes, or TEM modes. For $m = l = 0$, or when $n >> (m,l)$, the difference in wavenumbers between the nth and the $n^{th} + 1$ mode is:

$$\Delta \sigma_n = \sigma_{n+1,m,l} - \sigma_{n,m,l} = (2 \ d)^{-1}. \qquad 6.4.10$$

This result can be recognized to be the free spectral range of the Fabry-Perot defined in Equation 2.1.10b). Note that n, of Equation 6.4.8a), can also be identified with the order of Equation 2.1.8), for $\mu = 1$. For $n \gg (m,l)$ and $(m,l) \neq 0$, Equation 6.4.8a) can be expanded, to first order, into:

$$\sigma \simeq n\,(2d)^{-1} + d\,(m^2 + l^2)\,(16\,a^2 n)^{-1}\,. \qquad 6.4.8b$$

The results of Equation 6.4.10) need not be repeated here; however, the wavenumber difference for modes which differ by unity in m (or l) is:

$$\Delta\sigma_{m;l} = \sigma_{n,m+1,l} - \sigma_{n,m,l} = \sigma_{n,m,l+1} - \sigma_{n,m,l}$$

$$= d\,(m+1/2)\,(8na^2)^{-1} = \Delta\sigma_n\,(8N)^{-1}\,(m+1/2)\,, \qquad 6.4.11$$

$$N = a^2\,\sigma\,d^{-1}\,. \qquad 6.4.12$$

N can be recognized as the Fresnel number, that is, a measure of the diffraction spread of a plane wave of transverse dimension $2a$ relative to the geometrical angle spread for apertures, or mirrors, of dimension $2a$ separated by a distance d (Svelto, 1982). Therefore, large Fresnel numbers indicate a small diffraction spread relative to the geometrical spread. With the derivations of Equations 6.4.10) and 6.4.11) it is now possible, within the approximations used, to find the properties of a cavity when used as a classical Fabry-Perot etalon and as a laser. Since the present development has been made for a rectangular cavity, imagine a typical high-resolution Fabry-Perot 13.3 cm to the side (i.e., the equivalent of a 15 cm diameter etalon) with a 2.5 cm spacer and a laser cavity 1 m long with useful mirrors of about 0.1 cm to the side. The Fresnel numbers for these two examples are $\sim 1\,415\,000$ for the Fabry-Perot and 2 for the laser cavity. The separation between longitudinal modes is, from Equation 6.4.10):

$$\Delta\sigma_n\,(\mathrm{F\ P}) = 2 \times 10^{-1}\ \mathrm{K}\,, \qquad 6.4.13$$

$$\Delta\sigma_n\,(\mathrm{Laser}) = 5 \times 10^{-3}\ \mathrm{K}\,, \qquad 6.4.14$$

while the separation between the transverse (TEM) modes is given by:

$$\Delta\sigma_{m;l}\,(\mathrm{F\ P}) = 1.77 \times 10^{-8}\,(m+1/2)\,, \qquad 6.4.15$$

$$\Delta\sigma_{m;l}\,(\mathrm{Laser}) = 3.125 \times 10^{-4}\,(m+1/2)\,. \qquad 6.4.16$$

The results for the laser cavity show the TEM mode separation, i.e., Equation 6.4.16), to be a sensible fraction of the longitudinal mode separation, or free spectral range, of Equation 6.4.14) and this is schematically illustrated in Figure 6.8b. On the other hand, for the high-resolution Fabry-Perot, the transverse intermode separation is seven orders of magnitude smaller than the free spectral range and, thus, unresolvable. This result could have been directly derived from Equation 6.4.8), for the limiting case of $an \gg d$, or:

$$\lim_{an \gg d} \sigma_{n,m,l} = \sigma_{n,0,0} \; , \qquad\qquad 6.4.8c$$

that is, the wavenumber of the transverse modes is undistinguishable from the wavenumber of the longitudinal modes. Indeed, this finding about the high-resolution Fabry-Perot etalon can be used to justify the one-dimensional (bound string) approximation used in the classical derivation of the Fabry-Perot as given in Chapter 2.

The above discussion has provided a semi-quantitative description of the spectral position of the multiple modes in a cavity. The intensity distribution of these modes can also be discussed in terms of wave equations, neglecting diffraction. That is, the results derived here are applicable to optical systems of large aperture, i.e., apertures that intercept a negligible portion of the beam. A field component u of a coherent beam satisfies the scalar equation (Kogelnik and Li, 1966):

$$\nabla^2 u + k^2 u = 0 \; , \qquad\qquad 6.4.17$$

where $k = 2 \pi \lambda^{-1} = 2 \pi \sigma$ is the propagating constant of the medium. In this section, the method used by Kogelnik and Li (1966) will be closely followed to the extent necessary for the purposes of the discussion. Using the same coordinate representation of Figure 6.8a, the function for the light beam traveling in the z-direction can be represented by:

$$u = \Psi(x,y,z) \exp(-i \, k \, z) \; , \qquad\qquad 6.4.18$$

where $\Psi(x,y,z)$ is a slowly varying function which represents the differences between a laser coherent beam and a plane wave. Replacing Equation 6.4.18) into Equation 6.4.17) and assuming that $\Psi(x,y,z)$ varies slowly enough such that its second derivative with respect to z can be ignored, we have:

$$\frac{\partial^2 \Psi(x,y,z)}{\partial x^2} + \frac{\partial^2 \Psi(x,y,z)}{\partial y^2} - 2 \, i \, k \, \frac{\partial \Psi(x,y,z)}{\partial z} = 0 \; . \qquad 6.4.19$$

This equation has a solution similar to the Schrödinger time-dependent equation, or:

$$\Psi(x,y,z) = \exp\left\{ -\{ P(z) + [2 \, q(z)]^{-1} \, k \, r^2 \} \right\} \; , \qquad 6.4.20$$

$$r^2 = x^2 + y^2 \; . \qquad\qquad 6.4.21$$

$P(z)$ represents a complex phase shift associated with the propagation of the beam, while $q(z)$ is a complex parameter that describes the variation of the intensity of the Gaussian beam of Equation 6.4.21). Kogelnik and Li (1966) introduced two real parameters, R and w, which are related to q by:

$$[q(z)]^{-1} = [R(z)]^{-1} - i \, (\pi \, w^2(z) \, \sigma)^{-1} \; . \qquad 6.4.22$$

Replacement of this definition of $q(z)$ into Equation 6.4.20), shows the physical

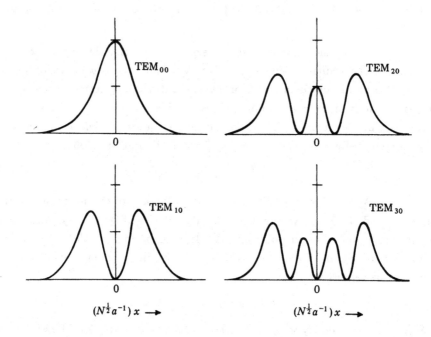

FIGURE 6.9. Intensity distribution of the TEM modes of a cavity as a function of $(N^{1/2}a^{-1})\,z$.

significance of $w(z)$ to be the HWHH of the beam at the e^{-1} height of the maximum of the Gaussian beam and $R(z)$ is the radius of curvature of the wavefront that intersects the axis at z. Equation 6.4.20) is not the only solution of Equation 6.4.19). Solutions of the following form also exist:

$$\Psi(x,y,z) = g\left(\frac{x}{w(z)}\right) h\left(\frac{y}{w(z)}\right)$$

$$\times \exp\{-i \left[P(z) + (2\,q(z))^{-1}k\ r^2 \right]\}\,. \qquad 6.4.23$$

For real g and h, insertion of this solution into Equation 6.4.19) gives the differential equation for Hermite polynomials, H_l (Hochstrasser, 1964); thus:

$$g\left(\frac{x}{w(z)}\right) h\left(\frac{y}{w(z)}\right) = H_m\left(2^{1/2}\frac{x}{w(z)}\right) H_l\left(2^{1/2}\frac{y}{w(z)}\right)\,, \qquad 6.4.24$$

$$\Psi(x,y,z) = H_m\left(2^{1/2}\frac{x}{w(z)}\right) H_l\left(2^{1/2}\frac{y}{w(z)}\right)$$

$$\times \exp\{-i \; [\; P(z) + (2 \; q(z))^{-1} k \; r^2 \;]\} \;. \qquad\qquad 6.4.25$$

Since $H_0(z) = 1$, the solution shown in Equation 6.4.20) is but a special case of Equation 6.4.25). Therefore, the intensities of the mode patterns are given as products of Hermite and Gaussian functions. The Hermite polynomials in m and l can be recognized to be the functions associated with the amplitudes of the TEM modes, which modulate the longitudinal (TEM_{00}) modes' amplitudes. Some of these patterns are illustrated in Figure 6.9. Photographs of these same patterns from a laser are shown in Kogelnik and Rigrod (1962) and Kogelnik and Li (1966). Although the present solutions are applicable to rectangular symmetry mirrors, extension to circular symmetry can be easily made. The solution is like changing from a square membrane to a circular membrane. The results for the latter and the relationships between the modes for the two symmetries are illustrated in Kauzmann (1957).

Further properties of the quantities $q(z)$ and $R(z)$ can be found by replacing Equation 6.4.20) into Equation 6.4.19) and comparing terms of equal power in r. The findings are (Kogelnik and Li, 1966):

$$\frac{\partial \; q(z)}{\partial \; z} = 1 \;, \qquad\qquad 6.4.26$$

$$\frac{\partial \; P(z)}{\partial \; z} = -i \; [\; q(z) \;]^{-1} \;, \qquad\qquad 6.4.27$$

and integration of Equation 6.4.26) readily yields:

$$q(z_2) = q(z_1) + z \;. \qquad\qquad 6.4.28$$

Returning to Equation 6.4.22), the beam width $w(z)$ has a minimum diameter at the beam waist, where the phase front is plane. When this minimum width is labeled w_0, we obtain:

$$q_0 = i \; \pi \; \sigma \; w_0^2 \;, \qquad\qquad 6.4.29$$

while, at a distance z away from the beam waist the $q(z)$ parameter becomes, from Equation 6.4.28):

$$q(z) = q_0 + z = i \; \pi \; \sigma \; w_0^2 + z \;. \qquad\qquad 6.4.30$$

Combining Equations 6.4.22) and 6.4.30), the following is obtained:

$$[\; w(z) \;]^2 = w_0^2 \; \{1 + [\; z \; (\; \pi \sigma w_0^2 \;)^{-1} \;]^2 \} \;, \qquad\qquad 6.4.31$$

$$R(z) = z \; [\; 1 + (\; \pi \sigma w_0^2 z^{-1} \;)^2 \;] \;. \qquad\qquad 6.4.32$$

The beam contour $w(z)$ is, therefore, a parabola with asymptotes inclined at an angle Θ to the axis, given by:

$$\Theta = (\; \pi \; \sigma \; w_0 \;)^{-1} \;, \qquad\qquad 6.4.33$$

or the far field diffraction angle of the fundamental mode. The ratio of Equation 6.4.31) over Equation 6.4.32) gives:

$$\pi \, \sigma \, [w\,(z)]^2 \, [R\,(z)]^{-1} = z \, (\pi \, \sigma \, w_0^2 \,)^{-1} \, , \qquad\qquad 6.4.34$$

which, in turn, allows the expression of w_0 and z in terms of $w\,(z)$ and $R\,(z)$:

$$w_0^2 = [w\,(z)]^2 \left\{ 1 + \left(\pi \, \sigma \, [w\,(z)]^2 \, [R\,(z)]^{-1} \right)^2 \right\}^{-1} \, , \qquad\qquad 6.4.35$$

$$z = R\,(z) \left\{ 1 + \left(\pi^{-1} \sigma^{-1} [w\,(z)]^{-2} R\,(z) \right)^2 \right\}^{-1} \, . \qquad\qquad 6.4.36$$

The complex phase shift can be determined at a distance, z, away from the waist, by replacing Equation 6.4.28) into Equation 6.4.27), or:

$$\frac{\partial\,P\,(z)}{\partial\,z} = -i\,(\,z\,+\,i\,\pi\,\sigma\,w_0^2\,)^{-1}\,, \qquad\qquad 6.4.37$$

$$i\,P\,(z) = \ln[\,1\,-\,i\,z\,(\pi\,\sigma\,w_0^2\,)^{-1}\,]$$

$$= \ln[1+z^2(\pi\,\sigma\,w_0^2\,)^{-2}\,]^{1/2} - i\,\tan^{-1}[z\,(\pi\,\sigma\,w_0^2\,)^{-1}\,]$$

$$= \ln[\,w\,(z)\,w_0^{-1}\,]\,-\,i\,\Phi(z)\,. \qquad\qquad 6.4.38$$

Replacing the above result and the definition of $q\,(z)$, given in Equation 6.4.22), into Equation 6.4.20), the solution of u in Equation 6.4.18) is found to be:

$$u = w_o\,[w\,(z)]^{-1} \exp \left\{ -i\,[kz - \Phi(z)] \right.$$

$$\left. -\,r^2 \left\{ [\,w\,(z)\,]^{-2} + i\,k\,2^{-1}\,[R\,(z)]^{-1} \right\} \right\}. \qquad\qquad 6.4.39$$

The solution for the beam including higher, or transverse, modes is obtained by using Equation 6.4.25), rather than Equation 6.4.20), in the above development. Note that the phase $\Phi(z)$ [defined in Equation 6.4.38)] is now given by (Kogelnik and Li, 1966):

$$\Phi(m\,,l\,;z) = (m + l + 1)\,\tan^{-1}[z\,(\pi\,\sigma\,w_0^2\,)^{-1}\,]$$

$$= (\,m\,+\,l\,+\,1\,)\,\Phi(z)\,, \qquad\qquad 6.4.40$$

that is, the phase velocity increases with increasing mode, leading to different wavenumbers for the various modes of oscillation.

A mirror of finite curvature, such as would be employed in a laser cavity, can be thought of as a thin lens equivalent which will transform a spherical wave with radius $R\,(z_1)$, immediately in front of the lens, to another spherical wave of radius $R\,(z_2)$, immediately in back of the lens, by:

$$[R(z_2)]^{-1} = [R(z_1)]^{-1} - f^{-1} , \qquad 6.4.41$$

where f is the focal length of the lens (mirror). Since the width of the beam is the same immediately in front of and in back of the lens, then we have:

$$[q(z_2)]^{-1} = [q(z_1)]^{-1} - f^{-1} . \qquad 6.4.42$$

A laser cavity, with equal curvature mirrors separated by a distance d, can be considered to be equivalent to a large sequence of (thin) lenses of the same focal length as the mirrors and separated from each other by the distance d. When these lenses (mirrors) are very large in comparison to the beam width, the parameter $q(z_1)$ in front of one of these lenses will be transformed into $q(z_2)$ immediately in front of the next lens and can be deduced from Equations 6.4.28) and 6.4.42) to be equal to:

$$[q(z_2)]^{-1} = [q(z_1) + d]^{-1} - f^{-1} . \qquad 6.4.43a$$

The symmetry of the system requires that $q(z_2) = q(z_1) = q(d)$ and, thus, the above equation becomes:

$$[q(d)]^{-2} + [f \ q(d)]^{-1} + [f \ d]^{-1} = 0 , \qquad 6.4.43b$$

which has solutions given by:

$$[q(d)]^{-1} = -2^{-1} f^{-1} \pm i \ (f^{-1} d^{-1} - 4^{-1} f^{-2})^{1/2} . \qquad 6.4.44$$

By comparison with Equation 6.4.22), the real parameters R and w can be obtained from the above expression. $R(d)$ is then equal to the radius of curvature of the mirrors (denoted by R_m), i.e., $2f$ and $w(d)$ can be deduced from the root that gives a real beam width. Therefore:

$$[w(d)]^2 = R_m \ \pi^{-1} \sigma^{-1} \ (2 \ R_m \ d^{-1} - 1)^{-1/2} , \qquad 6.4.45a$$

is the beam width at the mirror, while the beam waist can be obtained with the assistance of Equation 6.4.34) for $z = 2^{-1} d$, i.e.:

$$w_0^2 = (2\pi\sigma)^{-1} [d \ (2R_m - d)]^{1/2} . \qquad 6.4.46a$$

The above equations make it possible to calculate the beam parameters in terms of the cavity constants.

Since the beam parameters $R(z)$ and $w(z)$ describe the modes of all orders, the difference among orders is due to the phase shift given in Equation 6.4.40). Resonances in the cavity will then occur when the phase shift in one round trip is a multiple of 2π. In terms of Equations 6.4.39) and 6.4.40), the following is obtained for the phase shift, from one mirror to the other:

$$kd - 2(m+l+1) \ \Phi(d) = (v+1) , \qquad 6.4.47$$

where v is the number of nodes in the standing wave, that is, $v + 1$ one-half

wavelengths. From the earlier definitions of k, $\Phi(z)$, Equations 6.4.10) and 6.4.46),
as well as the usual relationships among inverse trigonometric functions (Gradshteyn
and Ryzhik, 1980), the resonant wavenumber for a given mode is:

$$\sigma \ (\Delta\sigma_n) = v + 1 + \pi^{-1} \ (m + l + 1) \ \cos^{-1}(dR^{-1} - 1) \ . \qquad 6.4.48a$$

For the special case of a confocal etalon, where $R = d$, Equations 6.4.45a), 6.4.46a)
and 6.4.48a) become:

$$[\ w \ (z) \ ]^2 = R \ (\ \sigma \ \pi \ )^{-1} \ , \qquad 6.4.45b$$

$$w_0^2 = d \ (\ 2 \ \pi \ \sigma \ )^{-1} \ , \qquad 6.4.46b$$

$$\sigma \ (\ \Delta\sigma_n \ )^{-1} = v + 1 + 2^{-1}(\ m \ + \ l \ + 1) \ , \qquad 6.4.48b$$

where $\Delta\sigma_n$ has been given in Equation 6.4.10). The result from the last expression
is the same as that obtained by Boyd and Gordon (1961). Kogelnik and Li
(1966) have also discussed the general case of unequal curvature mirrors and the
reader is referred to them for details.

Under the approximations made in the derivation of Equation 6.4.24), and its
special case of Equation 6.4.20), it is possible to separate the modes associated with
the x and y coordinates. Further examination of Equation 6.4.24) shows that x
(and y) appears in the form $x \ / \ w \ (z)$ in both the Gaussian and the Hermite polynomials. For a confocal etalon, the value of $w \ (z)$ can be determined in terms of d
with the help of Equations 6.4.31) and 6.4.46b), or:

$$w \ (z) = (d \ 2^{-1} \ \pi^{-1} \ \sigma^{-1})^{1/2} [\ 1 + (\ 2 \ z \ d^{-1} \ )^2 \ ]^{1/2} \ . \qquad 6.4.49$$

From this expression and Equation 6.4.12), it is possible to express the ratio of the
coordinate to the width as:

$$x \ [\ w \ (z) \ ]^{-1} = (\ 2 \ \pi \ )^{1/2}[1 + (2zd^{-1})^2 \ ]^{-1/2} \ x \ a^{-1} \ N^{1/2} \ , \qquad 6.4.50$$

where N is the Fresnel number. The ability to describe Equation 6.4.24) in terms of
the etalon cavity parameters, shows the width of the function $\Psi(x,y,z)$, for $m =
l = 0$, to decrease with increasing values of N (for arbitrary values of x or y).
However, when the following dimensionless quantity is defined:

$$\eta_x = (\ N^{1/2} \ a^{-1} \ ) \ x \ , \qquad 6.4.51$$

the function $\Psi(\eta_x,\eta_y,z)$ becomes a unique representation of the modes, regardless of
the value adopted for N, and this is the reason Figure 6.9 has been plotted as a function of η_x. With the help of this substitution, it can be seen that a cavity with large
values of N will support many transverse modes, while a cavity with low values of N
will tend to support only those modes which have appreciable values of the function
$\Psi(x,y,z)$. Mode selection can be achieved by masking down the mirrors to obtain a
low Fresnel number. However, low Fresnel numbers mean that diffraction effects are

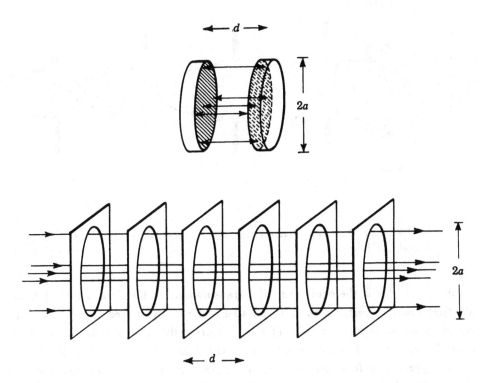

FIGURE 6.10. Fabry-Perot etalon and its transmission medium equivalent used in the study of diffraction effects. After Fox and Li (1961).

significant and cannot be ignored any longer.

In a laser, using a finite-size etalon as the cavity, a wavefront leaving one mirror and traveling towards the other will be amplified as given by Equation 6.4.4). However, at the same time, it will lose power due to scattering by inhomogeneities in the medium. When this wavefront arrives at the second mirror, further power will be lost by radiation around the edges of the (finite-size) mirror and by the less than perfect reflectivity of the mirrors (planned or otherwise). Therefore, for lasing to occur, the total loses must be smaller than the power gained from the active medium. The losses caused by the existence of diffraction play an important role in the operation of the laser, in the sense that these losses determine the distribution of the energy in the etalon cavity during operation of the laser. The recurring power loss from the edge of the wavefront will cause marked departures from a uniform amplitude and phase across the mirror (Fox and Li, 1961).

The approach taken by Fox and Li (1961; Kogelnik and Li, 1966) to estimate the diffraction losses, was to consider the properties of the propagating wave which is reflected back and forth between the etalon mirrors. As these authors noted, for a

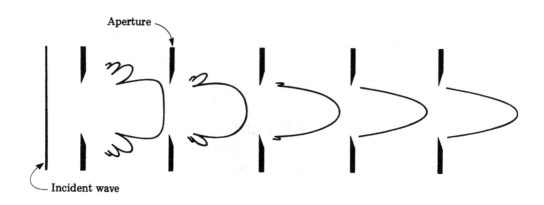

FIGURE 6.11. Schematic representation of the steady-state approach used by Fox and Li (1961).

plane etalon, this is the equivalent of a transmission medium consisting of a series of identical coaxial apertures cut into parallel and equally spaced (perfectly) absorbing partitions of infinite extent. This is schematically illustrated in Figure 6.10 for a plane Fabry-Perot. For spherical mirror cavities, the apertures are replaced by lenses of the appropriate focal length (Kogelnik and Li, 1966). The aperture example given in the figure is, then, the special case of zero optical power lenses. Using the scalar approximation, that is, a field which is nearly transverse and uniformly polarized (Svelto, 1982), Fox and Li (1961) assumed an arbitrary field distribution of the wavefront at the first mirror (or aperture). Then, they proceeded to compute the field distribution due to diffraction at the second mirror, using Kirchhoff's diffraction integral (Born and Wolf, 1964). This new field distribution is then employed to compute the field after the second transit, etc. This procedure is continued for a large number of transits until the distribution reaches a steady state, as schematically shown in Figure 6.11. Mathematically, using Fox and Li's (1961) development, the field u_p due to an illuminated aperture A with field u_a is:

$$u_p = i\,(4\pi)^{-1}k \int_A u_a \; r^{-1} \exp(-ikr)\,(1+\cos\Theta)\,ds$$

$$= \int_A \mathbf{K}\,u_a\,ds\;, \qquad\qquad 6.4.52\mathrm{a}$$

where k is the propagation constant (previously defined), r is the distance from a point in the aperture to the point of observation, and Θ is the angle r makes with the normal to the aperture. When an initial wave of distribution u_p is launched at one of the mirrors and allowed to reflect back and forth in the etalon, after j transits the field distribution will be given by Equation 6.4.52a) with u_p now replaced by

u_{j+1} and u_a by u_j. After many transits, the field distribution will suffer negligible change from reflection to reflection and achieves a steady state. At this point the fields in the mirrors are identical, save a (complex) constant γ, i.e.:

$$u_j = \gamma^{-j} v ,$$
6.4.53

where v is the unvarying distribution and γ is independent of position coordinates. Substituting the results of Equation 6.4.53) into Equation 6.452a) gives:

$$v = \gamma \int_A \mathbf{K} \, v \, ds .$$
6.4.54

The distribution v can be considered to be the normal mode of the etalon at the mirror surface and the logarithm of γ, which specifies the attenuation and phase shift the wave undergoes during each transit, can be regarded as the propagation constant associated with this normal mode.

For a spherical etalon, with mirrors of radii R_1 and R_2 separated by a distance d, Equation 6.4.52a) can be expressed in cylindrical coordinates (Fox and Li, 1961), as indicated in Figure 6.12:

$$u_{j+1}(r_2,\phi_2) = i \, 2^{-1} \, \sigma \int_0^a \int_0^{2\pi} u_j(r_1,\phi_1)$$

$$\times \, r^{-1}\exp(-ikr) \, (1+d_1 r^{-1}) r_1 \, d\phi_1 \, dr_1 ,$$
6.4.52b

where d_1 and r are defined by the following (Fox and Li, 1961; Kogelnik and Li, 1966):

$$d_1 = d - R_1 \left[1-(1-r_1^2 \, R_1^{-2})^{1/2}\right]$$

$$- R_2 \left[1-(1-r_2^2 \, R_2^{-2})^{1/2}\right] ,$$
6.4.55a

$$r = \left[d_1^2 + r_1^2 + r_2^2 - 2 \, r_1 r_2 \cos(\phi_1-\phi_2) \right]^{1/2} .$$
6.4.56a

These expressions can be simplified when $d >> a$ and $R_i >> r_i$, i.e.:

$$d_1 \simeq d - r_1^2 \, (2R_1)^{-1} - r_2^2 \, (2R_2)^{-1} ,$$
6.4.55b

$$r \simeq d_1 + r_1^2 \, (2d_1)^{-1} + r_2^2 \, (2d_2)^{-1} - r_1 r_2 d^{-1} \cos(\phi_1-\phi_2)$$

$$\simeq d + r_1^2 \, (2d)^{-1}(1-dR_1^{-1}) + r_2^2 \, (2d)^{-1}(1-dR_2^{-1})$$

$$- r_1 r_2 \, d^{-1} \cos(\phi_1 - \phi_2)$$

$$\simeq d + g_1 r_1^2 \, (2d)^{-1} + g_2 r_2^2 \, (2d)^{-1} - r_1 r_2 d^{-1} \cos(\phi_1-\phi_2) .$$
6.4.56b

Replacing the above results in Equation 6.4.52b), we obtain:

$$u_{j+1}(r_2,\phi_2) = i \, \exp(-ikd) \, \sigma \, d^{-1} \int_0^a \int_0^{2\pi} u_j(r_1,\phi_1)$$

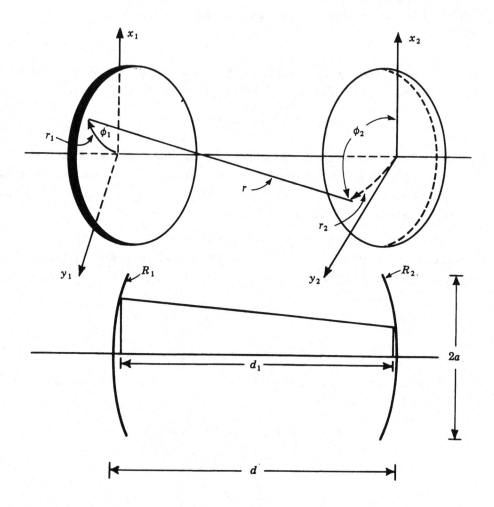

FIGURE 6.12. Geometry and nomenclature used in the determination of diffraction losses of a spherical etalon. After Fox and Li (1961).

$$\times \exp\{-i \; k \, (2d)^{-1}[g_1 r_1^{\,2} + g_2 r_2^{\,2} - 2r_1 r_2 \cos(\phi_1 - \phi_2) \,] \, \}$$

$$\times r_1 \, d\phi_1 \, dr_1 \, , \qquad\qquad\qquad\qquad 6.4.52c$$

which, as stated earlier, is valid for $d \; a^{-1} >> 1$, or $(da^{-1})^2 >> (a^2 d^{-1} \sigma) = N$. Incorporation of the steady state conditions, given in Equation 6.4.54), into Equation 6.4.52c) gives:

$$v \, (r_2, \phi_2) = \gamma_1 \int_0^a \int_0^{2\pi} \mathbf{K}\!\left(r_2, \phi_2; \; r_1, \phi_1\right) v \, (r_1, \phi_1) \, r_1 \, d\phi_1 \, dr_1 \, , \qquad 6.4.57$$

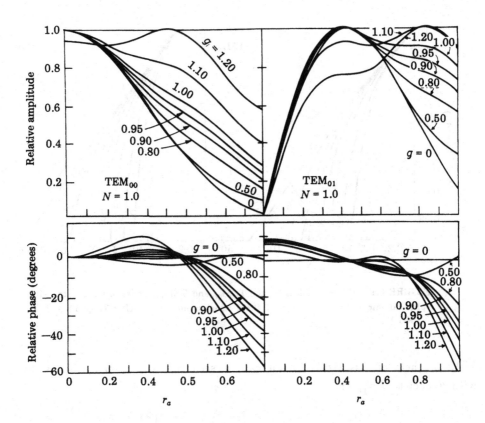

FIGURE 6.13. Field distributions for the TEM_{00} and TEM_{01} modes of a spherical etalon in the presence of diffraction, with the Fresnel number as a parameter. After Kogelnik and Li (1966).

where γ_1 and the kernel $\mathbf{K}$ are defined by the following expressions:

$$\gamma_1 = \gamma \exp(-i\,k\,d) , \tag{6.4.58}$$

$$\mathbf{K}(r_2, \phi_2; r_1, \phi_1) = i\,\sigma\,d$$

$$\times \exp\{-ik\,(2d)^{-1}[\,g_1 r_1^{\,2} + g_2 r_2^{\,2} + 2r_1 r_2\cos(\phi_1 - \phi_2)\,]\} . \tag{6.4.59}$$

Fox and Li (1961) make use of the following identity to simplify the solution:

$$\exp[\,i\,n\,(2^{-1}\pi - \phi_2)\,]\,J_n(k\,r_1 r_2 d^{-1}) = (2\,\pi)^{-1}$$

$$\times \int_0^{2\pi} \exp[\,i\,k\,r_1\,r_2 d^{-1}\cos(\phi_1 - \phi_2) - i\,n\,\phi_1\,]\,d\phi_1 , \tag{6.4.60}$$

where $J_n(x)$ is a Bessel function of the first kind of integer order n (Olver, 1964).

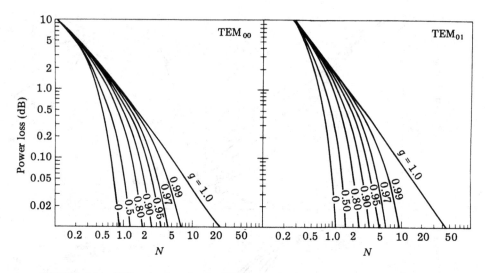

FIGURE 6.14. Diffraction loss for the TEM_{00} and TEM_{01} modes of a resonator with spherical mirrors as a function of the Fresnel number N. After Kogelnik and Li (1966).

With the help of this expression, it can be seen that Equation 6.4.57) is integrable with respect to $d\,\phi_1$, when:

$$ v\left(r_i,\phi_i\right) = v_n\left(r_i,\phi_i\right) = S_n\left(r_i\right)\exp(-i\;n\;\phi_1)\,. \qquad 6.4.61 $$

Therefore the function $S_n\left(r_1\right)$ will satisfy the reduced equation and orthogonality conditions given below:

$$ S_n\left(r_2\right)r_2^{1/2} = \gamma_n\int_0^a \mathbf{K}_n\left(r_2,r_1\right)S_n\left(r_1\right)r_1^{1/2}\,dr_1\,, \qquad 6.4.62 $$

$$ \mathbf{K}_n\left(r_2,r_1\right) = i^{\,n+1}\,d^{-1}\,k\;J_n\left(kr_1r_2d^{-1}\right)\left(r_1r_2\right)^{1/2} $$

$$ \times\exp\left[-i\;k\;(2d)^{-1}\left(g_1r_1^2 + g_2r_2^2\right)\right]\,, \qquad 6.4.63 $$

$$ \int_0^a S_n\left(x\right)S_m\left(x\right)\,dx = 0 \qquad n\neq 0\,. \qquad 6.4.64 $$

The solutions of Equation 6.4.57) for $v\left(r,\phi\right)$ and the eigenvalues γ for the steady state conditions have been obtained by numerical methods (Fox and Li, 1961) and by analytical approximations (Boyd and Gordon, 1961). Figure 6.13 illustrates the results of a spherical etalon, for the TEM_{00} and TEM_{01} modes with a Fresnel number, N, equal to 1. In this figure, the distance on the mirror surface (away from the axis) is denoted by the dimensionless quantity r_a, given as the ratio of the

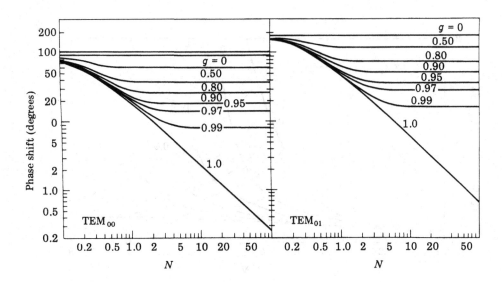

FIGURE 6.15. Phase shift for the TEM_{00} and TEM_{01} modes of a resonator with spherical mirrors as a function of the Fresnel number N. After Kogelnik and Li (1966).

true distance over the mirror radius, a. Note the similarity in shapes with the corresponding modes for negligible diffraction illustrated in Figure 6.9. The diffraction loss, in terms of the fractional energy loss per transit, is given by (Kogelnik and Li, 1966):

$$\alpha = 1.0 - \left| \gamma \right|^2 , \qquad\qquad 6.4.65$$

which is illustrated in Figure 6.14 for the TEM_{00} and the TEM_{01} modes, while the phase shift suffered by the wave per transit is:

$$\beta = \arg(\gamma) . \qquad\qquad 6.4.66$$

For $\gamma = x + i\, y$, the argument is given by $\tan^{-1}(y/x)$. The phase shift is shown in Figure 6.15 for the same modes as Figure 6.14. In view of the above phase shift, the resonant frequency of Equation 6.4.48b) is changed to (Kogelnik and Li, 1966):

$$\sigma \left(\Delta\sigma_n \right)^{-1} = v + 1 + 2^{-1} \left(m + l + 1 \right) + \beta\, \pi^{-1} . \qquad\qquad 6.4.48c$$

As mentioned earlier in the derivations, mode selection can be achieved by suitably masking down the aperture of the cavity mirrors. With the introduction of diffraction, a quantitative measure of these masking effects can be given, at least for the two modes illustrated in Figure 6.14. The effects caused by masking the cavity mirrors are given in Figure 6.16 as the ratio of the TEM_{00} mode losses over the TEM_{01}

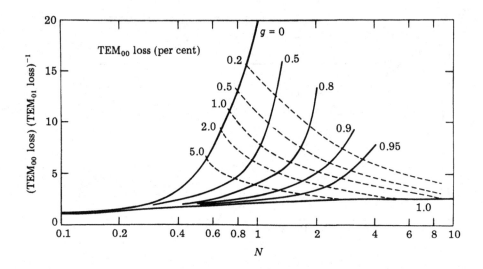

FIGURE 6.16. Ratio of the losses per transit of the TEM_{00} mode over the TEM_{01} mode as a function of the Fresnel number, N. The dashed lines are contours of constant loss per transit for the TEM_{00} mode. After Li (1965).

mode losses per transit as a function of the Fresnel number N. As can be seen in the figure, for a confocal cavity, or $g = 0$, Fresnel numbers near unity tend to favor the TEM_{00} mode over the TEM_{01} mode by a factor of about 20. The contours of the absolute loss of the TEM_{00} mode are also shown to estimate the actual power loss associated with masking.

Presume that suitable masking allows a laser to oscillate in the TEM_{00} mode, i.e., $m = l = 0$. Under these circumstances, the laser can still oscillate on several longitudinal modes, separated in wavenumber by integer values of $\Delta\sigma_n$, as defined in Equation 6.4.10). There are a number of techniques to isolate a single longitudinal mode. The simplest one is to reduce the cavity length, such that only one of the modes falls within the critical population inversion region. However, by decreasing the cavity length, the volume of the active medium also decreases which results in lower power output. As expected, for those media where the line width is very broad, this technique is not applicable. A very useful scheme is to introduce a passive Fabry-Perot inside the cavity, as illustrated in Figure 6.17. For some angle Θ, the Fabry-Perot will have maximum transmission for the wavenumber associated with a given longitudinal mode. As has been shown earlier (cf. Figures 2.2 and 2.3), when an etalon has maximum transmission for a given wavenumber, other wavenumbers (separated from the original line other than by multiples of the etalon free spectral

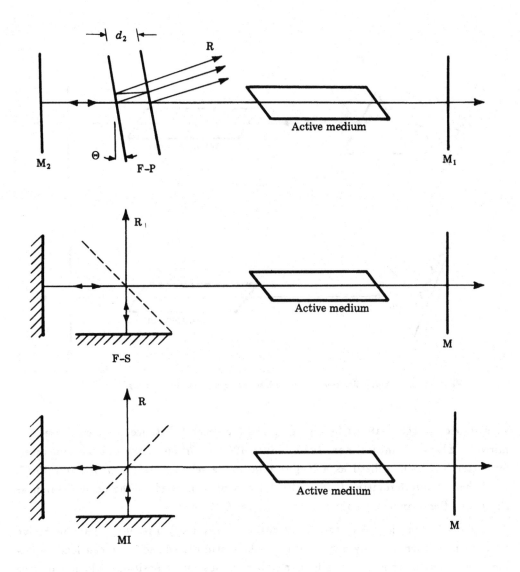

FIGURE 6.17. Selection of a longitudinal mode with a Fabry-Perot etalon (top), Fox-Smith (center) and a Michelson interferometer (bottom) in the cavity.

range) are reflected. Therefore, an etalon inside the cavity and inclined with respect to the beam will reflect the undesired wavenumbers outside the cavity, such that this radiation is no longer able to contribute to the stimulation of radiation. This is the same effect Brewster angle windows have on the perpendicular polarization component of the radiation. In practice, the mirror separation of the Fabry-Perot is selected such that the etalon free spectral range is large enough for only one passband to fall within the line width. Thus, only one mode of the cavity can be stimulated to lase. Tuning of the etalon is possible by either changing the inclination of the etalon

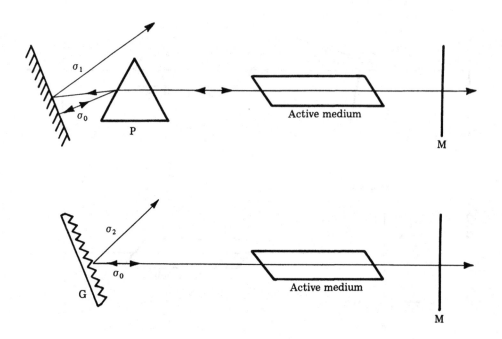

FIGURE 6.18. Single line selection for a laser using a prism or a grating.

with respect to the beam or by changing the (optical) distance between the etalon
mirrors. The Fox-Smith (Fox, 1970; Smith, 1965) variant of the Fabry-Perot interferometer has been utilized as a longitudinal mode selector, as given in Figure 6.17.
A Michelson interferometer (1881, 1882) can also be used in the same fashion as
the Fabry-Perot and this is illustrated in Figure 6.17 as well.

Thus far in the presentation, it has been explicitly assumed that the active
medium in a laser has only one transition where stimulated radiation can lead to lasing. In practice, there is usually a number of possible transitions which can lase
simultaneously. When single line operation of such a laser is desired, removal of the
radiation that stimulates the undesired transitions is again the solution. When the
wavenumber separation between lines is sufficiently large, the reflectivity of the cavity mirrors can be made to be wavenumber-dependent such that only the desired
wavenumber will oscillate while all others escape the cavity. This can easily be
accomplished with multilayer dielectric mirrors. Single line operation can also be
achieved by placing a dispersive element in the cavity as shown in Figure 6.18.
These devices operate by sending the radiation outside the cavity, in the same fashion
as the Fabry-Perot in the cavity example given earlier, and the explanation needs not
be repeated here. The use of these dispersive devices allows the operation of the laser
at any given laser line by simple rotation of the prism or grating. The use of

dispersive devices becomes mandatory in dye lasers where stimulation of radiation over a wide wavenumber range is possible.

In the derivations of the properties of the laser, it has been found that the succession of many diffraction stages in a Fabry-Perot leads to the separation of the modes of the cavity. Koppelmann (1969) has illustrated the reverse process, namely how the Fabry-Perot interference effects can be derived from the properties of a cavity. His results show that, when the mirror diameter (a) obeys the relationship $a \gg \sigma^{-1}$ and for not too a small Fresnel number, 5 to 10 neighboring modes of high m (or l) are the modes which have appreciable contributions to the resultant field, whose intensity is the classical formula for the Fabry-Perot. Therefore, the Fabry-Perot fringes are not the fundamental modes of the etalon.

Chapter 7

Practical Fabry-Perot interferometers

1. Introduction

In the previous chapters the mathematical framework necessary for understanding the behavior and properties of a Fabry - Perot device has been presented. On the basis of the mathematical treatment, it is possible to select a Fabry - Perot etalon instrument suitable for making a desired measurement to some prescribed uncertainties in the least possible time. However, in the practical world, all the desired characteristics of this (ideal) Fabry - Perot may not be possible, or available, or the cost may be prohibitive. In the following text, the ability to design a given Fabry - Perot will be tempered with the art of constructing one which will approximate the desired result, yet be within the given constraints.

Previously, an etalon had been described simply as two mirrors, with a specified reflective power, which were parallel to each other. In practice, an etalon consists of a physical substrate (or substrates) that supports the previously disembodied mirrors, some means to keep an arbitrary distance, or gap, between the mirrors, as well as a mechanical arrangement that supports all of the above. In addition, the last usually has means to ensure parallelism of the mirrors and, sometimes, the capability of changing the separation of the mirrors in a known way, yet holding them parallel. Also the etalon has to be separated from external influences that may affect its operation (such as mechanical vibrations, changes in the ambient pressure, environmental temperature variations and chemical substances which affect the mirrors).

For other than visual measurements, a Fabry - Perot instrument will require a lens to project the fringes on a plane where they will be recorded. In photographic applications, this plane is where the sensitive material is located, while for photoelectric recording applications a selection of part of a fringe (or fringes) is made here, ahead of the detector. Thus, in general, a detector-head section needs to be included as part of the instrument. The lack of selectivity of a single Fabry - Perot etalon instrument requires isolation, or filtering, of the desired spectral feature to be measured. This filtering may be simple, or it may consist of other etalons as discussed in Chapter 4. Since Fabry - Perots normally are used in the transmission mode, this configuration will be discussed in some detail. Because of the symmetry of the etalon, the discussion is also applicable to the reflection mode configuration.

2. Practical etalons

2.1. Mirrors and their substrates

Although the final quality of a Fabry - Perot etalon is intrinsically dependent on the substrate-mirror combination, some of the properties of the mirrors and the substrates can be treated separately. For instance, the substrate material must be transparent to radiation at the spectral region where the etalon will be used. Substrates, such as lithium fluoride (LiF) (Abjean and Johannin-Gilles, 1970) and magnesium fluoride (MgF_2) (Bidaeu-Mehu *et al.,* 1976), have been used in the far ultraviolet region, fused quartz (Meggers and Peters, 1918) in the near ultraviolet, visible and near infrared regions, glass in the last two regions (Bouloch, 1893; Fabry and Perot, 1896, 1898a; Perot and Fabry, 1898a), rock salt ($NaCl$) (Greenler, 1957) and zinc selenide ($ZnSe$) (Andriese *et al.,* 1979) in the infrared region and metallic grids in the mm and radiofrequency regions (Korolev and Gridnev, 1964; Lecullier and Chanin, 1976). For a given spectral region, the material with the smallest thermal expansion coefficient is to be preferred, as it simplifies the control of the ambient temperature, leading to better stability of the etalon. The use of fused quartz is nearly universal in the visible region, because this material combines a low thermal expansion coefficient with high optical transparency.

A most important property of the substrate material is its ability to be polished to a high degree of flatness, approaching the ideal perfectly flat surface. This property must be coupled with high mechanical stability in order to retain the highly polished surface figure indefinitely, regardless of the position of operation. Marioge (1971) has studied the behavior of substrates used in Fabry - Perot etalons under mechanical stress and defined this behavior. The substrate is usually made mechanically stiff by fabricating on a massive size, the empirical rule being a diameter-to-thickness ratio of at least 4 to 1 for fused quartz. This empirical rule is supported by Marioge's investigations. The polished substrates are usually known as 'flats', and it is common practice to specify their mechanical position of operation when purchasing them as the final surface figure is dependent, to a small degree, on this factor. As discussed in Chapter 4, the surface finesse factor is quite critical for multiple-etalon operation and every small gain in the final surface figure is useful.

Another property, directly associated with the substrates, is homogeneity, not only in such obvious things as being free of foreign embedded material, but in the index of refraction of the material as well, since the etalon-delayed beams must pass through this medium essentially undisturbed. A simple estimate of the tolerable index of refraction inhomogeneities of a given substrate can be made easily. If these inhomogeneities are presumed to behave as a wedge prism, which deviates the light by a prescribed (tolerable) amount, the approximate solution is:

$$\Delta\mu \; \mu^{-1} = \delta \; \mu \; (\; 2 \; n_0 \; \Delta n \;)^{-1/2} \; , \qquad\qquad 7.2.1$$

where μ is the index of refraction of the substrate, δ is the prescribed tolerance (in orders), n_0 is the value of the central order [cf. Equation 2.1.9)], and Δn is the number of orders away from the central order where the tolerance is desired. As would be expected, the tolerances become smaller as the operating order becomes larger or when off-axis measurements are desired, since the angular separation between fringes become smaller in both cases. As an example, consider a photographic measurement of wavelength, or Doppler shift, with an etalon operating at $n_0 = 50\,000$ and where an arbitrary tolerance of 10^{-2} orders is acceptable. This tolerance is equivalent to 2 mK or 30 m/s precision (at 20 000 K) respectively. The degree of acceptable inhomogeneity is found from Equation 7.2.1) to be 5×10^{-5}, when the measurement is made one order away from the central order. If the same measurement is to be made at 0.1 order and 10 orders away, the acceptable inhomogeneity is 1.5×10^{-4} and 1.5×10^{-5} respectively. The tolerance results derived above apply to the total thickness of the substrate, that is, as the flats become larger in diameter (and thicker) the tolerance per unit thickness becomes stiffer. Note that for an otherwise-perfect etalon of 0.92 reflectivity, a tolerance of 10^{-2} orders is slightly less than one-third of the instrumental width a^* of Equation 2.1.12a).

The degree of flatness to which a polished substrate can be obtained commercially seems to be limited to about $\lambda/200$ ($\lambda = 632.8$ nm) in the visible. Flats to this tolerance are offered with diameters up to 15 cm. This form of specifying flatness, in terms of λ/m, is quite useful as it is already in the proper units as utilized. However, note this specification is wavelength-dependent. The flats are usually matched sets, where the errors of polishing of one flat tend to cancel those of the other (Rayleigh, 1906). The actual surface finesse attainable from such flats depends on the specifications of the surface defects. As discussed in Chapter 2, Section 3, the finesse is quite sensitive to the type of defects present. When the dominant effect is curvature of the substrate, the surface finesse is given in Equation 2.3.5) to be equal to $m/2$, while for predominant micropolish defects, the surface finesse is, by Equation 2.3.3), equal to $m/4.7$. Using the previous flatness figure of $\lambda/200$, or $m = 200$, the resultant surface finesse is either 40 or 100 depending on the type of polishing error present. According to the LRP criterion, as given in Figures 2.8 and 2.9, optimum operation of an etalon would occur for a reflective finesse equal to the surface finesse. For the example given above, this corresponds to reflectivities of 0.97 and 0.92 for spherical curvature and micropolish surface defects respectively. Note that the magnitude of spherical curvature defects is aperture-dependent, while micropolish defects are not. Thus, no appreciable improvement is to be expected by masking down an etalon characterized by micropolish surface defects.

As shown by Pelletier *et al.* (1964), Koppelmann and Schreck (1969) and Hodgkinson (1972), to name a few, the final degree of flatness of the substrates can be further improved by judicious coating of the substrate prior to mirror deposition, or as part of the mirror deposition process proper. For the usual Fabry-Perot, the mirrors are applied onto two separate flat substrates in order to allow for flexibility of operation, such as changes of the free spectral range. When this is not the case, the mirrors can be coated on both sides of a single substrate and thus a monolithic, or 'solid', Fabry-Perot etalon is obtained (Jobin, 1898; Fabry and Perot, 1899).

The mirrors to be deposited on the substrates must satisfy many requirements. The first of these is the attainment of some specified reflectivity, over an arbitrary region of the spectrum, with the least amount of light loss. In addition, this mirror must be durable, should not permanently affect the substrates and yet be removed easily when the need arises. Historically, silver coatings were the first to be used (Fabry, 1923), as they satisfy most of the requirements set for semi-transparent mirrors, except possibly those of transparency or light loss. As would be expected, the reflectivity of silver mirrors increases as the thickness of the deposited silver increases; however, the transmission of this layer decreases very fast due to absorption by the metallic layer (Tolansky, 1946, 1947, 1955), making the $[\,1-A\,(1-R\,)^{-1}]^2$ term of Equation 2.1.7b) of considerable importance. For most applications, a reflectivity near 0.75 represents a reasonable compromise between the transparency of the coatings and the reflective finesse (Tolansky, 1947). The reflective power of silver is quite high over the near infrared and visible spectra, and starts decreasing in the near ultraviolet due to an absorption band of the metal. In this last region the metal of preference is aluminium (Burridge *et al.*, 1953; Bates and Bradley, 1967), even though its transmission characteristics are far from ideal. Although other metals, such as gold (Babcock, 1923) and tellurium (Greenler, 1957), have been used, the previously mentioned silver and aluminium mirrors remain the most useful pair (Tolansky, 1955). The practical deposition of metallic coatings is by evaporation and sputtering (Fabry, 1923; Tolansky, 1947), and the reader is referred to these authors for details on the coating techniques. The metallic coatings, as well as other coatings, tend to adopt the shape of the surface where they are deposited, thus explaining the need for high-quality substrates.

The introduction of multilayer dielectric coatings nearly eliminated the absorption-limiting factor of metallic coatings. The ability to obtain very high reflectivities, with very small transmission losses, allows for potentially high etalon finesses. As discussed earlier in Chapter 2, high etalon finesse is limited by the surface finesse to the extent that the latter is described as the limiting finesse. As these multilayer coatings, as implied by their name, require the deposition of more than 5 layers (Born and Wolf, 1964), the potential to alter the apparent surface quality by accident or by planning becomes available (Pelletier *et al.*, 1964). Although the basics of multilayer stacks are well known (Born and Wolf, 1964), the field of multilayer coatings is better discussed elsewhere (Dobrowolski, 1978). For the purposes of the present discussion, a multilayer dielectric coating consists of a stack of alternately high and low index of refraction layers each one-quarter wavelength thick. For such stacks, the reflectivity will be high over a limited range about the specific λ where the layers have the $\lambda/4$ optical thickness. This wavelength sensitivity of dielectric mirrors limits the range of their spectral usefulness to a region about 70 nm about the design wavelength. This range of usefulness can be extended by coating more than one stack with a different design center wavelength, or by a thicker stack of varying thickness layers such that the final effect is a broadband coating ($\sim$ 300 nm bandwidth) (Baumeister and Stone, 1956; Netterfield *et al.*, 1980). The presence of many layers has the enhanced possibility of changing the effective surface quality of the finished product (Giacomo, 1958).

Since the deposition of multilayer coatings is by evaporation of suitable materials, the resultant stack then is composed of amorphous material. When deposition occurs onto a relatively cool substrate, a 'soft' coating is obtained, which has the advantage of ease of removal but can be scratched easily. Hot substrate deposition gives a 'hard' coating which adheres tenaciously to the substrate, due to the partial melting and crystallization of the layer material(s), and which also has the ability to stress the substrate upon cooling, since the thermal expansion coefficient of the layers is unlike that of the substrate. A possible side-effect of hot deposition of these 'hard' coatings is to release internal strains of the substrate with the ensuing likelihood of degrading the surface quality.

The dielectric materials more commonly used in the visible are zinc sulfide, ZnS , and cryolite, Na_3AlF_6 (sodium fluoaluminate), for the high and low index of refraction layers respectively. Cryolite is hydrophilic, which requires some care in the handling of the mirrors, since the swelling by moisture of coatings containing cryolite is not fully reversible upon drying. A protective coating (usually thorium oxyfluoride, $ThOF_2$, or thorium fluoride, ThF_4) over a multilayer coating decreases problems with humidity. This protective coating is more effective when the mirror coating is smaller than the substrate, because of the edge-sealing effect by the over-coat on the mirror stack. Metallic films can be 'enhanced' by coating with dielectric materials, although this technique seems to be more useful with interference filters (Turner, 1950). Other materials used for dielectric multilayer coatings have been: aluminium fluoride, AlF_3; antimony oxide, Sb_2O_3; antimony trisulfide, Sb_2S_3; cadmium sulfide, CdS; calcium fluoride, CaF_2; cesium monoiodide, CsI; cesium mono-bromide, $CsBr$; cerium dioxide, CeO_2; lead sulfide, PbS; lead fluoride, PbF_2; lead chloride, $PbCl_2$; lead selenide, $PbSe$; lead telluride, $PbTe$; indium antimonide, $InSb$; magnesium fluoride, MgF_2; potassium bromide, KBr; rubidium iodide, RbI; silicon monoxide, SiO; silicon dioxide, SiO_2; sodium chloride, $NaCl$; stannous oxide, SnO; titanium dioxide, TiO_2; molybdenum trioxide, MoO_3; tungsten trioxide, WO_3; silicon, Si; germanium, Ge; and tellurium, Te. Some of these are more useful in regions like the ultraviolet and the infrared, rather than the visible (Heavens, 1960).

As discussed earlier, the process of deposition of dielectric multilayer coatings leaves an amorphous material stack, which will have a finite amount of both forward and backward scattering of the incoming illumination (Giacomo, 1956a, b). Similar behavior was observed also by Fabry and Perot (1897) for silvered surfaces. This scattering proceeds from the combination of the scattering from the microstructure of the layers' surface and from the bulk amorphous material. The main effect of scattering in multilayer mirror coatings is similar to that observed with metallic coatings, namely a loss in the transmission of the etalon, but of a much smaller magnitude. The associated coefficient A is small, about 0.005, and its effect does not become appreciably important until reflectivities near unity are desired. Since this light scattering redistributes the light, its more important effect is a reduction of the contrast an etalon can deliver. The use of 'hard' coatings diminishes the scattering problem. Although, from the discussion of Chapters 2 and 5, it is possible to determine the required reflectivities for a given etalon, the article by Davis (1963) on the selection of dielectric coatings should be consulted, since it provides practical advice.

As briefly mentioned earlier, the coating process places a stress on the coatings and substrates (Ennos, 1966), which will degrade the quality of the etalon mirrors (Killeen *et al.,* 1982). Ennos (1966) found certain combinations of materials to have the least residual stress when used as multilayer coatings. $ThOF_2$ - ZnS coatings had the best properties of the many coatings he investigated.

2.2. Etalon internal reflections and scatter

The requirement of etalon mirror substrates, each with finite thickness, has some disadvantages. The most obvious of these is the appearance of extraneous etalons, which are formed by the reflections between the mirror-coated sides and the backsides of each flat, due to the finite reflectivity of the backside-air interface. These extraneous etalons are rendered ineffective either by reducing the backside reflectivity with anti-reflection coatings (Killeen *et al.*, 1982) or, more commonly, by simply fabricating each of the substrates with varying thickness, i.e., a wedge, and thus forcing the extraneous etalons to be out of adjustment (Tolansky, 1944). Needless to say, when the substrate is wedge-shaped and the backside is anti-reflection coated, the effects of the extraneous etalons can be made arbitrarily small. As would be expected, the present and following discussion is only applicable to an etalon made up of two substrates since, for a solid etalon, consisting of one substrate coated on both sides, the present effects are non-existent.

Although the extraneous etalons formed by the presence of slightly reflective substrate backsides can be reduced to a negligible level, by having a wedge-shaped thickness substrate, the back surface still is present with a small, but finite, reflectivity. As discussed by Mrozowski (1941), Tolansky (1944), Dufour (1951) and Georgelin (1970b), the effects of the substrate backside reflectivity can be separated into two categories, depending on the location of the backside with respect to the incoming light into the etalon. This is shown schematically in Figure 7.1 for an etalon consisting of two substrates with an arbitrary thickness wedge characterized by the angle Φ. Figure 7.1a illustrates the effects of reflection on the etalon when the substrate has its backside closer to the source. In this example, as well as in the following discussion, refraction by the substrate material will be ignored. In the absence of reflection at surface A, a beam from source S will be partially transmitted, indicated by SO, and partially reflected, indicated by SBP, in the same fashion as described in Chapter 2, Section 1. However, in the presence of reflection at surface A, a new beam, indicated by $SBAQ$, has appeared, making an angle of 2Φ with respect to SO. The final effect of the reflection from A is to increase the illumination at an angle 2Φ away from the original beam. Note that this illumination adds to the illumination already present in that direction and it contributes to the one ring pattern formed by the etalon. For an extended source, this is of little consequence, as the final illumination is evenly spread, but for a collimated source, such as that from a spectrograph slit (Tolansky, 1944), this redistribution of the illumination acts as if there were a secondary slit apart from the main slit by the angle 2Φ, giving rise to the appearance of a weak secondary fringe pattern concentric with the main fringe pattern, but with maximum brightness at 2Φ away from the main pattern. When this secondary reflection occurs at an air-glass interface, with a reflectivity of about 0.04, the increase in brightness incurred is about 3 %.

Reflections from the etalon substrate with its backside farthest from the source are illustrated in Figure 7.1b. Again, one finds a secondary beam $S'Q$ forming an angle 2Φ away from SO, except that for each incoming ray SO there is a large number of (delayed) rays leaving the etalon, which will be partially reflected and

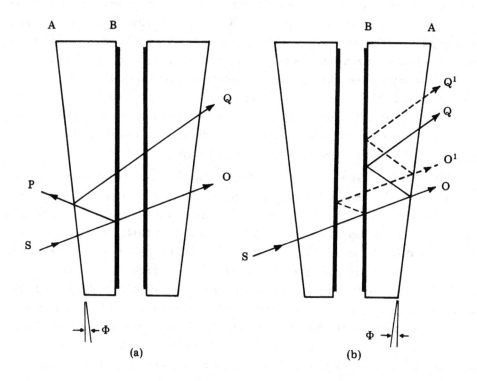

FIGURE 7.1. Formation of ghost images in an etalon with finite thickness substrates: a) substrate backside nearest to the source; b) substrate backside farthest from the source.

shifted in direction, as is the main ray. Since the rays, reflected by the substrate backside, have the same properties as those rays associated with the main ray (except intensity and direction), the resultant effect is a set of fringes displaced by an angle 2 Φ from the main fringe pattern (Mrozowski, 1941; Tolansky, 1944). For an uncoated glass substrate, the relative intensity of the displaced ghost fringe pattern is about 3 % of the main pattern and, as has been seen earlier, the angular position relative to the main pattern is dependent on only the wedge angle of the substrate. Note that this ghost pattern is a property of the etalon and it appears regardless of the source geometrical parameters.

Similar effects to those described above for the etalon substrate also occur when an interference filter is used as a pre-filter for an etalon spectrometer (Georgelin, 1970b). However, in this case, the reflectivity of the interference filter is much higher than a simple glass-air interface, and the relative brightness of the ghosts are a larger proportion of the main fringe pattern. Also, since interference filters are usually inclined with respect to the optical axis (in order to tune them for maximum transmission, cf. Chapter 2, Section 5.2), the angular location of the ghost patterns is no longer a constant of the etalon, as it was for the substrate backside reflections. As would be expected, the presence of ghosts is found also in multi-etalon devices (Mack *et al.*, 1963), where the etalon coated surfaces serve the same function as the etalon substrate backsides and interference filter just discussed.

As Tolansky (1944) has shown, the contamination of the desired spectrum by the ghost pattern can be removed, for the specific case of an etalon in front of an spectrograph, by using collimated light and properly wedged etalon substrates, such that the ghosts are projected away from the slit. The same method can be used for a Fabry-Perot spectrometer in the Jacquinot and Dufour (1948) single aperture configuration and its multiple-etalon extension, or PEPSIOS device discussed in Chapter 4, Section 3.

When etalon devices are used for imaging purposes, usually with interference filters as pre-filters, the ghosts cannot be eliminated, and one must deal with their presence. In photographic applications, the simplest solution is not to use that section of the record where the ghosts have been projected. By proper positioning of the etalon substrates wedges and other reflections, such as those caused by interference filters in the system, the ghosts can be located at an arbitrary section of the record where they will be ignored. For those imaging applications where radial symmetry detectors are used (Killeen *et al.*, 1983), and the presence of off-center ghosts cannot be tolerated, the solution has been to have no wedge in the substrate so that the ghosts are concentric with the main pattern, and to anti-reflection coat the etalon backsides to make the ghosts as weak as possible. This approach simply minimizes the contribution of the ghost images as long as they must be tolerated. Note that simple masking of radial symmetry detectors will accomplish the same effect as ignoring part of a photographic record, and then the usual wedged-substrate etalons can be used.

For the specific cases of an etalon illuminating a spectrograph, or a single aperture Jacquinot and Dufour (1948) etalon spectrometer, the etalon substrate wedge should be made large enough so that the ghosts are displaced by a nominal two to three times the angular spread of the fringe pattern actually used. As the reader is aware, the latter depends on the central order of operation, the reflectivity of the coatings and the width of the (line) source under investigation. The following example will give an idea of the wedge necessary to avoid significant ghost contamination in a single-aperture Fabry-Perot spectrometer. Assume an etalon operating order of 50,000 where the scanning aperture will examine an (arbitrary) 0.1 order. The half-angle associated with this aperture then is found to be, from Equation 2.1.8), about 7 minutes of arc. Thus, if the wedge angle of the substrate thickness is made to be 30 minutes of arc, the ghost images will be displaced from the main image by approximately twice the diameter of the aperture. On the other hand, if the

FIGURE 7.2. The original interferometer built by Jobin for Fabry and Perot. From
Benoit *et al.* (1913). Courtesy of P. Giacomo, Bureau International des Poids et
Mesures.

FIGURE 7.3. Fabry and Perot's étalon. Fabry and Perot (1902a).

operating order was to be 5,000, rather than 50,000, the wedge angle of the substrate thickness would need to be one and one-half degrees to give the same separation.

As Tolansky (1944) noted, scattered light by reflections in instrumentation containing etalons can undo the efforts made in removing the etalon ghosts. Etalon windows, lenses, lens cells, etc., are usually the main culprits. A little care in the design stage, anti-reflection coating of transmission elements, and careful baffling of the optics and the light sources allow attainment of near-ideal Fabry-Perot devices.

3. Spacers and etalon support

In the same manner as the substrates and mirrors previously described, the etalon spacers and the etalon support are closely related and will be discussed together. In the limiting cases the spacers can become the etalon support, and the support can become the spacing as well. Historically, (Jobin, 1898) there existed no spacer as such, and this is shown in Figure 7.2 for the original interferometer of Fabry and Perot. Physical spacers appeared when Fabry and Perot (1897) required an

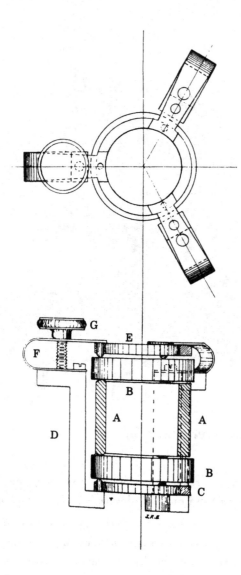

FIGURE 7.4. Rayleigh's tube spacer etalon. Rayleigh (1908).

étalon or interference gauge (Rayleigh, 1906) and this interference gauge is given in Figure 7.3. Although not shown in the figure, the spacing between the silvered glass plates is defined by three steel rods. The adjustable pressure springs, shown in both front and back, were used to compress slightly the steel rods in order to reach near-perfect parallelism and stabilization of the overall assembly. To this day, the use of adjustable pressure devices is still commonplace. Other spacers used have been bicycle ball bearings (Fabry, 1923). Ball bearings are made to close tolerances of 2λ - 5λ (Fabry, 1923; Tolansky, 1947) and presently are available to $\lambda / 8$ tolerances. Bent tungsten wires (Mewe and de Vries, 1965) have also been used as spacers.

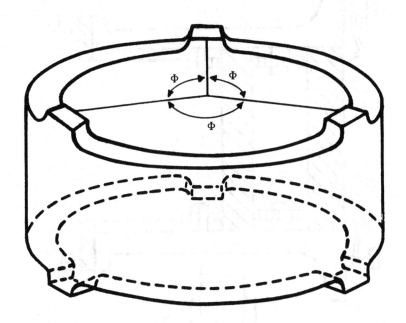

FIGURE 7.5. Contemporary etalon spacer. After Phelps (1965).

Later etalons used invar spheres and shaped invar rods as spacers (Rayleigh, 1908; Pfund, 1908) because of the low thermal coefficient of expansion of this alloy. Also at that time, Rayleigh (1908) introduced the use of (glass) tubing spacers with three equally spaced and parallel protuberances at the ends. Rayleigh's etalon, with the tubing spacer, is given in Figure 7.4. Note again the spring assembly used to compress the spacer to attain parallelism. This design by Rayleigh is the most commonly used spacer configuration for etalons (Meissner, 1941, 1942; Phelps, 1965). A modern version of this type spacer is illustrated in Figure 7.5. Present day Rayleigh-type spacers use Zerodur as the basic material because of its extremely low thermal coefficient of expansion (Killeen *et al.*, 1980). Figures 7.6 and 7.7 illustrate two contemporary versions of the etalon supports.

For those etalons where it is necessary to have access to the internal cavity, the design of Benoit *et al.* (1913) in Figure 7.8 is quite useful. The collection of etalons used by these workers in the determination of the meter are given in Figure 7.9. The use of quartz as a spacer material was reported by Meggers and Peters (1918) in

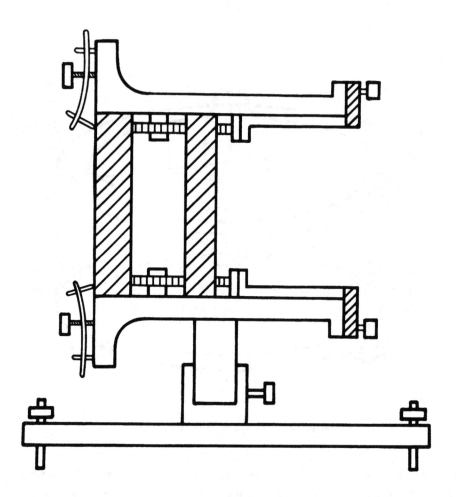

FIGURE 7.6. Laboratory etalon. After Tolansky (1947).

their original determination of the index of refraction of air. A very sturdy etalon design evolved from the work of Sears and Barrell (1933). A schematic diagram of this design is given in Figure 7.10, where the screw and wire assemblies found between the collars are used to compress the portion between the collars. The mirrors are wrung to the ends of the invar assembly, after the ends of the latter are coated with a harder metal than invar which will allow lapping and polishing of surfaces suitable for wringing. This design is a good example of the spacer becoming the etalon support. The stability gained by wringing the mirrors to the substrates was quickly appreciated (Jackson and Kuhn, 1935; C. V. Jackson, 1936), and has been

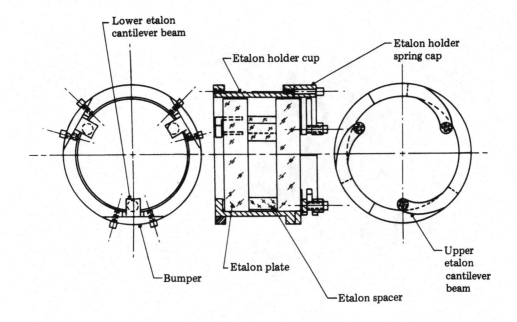

FIGURE 7.7. Modern etalon assembly. Hays *et al.* (1981).

carried farther into optically contacted etalons which offer rigidities approximating those of the solid etalon originally used by Fabry and Perot (1899), yet are simpler to fabricate. Also, in this design, the difficulty of obtaining a uniform optical thickness material is avoided, since the spacer material is now air or vacuum. An example of this design (Bates *et al.*, 1973) is illustrated in Figure 7.11.

Although, for most applications, a fixed spacer is satisfactory, there are some spectroscopic investigations where it becomes necessary to change the etalon gap by many small increments. In these cases, an infinitely variable etalon gap, such as that shown in the interferometer of Figure 7.2, is eminently suited. Chabbal and Jacquinot (1961) have designed a very stable variable-spacing support and spacer, which is given in Figure 7.12. Another solution for the infinitely variable spacer is the use of air bearings (Barrett and Steinberg, 1972).

Previous to Jacquinot and Dufour's (1948) introduction of photoelectric recording, the majority of the measurements with Fabry - Perot interferometers were made photographically. The use of this technology requires long-term stability of the etalons, a necessity encountered because of the relatively long exposure times used in the measurements. Because of the spatial resolution of photographic materials, the angular location, thus its projection, of a given Fabry - Perot fringe is not as important as is its stability. Only when more than one etalon was used, it did become

FIGURE 7.8. Open etalon. From Benoit *et al.* (1913). Courtesy of P. Giacomo, Bureau International des Poids et Mesures.

necessary to control the angular location of a given fringe, and this was attained by mechanically tilting the etalon axis with respect to the instrument axis (Houston, 1927), or more elegantly, by sealing the etalon and changing the index of refraction of the medium between the mirrors by changing the air pressure (Jackson and Kuhn, 1938a). This is the reverse of the technique used by Meggers and Peters (1918) to determine the index of refraction of air.

4. Scanning

The Jacquinot and Dufour (1948) recording method requires the fringes to move relative to the scanning aperture in order to scan the spectrum. This relative movement requires both the aperture and the fringe(s) to have a common center, namely the axis of the instrument. Using any other geometry, such as the aperture scanning the fringes like a microdensitometer scans a photographic plate, negates the gains obtained by the Jacquinot and Dufour (1948) scheme. As discussed earlier, in Chapter 2, the aperture must be radially symmetric, i.e., an annulus or a number of properly spaced annuli (illustrated in Figure 7.13), of specified width for the most efficient operation of the Fabry - Perot. The possible means to achieve this relative

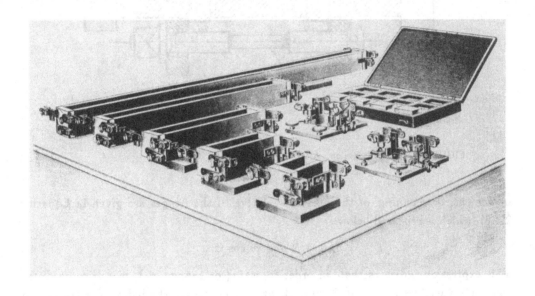

FIGURE 7.9. Etalons used by Benoit, Fabry and Perot for the determination of wavelength in terms of the standard meter. From Benoit *et al.* (1913). Courtesy of P. Giacomo, Bureau International des Poids et Mesures.

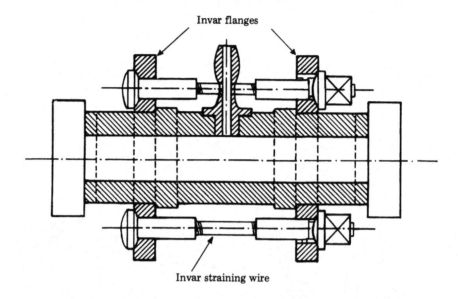

FIGURE 7.10. Sturdy etalon. After Sears and Barrell (1933).

movement, or scanning, of the aperture relative to the fringes are given in Equation 2.1.8), which is repeated below:

$$n = 2 \, \mu \, d \, \sigma \, \cos \Theta \, .   \qquad 7.4.1$$

This expression indicates that, in order to scan (or change n), one or more of the right side variables μ, d and Θ must change. A change in the index of refraction, μ, of the medium between the mirrors may be considered to be a change in the optical thickness, μd, of the etalon, or in the effective wavenumber of the radiation, $\mu \sigma$. As mentioned earlier, these two approaches are equivalent and, as would be expected, this method of scanning is called index of refraction scanning. Changing the physical distance between the mirrors is described as mechanical scanning, while changing the angle Θ is called spatial scanning. Index of refraction and mechanical scanning imply a stationary aperture, while spatial scanning has an effective moving aperture. With the usual optical configuration of a Fabry - Perot spectrometer, moving the aperture amounts to examining different parts of the source depending on the (angular) position of the aperture. Unless this scanning in space is properly accounted for (Shepherd *et al.*, 1965), the results obtained may not be those originally intended.

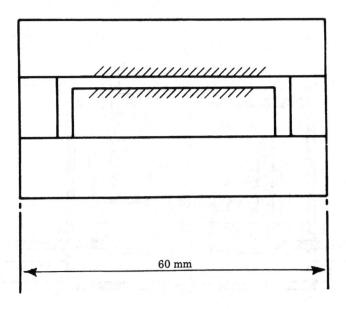

FIGURE 7.11. Optically contacted etalon. After Bates *et al.* (1973).

4.1. Index of refraction scanning

The solution presented by Jacquinot and Dufour (1948), i.e., changing the index of refraction of the medium between the Fabry - Perot mirrors to move the projected fringes relative to a static aperture, satisfies most of the requirements for scanning, yet makes use of the previous etalon spacing and support techniques. Their practical solution was to place the etalon in an airtight vessel and then change the pressure within the vessel, and thus its index of refraction, by means of a drum-driven contact electrically controlling a leak. Improvements on the means to change the pressure in a linear fashion were developed by Rank and Shearer (1956) using a laminar flow leak, or near-supersonic leak, where the mass flow is nearly independent of back pressure. Biondi (1956) realized similar results using a needle valve. Baird (1958) and Cook (1960) used a closed system utilizing a piston, which has the advantage of allowing the reuse of the scanning medium. Shepherd (1960) improved on the piston by utilizing a sylphon bellows arrangement which provides an almost leak-free vessel (and a medium not contaminated with lubricant). Chantrel *et al.* (1971) reported the operation of index of refraction scanning in discrete steps, rather than in a continuous fashion, as done earlier, thus allowing true sampling of the etalon fringes. Figure 7.14 illustrates a compact pressurized etalon vessel (Nilson and Shepherd, 1961). The use of commercial current-to-pressure transmitters (i.e., electrically controlled pressure regulators) in a feedback-servo loop (Hernandez,

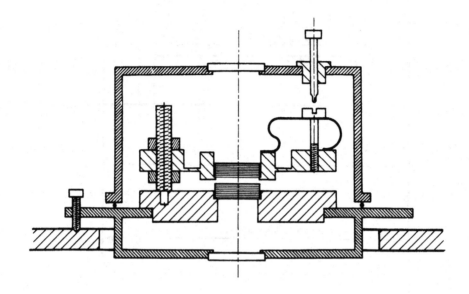

FIGURE 7.12. Variable-spacer etalon. After Chabbal and Jacquinot (1961).

1980), allows high linearity drive of the pressure either in a continuous or stepwise manner. A schematic configuration of such a device is given in Figure 7.15.

The limitations of using pressure scanning, to vary the index of refraction of a gas, are the linearity of the index of refraction change with pressure, the pressure variation necessary to obtain an arbitrary index of refraction change, and the temperature sensitivity of the index of refraction. Implicit in the earlier discussion, has been the existence of a change in the index of refraction directly proportional to the change in pressure. Since air is almost universally used for scanning, its properties bear investigation. From the American Institute of Physics Handbook (1963) formulas for the index of refraction of air, it can be shown that:

$$\left(\frac{\partial \mu}{\partial p}\right)_{\tau = 15\,^{\circ}C} = (\mu_0 - 1)\,(1.3149 \times 10^{-3} + 1.626 \times 10^{-9} p)\,, \qquad 7.4.2$$

where p is pressure in Torr units and μ_0 is the index of refraction of standard air (1 atmosphere, 15 $^{\circ}$C) at the wavenumber of interest. Using air as the refractive medium, a pressure change of about 0.1 atmosphere is necessary to obtain one free spectral range change, in the visible, with a 1 cm spacer etalon. From Equation 7.4.2) it is easy to see that the index of refraction change relative to the pressure change is non-linear to about 1 part per thousand, hence negligible except for the

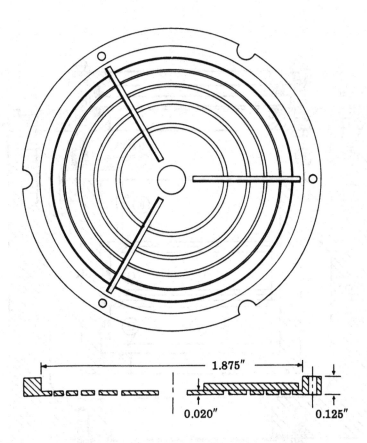

FIGURE 7.13. Multiple-annuli aperture. Courtesy of M. A. Biondi, University of Pittsburgh.

most exacting work. On the other hand, if 100 orders of this 1 cm etalon were to be scanned, or its equivalent of a one order scan of a 0.01 cm gap etalon, it takes nearly a ten atmosphere pressure change and the non-linearity is then near 0.5% which cannot be dismissed easily . Note that this non-linearity translates as a one-half order displacement for the 1 cm gap. The temperature dependence of the index of refraction of air also can be deduced to be:

$$\left(\frac{\partial \mu}{\partial \tau} \right)_{p\,=760} = (\mu_0 - 1)\ 3.8753 \times 10^{-3}\ (1 + 0.003661\tau\ )^{-2}$$

$$\approx (\mu_0 - 1)\ (\ 3.8753 \times 10^{-3} - 2.8374 \times 10^{-5}\tau\ )\ , \qquad 7.4.3$$

where τ is expressed in Celsius. Near room temperature, ~ 20 C, and for small pressure changes, the pressure and temperature sensitive terms in the right side of Equations 7.4.2) and 7.4.3) can be neglected. The remainder indicates the stronger sensitivity of the index of refraction to temperature changes relative to pressure changes.

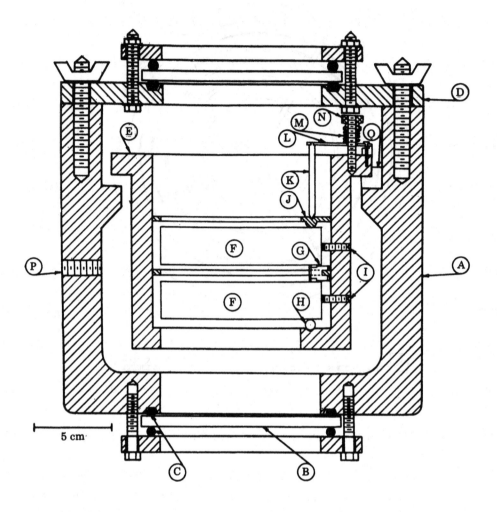

FIGURE 7.14. Compact etalon and pressure vessel. Nilson and Shepherd (1961).

Again, using room temperature as an example, a one degree change in this temperature is equivalent to ∼ 3 Torr change in the pressure, indicating the need for temperature regulation of a Fabry - Perot etalon with an air gap. For a 1 cm gap etalon,
the temperature regulation of the system needs to be ± 0.01 ^{o}C to have an uncertainty in the results comparable to that caused by the non-linearity of the index of
refraction change with pressure for a one order change. As mentioned earlier, nearly
a ten atmosphere change in air pressure was needed for a 0.01 cm gap etalon to
change one order, which complicates the construction of an airtight vessel. This pressure problem associated with small gaps can be ameliorated by using other gases
(Biondi, 1956; Meaburn, 1972) with a larger $\left(\partial \mu / \partial p \right)_T$ than air. A list of gases

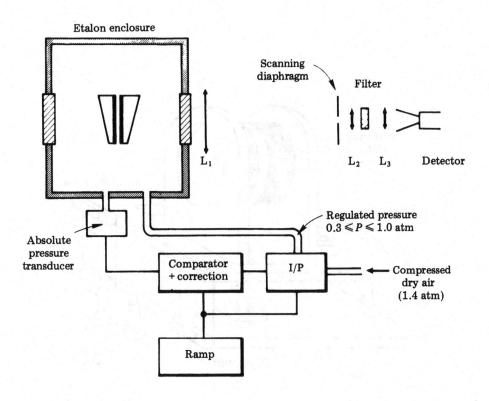

FIGURE 7.15. Schematic arrangement of a feedback loop pressure regulator for use with a Fabry - Perot etalon. Hernandez (1980).

useful for index-of-refraction scanning has been given by Gault and Shepherd (1973). When using gases as the scanning medium, there is a limit to the speed at which pressure can be changed, as adiabatic processes will lead to local heating (cooling), as well as turbulence with its accompanying inhomogeneities in the index-of-refraction of the gap. Since these effects are dependent, to a large extent, on the particular configuration of the pressure container and etalon assembly used, there exists no hard and fast rule in the maximum allowable pressure rate of change. For high resolution studies, where pressure changes are small ($\leqslant$ 50 Torr), the usual rule of thumb is not to exceed a one order change per minute.

Gagnè *et al.* (1966) have measured simultaneously the index of refraction with pressure of the gas used to scan their Fabry-Perot etalon, using a Kösters interferometer as a refractometer. With this method, the uncertainties of index-of-refraction scanning are avoided.

As the reader is aware, index-of-refraction scanning can be caused by both barometric and room temperature changes and this had to be avoided in

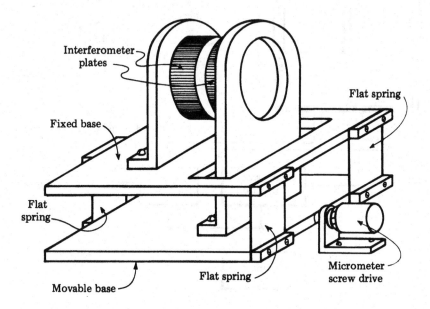

FIGURE 7.16. Mechanically scanned etalon. Greenler (1957).

photographic recording. The solution was similar to that described above, namely enclosing or sealing the etalon and controlling its temperature.

Index-of-refraction scanning has also been achieved by the use of electrooptic materials in the etalon cavity (Baird-Atomic, 1963; Del Piano and Quesada, 1965; Gunning, 1982). The index of refraction of electrooptic materials changes as a function of an applied electric field by (Gunning, 1982):

$$\Delta\mu = -2^{-1} \mu_0^3 \, r_{1,3} \, E \, , \qquad\qquad 7.4.4$$

where μ_0 is the index of refraction at zero field, $r_{1,3}$ is a constant describing Pockel's effect (Hartfield and Thompson, 1978) and E is the applied electric field. Some of the materials used with Fabry-Perot devices have been $NH_4H_2PO_4$ (ammonium di-hydrogen phosphate, or ADP), KH_2PO_4 (potassium di-hydrogen phosphate, or KDP), KD_2PO_4 (potassium di-deuterium phosphate, or KD*P) and $LiNbO_3$ (lithium niobate). Since these electrooptic materials are used as longitudinal modulators, i.e., the electric field is in the same direction as the light path, the electrodes must become part of the etalon mirror assembly. Gunning (1982) has

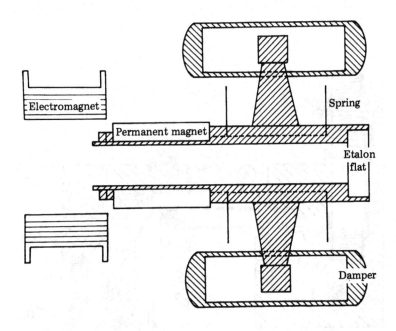

FIGURE 7.17. Electromagnetic scanned etalon. After Gobert (1958).

demonstrated a (lithium niobate) double etalon configuration for use in the infrared region of the spectrum.

4.2. Mechanical scanning

Mechanical scanning of a Fabry-Perot etalon is very attractive, as can be seen in Equation 7.4.1), since it is necessary only to change the etalon spacing by $\lambda/2$ to effect a one order change for any etalon spacing. This is, in particular, useful for very small gaps where index-of-refraction scanning would require inconveniently large pressure changes. Jacquinot and Dufour (1948) originally suggested the use of sliding ways (such as those shown in Figure 7.2) to effect the mirror separation but, because of the obvious difficulties, their use was abandoned for the more convenient index-of-refraction scanning. An improvement on the sliding ways has been the use of air bearings (Barrett and Steinberg, 1972). Note that the double-wedge etalon of Lummer (1901b) also could have been used to obtain mechanical scanning. Another suggestion by Jacquinot and Dufour was temperature scanning, which can take two forms. The first is to heat (cool) a solid etalon such that the thermal expansion of the material changes the separation of the mirrors, as done earlier by Burger and Van Cittert (1935). The second is to heat the material that defines the spacing between

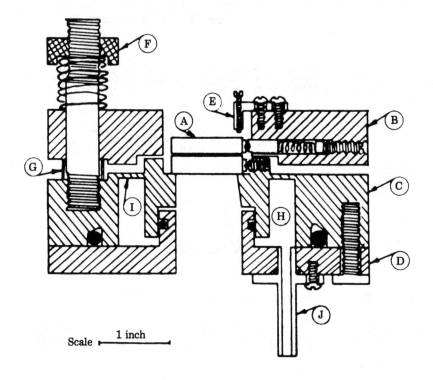

FIGURE 7.18. Pneumatically driven mechanically scanned etalon. Shepherd (1960).

the mirrors, which by simple thermal expansion will then provide the change in mirror separation (Roig, 1958). Magnetostrictive scanning was also mentioned by Jacquinot and Dufour (1948), Dupeyrat (1958) and Slater *et al.* (1965). The drawbacks of this method of scanning appear to be limited dynamic range and secondary thermal scanning caused by the ohmic heating of the electromagnet used.

Purely mechanical scanning, in the sense of springs, cams, etc., has been reported by Greenler (1957, 1958), Chabbal (1958c), Chabbal and Soulet (1958) and Bradley (1962b). Greenler's mechanically scanned etalon is given in Figure 7.16. Electromagnetic scanning has been used by Gobert (1958), Tolansky and Bradley (1960) and Bruce and Hill (1961). This system can be described as an audio transducer, or speaker, modified in such a way that light can pass through, as illustrated in Figure 7.17. Fabry and Perot (1899) used a rubber bellows and a head of water to distort the mirror holder to obtain mechanical scanning of the mirrors. A

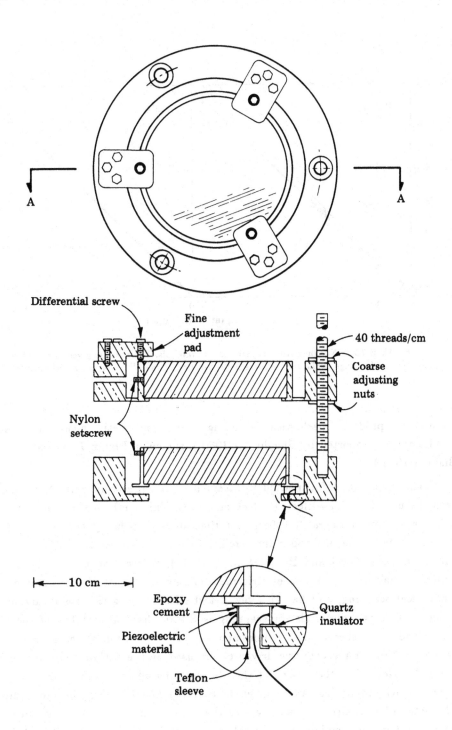

FIGURE 7.19. Piezoelectric scanned etalon. Hernandez and Mills (1973).

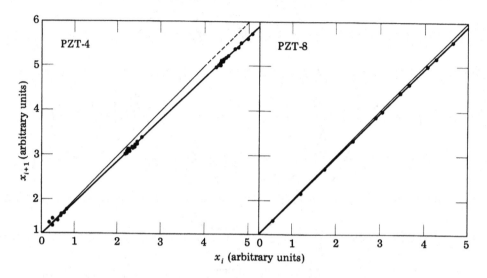

FIGURE 7.20. Nonlinearities of piezoelectric materials. The abscissa gives the position of the ith fringe, while the ordinate is the position of the $i^{th} + m$ fringe. Hernandez (1978).

method to produce mechanical scanning using pneumatic means to deform a diaphragm was reported by Shepherd (1960), shown in Figure 7.18, and later by Di Biagio (1974).

The most successful method to attain mechanical scanning has been associated with the use of the piezoelectric effect materials. This method of scanning was originally tested by Dupeyrat (1958) and then made viable by the investigations of Ramsay (1962, 1966), Cooper and Greig (1963), Mielenz *et al.* (1964), Gadsden and Williams (1966) and Bates *et al.* (1971), to mention a few. These studies indicated that the use of piezoelectric materials satisfies most of the requirements for mechanical scanning, and has very few drawbacks. Among the useful advantages of using piezoelectric materials for scanning is the mechanical stability of the finished etalon and the rather modest requirements for the operation of these devices (Hernandez, 1970). A piezoelectric driven etalon used by the author is illustrated in Figure 7.19. Note that the design of this device is based on Chabbal and Jacquinot's (1961) variable-spacing etalon given in Figure 7.12. However, the most important advantage of piezoelectric scanning was the realization of dynamic alignment of the etalon, rather than relying on the mechanical stability of the etalon itself (Ramsay, 1962, 1966; Gadsden and Williams, 1966; Hernandez and Mills, 1973; Hicks *et al.*, 1974), which will be discussed later. The presently-available piezoelectric materials are slightly non-linear and their behavior has been characterized fully (Hernandez,

FIGURE 7.21. A commercially available piezoelectric scanned etalon. Photograph courtesy of Burleigh Instruments, Inc., Fishers, New York.

1978, Basedow and Cocks, 1980). The behavior of two types of piezoelectric materials is shown in Figure 7.20. These non-linearities can be arbitrarily reduced (Hernandez, 1978) or other methods to monitor the etalon gap can be employed (Hicks *et al.,* 1974). A commercial piezoelectric scanned Fabry-Perot is shown in Figure 7.21.

4.3. Spatial scanning

As described briefly in the introduction to scanning methods, spatial scanning consists of varying the order number of Equation 7.4.1) by changing the value of the angle Θ. For the most efficient results this can be accomplished in two general ways, namely varying the effective radius of an aperture and its width (an annulus is presumed in this discussion) in a controlled manner at the projection plane of the etalon fringes, or having a number of detectors at varying positions on this plane. In all cases, the etalon gap is static and no scanning is performed by the etalon proper.

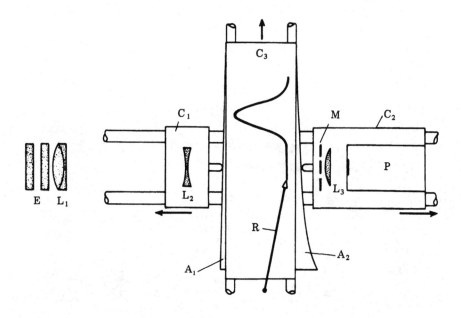

FIGURE 7.22. Spatial scanning, as proposed by Armstrong (1958).

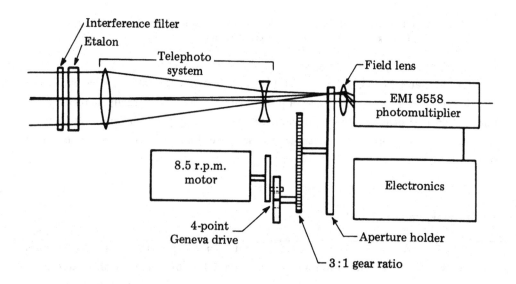

FIGURE 7.23. Stepped aperture device used to realize spatial scanning. Shepherd *et al.* (1965).

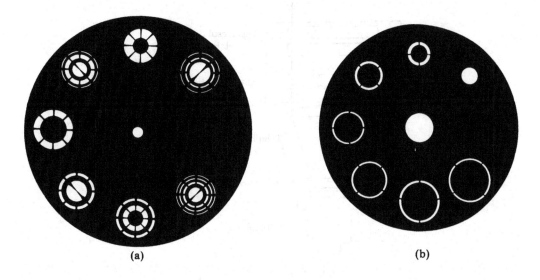

FIGURE 7.24. *a*) Hadamard coded and *b*) standard apertures. Shepherd *et al.*
(1978).

As expected, purely mechanical movement of the aperture presents a challenge, and
the solutions reported have side-stepped the associated problems. The original interferential grille of Jacquinot and Dufour (1948) could have been used as a spatial
scanning method, but it was used as a multi-slit device instead. Armstrong (1958)
proposed using a fixed annulus, upon which the etalon fringes were projected with a
variable-focal-length lens. Variation of the focal length of this lens would change the
size of the fringes and thus scan them across the annulus aperture. This is illustrated
in Figure 7.22. Another way to look at this process, i.e., from the point of view of the
source, is to say the angular dimension of the annulus is being changed. Bradley
(1962a) attained similar results using an intensifier tube as the variable 'focal
length' element. Shepherd *et al.* (1965) presented a stop and varifocus lens in combination with a diaphragm, as well as the use of stepped apertures, while Hoey *et al.*
(1970) utilized a variation of the latter by stepping increasingly larger apertures.
The stepping method and aperture shapes used by Shepherd *et al.* (1965, 1978) are
illustrated in Figures 7.23 and 7.24.

Hirschberg and coworkers (Hirschberg and Fried, 1970; Hirschberg *et al.*,
1971) introduced multiplexed apertures, while Neo and Shepherd (1972) utilized
Hadamard code apertures, and more general multiplexed spatial methods have been
discussed by Shepherd *et al.* (1978). The use of multiple apertures and multiple
detectors was successfully attained by Hirschberg and Platz (1965). These workers
used a number of mirrors mounted coaxially in the plane of projection of the fringes,

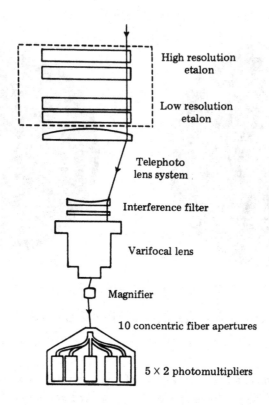

FIGURE 7.25. Multiple-aperture multiple-detector fiber optic device. Shepherd *et al.*
(1965).

and each of these mirrors directed the reflected light in separate directions where it
was focused onto separate detectors. That same year, Shepherd *et al.* (1965)
showed the use of two apertures with two detectors, as well as a multiple-aperture
multiple-detector device where the multiple aperture was defined by separate optical
fiber bundles, with the latter shown in Figure 7.25. Katzenstein (1965) used an axicon to separate selectively the different orders by simply moving the axicon in the
converging light beam used to project the fringes, and thus scanning the fringes across
an aperture. This is illustrated in Figure 7.26. Spatial scanning can also be attained
with the axicon by having a number of detectors in separate planes in the instrumental axis. Platisa (1974) has discussed the polarization effects the use of an axicon
introduces in an etalon system.

Multiple apertures with multiple detectors have also been demonstrated with
the use of Fresnel lenses (Hirschberg and Cooke, 1970), shaped multiple detectors
(Chaux *et al.*, 1976), single detector with shaped anodes (Abreu *et al.*, 1981; Killeen *et al.*, 1983) [illustrated in Figure 7.27] and finally by the use of devices with
a continuum of detectors, such as television tubes (Bradley *et al.*, 1964; Sivjee *et al.*,

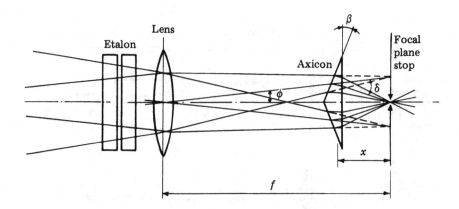

FIGURE 7.26. Axicon spatial scanning. After Katzenstein (1965).

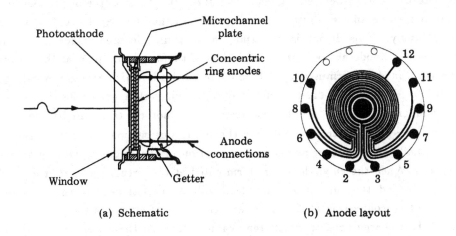

(a) Schematic   (b) Anode layout

FIGURE 7.27. Multiple anode photomultiplier detector. (Killeen *et al.*, 1983). Courtesy of T. Killeen, University of Michigan.

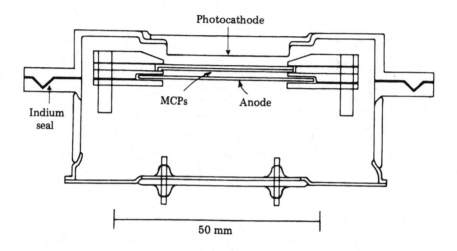

FIGURE 7.28. Imaging photon detector (IPD). Rees *et al.* (1981a).

1980), solid state array detectors (Chaux and Boquillon, 1979) and the imaging photon detector (IPD) devices (Rees *et al.,* 1981a). An example of this device is given in Figure 7.28 and, since its mode of operation is not self-explanatory, it will be described briefly. This device is essentially a proximity focused image intensifier, where a resistive anode has been installed in lieu of the usual phosphor at the output. This resistive anode is terminated to the outside via leads which are attached to the anode at 90° intervals. Consider one photon arriving at a given position in the cathode and giving rise to one photoelectron. The photoelectron will be converted into a large ($\sim 10^7$ electrons) pulse at the multiplier section and this pulse will reach the anode at a position reflecting the position of the original photoelectron at the cathode. When the anode leads are terminated at (a virtual) ground, the anode will discharge and the current at each lead will be partitioned by the resistance between the lead connection at the anode and the point of arrival of the pulse. Thus, in general, four charge packets of different value will reach these (virtual) grounds. When these virtual grounds are the input of amplifiers, which convert the charge packets into pulses whose height is proportional to charge, then it is rather straight-forward to convert the ratios of the pulse amplitudes into a coordinate position. In turn, this position is mapped onto a memory device where one event is added. Therefore, when the fringes from a Fabry - Perot are projected onto one of these position-sensitive photomultipliers, the end result from the memory device is, usually, a digital representation of the radiation reaching the cathode. These (digitized) fringes are then treated in the same manner as photographically recorded fringes after they have been examined with a microdensitometer. It is interesting that the improvements in spatial detectors have effectively resulted in the equivalent of a highly sensitive and linear-responding photographic plate.

The properties and use of photodetectors are better discussed elsewhere. The reader is referred to the articles of Lallemand (1962) and Young (1974) for details on these devices. For single detectors with shaped anodes, and reduction of the information obtained with them, the articles by Killeen *et al.* (1983) and Killeen and Hays (1984) are suggested.

5. Alignment and spacing control

Proper operation of a Fabry - Perot etalon presumes the mirrors to be parallel, or aligned. This process of alignment is dependent on the etalon gap and the available light sources, in the sense that the line width of the latter should be narrower than the desired free spectral range of the instrument. Fabry and Buisson (1919) and Fabry (1923) have described methods for the alignment of etalons, which are used to this day. For small etalon spacings ($\leqslant$ 10 cm), the following methods have survived the test of time. The first makes use of equal thickness, or Fizeau, fringes in the following manner. The etalon is illuminated with a collimated nearly monochromatic beam and the radiation transmitted through the etalon is focused on a screen with a small aperture. Looking through this aperture, possibly with the aid of a telescope, the following will be observed. When the mirrors are not parallel, as is usually the case, a number of nearly straight lines (or Fizeau's equal thickness fringes) will be observed crossing the mirrors. Application of a slight pressure to one side of the mirrors and perpendicular to the fringes will decrease the air wedge and the fringes will separate and become wider until, in the limit when the mirrors are parallel, the whole field will be evenly illuminated. Of course, if the pressure to the mirror is applied to the wrong end, the fringes will come closer together. In practice, when the mirrors are not ideally flat, upon reaching parallelism the field will not appear fully illuminated but will look splotchy, in particular when the width of the line used to do the alignment is very narrow. As expected, this behavior near parallelism is employed to determine the quality of the surfaces of the etalon, as will be discussed later. The limitation of this alignment method occurs when the spacing between the mirrors is large and no source narrow enough is available to give the equal thickness fringes. In this case, the second method (Perot and Fabry, 1901b; Fabry, 1923) is employed. Briefly, this method consists of illuminating the etalon with the collimated light from a very small pinhole which is, in turn, illuminated with a bright white light source and the brighter the better. Observing the light transmission through the etalon will show the multiple reflections of the pinhole in the mirror cavity. When the etalon is not parallel, these reflections will have the appearance of a comet's tail. Adjustment of the mirrors until this comet-tail appearance is no longer noticeable will force their parallelism. As would be expected, this method is more sensitive with the higher-reflectivity mirrors. In practice, this method is quite useful to bring any size gap near parallelism very quickly, and then the other methods, such as the Fizeau fringe method described earlier, are used to complete the alignment.

Another method, reported by Perot and Fabry (1901b), is alignment by fringes of superposition using two etalons. Briefly, when two etalons in series of equal, or integer ratio, spacing are illuminated by white light, they give rise to white light fringes. Assuming only one of these etalons is in alignment, the white light fringes would be observed at that part of the common aperture where the etalon gaps are equal, or integer ratio, to each other. Thus, changing the spacing of the out-of-alignment etalon until the full aperture shows the white light fringes, has transferred the alignment of one etalon to the other. Note that the etalon spacings are also fixed.

The most common method goes back to the original description of the etalon by Bouloch (1893) and Fabry and Perot (1899). When an etalon near parallelism, illuminated with an extended near-monochromatic source, is observed with the naked eye, it will show a set of brightly illuminated equal inclination, or Haidinger (1849), fringes, such as those shown in Figure 2.5. As the eye is moved about the etalon aperture, these fringes will change in (angular) diameter as the distance between the mirrors changes. As can be found from Equation 2.1.8), the fringes will appear to contract as the distance between the mirrors decreases and to expand when the distance increases. Again, as in the Fizeau fringe alignment, application of pressure at the apex of the etalon's wedge will reduce the variation of the fringe diameters until, when parallelism is attained, the fringes will show no further changes in size.

These alignment procedures have been extended for use with multiple etalons and multipassed etalons; see, for instance, Harley (1979).

Tolansky (1947) estimates the precision of obtainable parallelism to be $\lambda/100$ for both the equal-inclination and equal-thickness etalon alignment procedures, and $\lambda/50$ for the adjustment procedure with the brightly illuminated pinhole. The degree to which parallelism can be attained, with the equal-inclination fringe procedure, is dependent on the narrowness of the observed fringes, and the lower limit is therefore dictated by the etalon finesse of Equation 2.3.16). Roesler (1974) reports the precision to which alignment can be made in the equal-inclination alignment procedure to be equal to $\lambda/(4N_R)$, although a more conservative estimate would be $\lambda/(4N_e)$ based on the following argument. Etalons with finite surface defects will show, when carefully observed, a slight 'breathing' effect as the eye is moved about the etalon aperture regardless of how much adjustment is made, and this is the limit of the possible adjustment when surface defects are caused by microscopic random structure.

FIGURE 7.29. Etalon alignment aid. After Kinder and Torge (1971).

On the other hand, for an otherwise-aligned etalon with spherical curvature surface defects, the fringe pattern will show slightly different diameters at the center than at the edges of the etalon. The best alignment under the above-described conditions requires some judgement, based on experience, which may include masking out part of the etalon (Roesler, 1974); that is, the final alignment is indeed dependent on the etalon finesse rather than the reflective finesse.

Parallelism is usually reached by applying pressure to the etalon spacer or to the etalon support by means of springs, differential screws, etc., as has been shown in some of the etalons given earlier. Other means, such as gravity (Burnett and Lammer, 1969), also have been employed to apply this pressure on the spacer or the etalon support. Rayleigh (1908) cautioned that the use of pressure, as well as its point of application, to attain alignment by squeezing the spacer can create distortions on the mirror surfaces and degrade their quality, or conversely, improve the quality of the mirrors. Caplan (1975) has also discussed this topic.

In the conventional Haidinger fringe alignment procedure, the sensitivity of the parallelism adjustment decreases as the diameter of the fringes decreases with decreasing free spectral range. Aids, such as a (small aperture) telescope, will regain some of the lost sensitivity, but there exist other means to either regain or, for that matter, improve this alignment sensitivity, as well as to provide more convenient ways to align an etalon for parallelism. Molby (1936) reports using a small slit in front of a telescope while aligning an etalon. Moving the slit about the telescope aperture allows the entrance of rays from different parts of the etalon, thus, the local alignment can be determined and improved if necessary. Kinder and Torge (1971) report a similar device ascribed to Hansen, and it is illustrated in Figure 7.29. The prism devices used by Rogers (1969) and Khodinskii (1980) are given in Figures 7.30 and 7.31. All of these devices satisfy the requirement of observing a small section of the etalon at one time in order to determine the local adjustment. The most

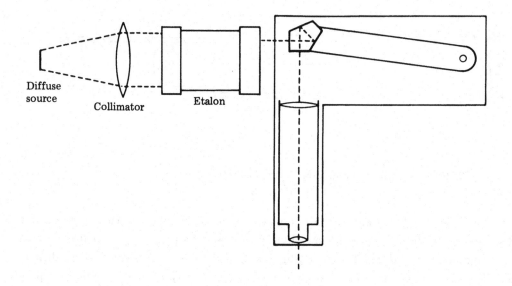

FIGURE 7.30. Prism based etalon alignment aid. Rogers (1969).

recent method of etalon alignment has been reported by Colombeau and Froehly (1976). These authors use the speckle pattern of coherent source projected etalon fringes to reach both best parallelism and best focus of the instrument, with a sharp symmetric speckle pattern indicating the best alignment and focus point.

The methods for attaining parallelism thus far discussed are passive methods since, after alignment is reached, the etalon spacer and support combination are supposed to remain stable and unchangeable while the particular measurement, where the etalon is employed, is in progress. In practice, an experiment is routinely interrupted and the alignment is either checked visually, or by some means built into the experiment, such as the measurement of a laboratory reference source. Depending on the results of these tests, the instrument may be realigned.

Ramsay (1962) showed that an etalon could be kept in automatic alignment over a prolonged term. This was accomplished by electronically forcing the etalon to have maximum transmission for white light fringes derived from an auxiliary source. In order to do this, light from this auxiliary source is forced to pass through the etalon twice by means of reflections. In effect, the method of Fabry and Perot (1899) and Perot and Fabry (1901b) to find, and force, the equality of the spacing of two etalons is being utilized. By producing a small oscillation in the etalon gap and detecting the resultant variation in the transmitted illumination from the white light source, it is possible by electronic means (phase sensitive detection), to have a

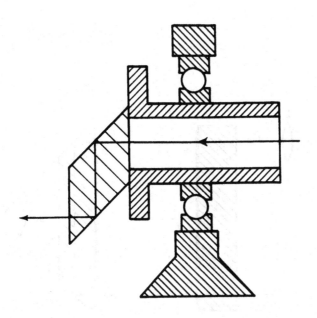

FIGURE 7.31. Prism alignment aid. After Khodinskii (1980).

servo loop that will change the etalon gap until the etalon is locked to remain at the maximum of the white light transmission. When this is done in two orthogonal dimensions, then, by definition, the etalon gap is parallel. This method of alignment is schematically illustrated in Figure 7.32. The oscillation and movement of the mirrors was derived from piezoelectric material stacks (Ramsay, 1962; Kobler, 1963), although any other means of controllable scanning can be used as long as independent variation of the mirrors at a minimum of two (orthogonal) positions is available. This method of alignment keeps the etalon mirrors parallel, but has no control on their spacing, since, as long as the mirrors are parallel, the alignment method is satisfied regardless of the size of the etalon gap. Ramsay (1966) included a third gap, or reference interferometer, as part of the system. With this reference interferometer as part of the illumination path, the etalon is forced to have the same gap as the reference etalon, or an integer ratio thereof. During operation of the etalon, the gap parallelism is controlled by the system described earlier, and the etalon spacing is independently controlled by the reference etalon. If the gap of the reference etalon is changed, the etalon gap must follow and thus alignment also can be attained in a dynamic fashion. This alignment system is illustrated schematically in Figure 7.33.

The use of line spectral sources, rather than the continuum source(s) of Ramsay (1962), makes it possible to both fix the gap and force the etalon to be parallel, as well as to be referenced to a basic standard. This has been done by Hernandez and

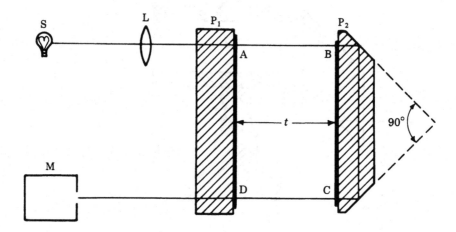

FIGURE 7.32. White light etalon parallelism alignment procedure. Ramsay (1962).

Mills (1973) for a dynamically scanned Fabry - Perot spectrometer of the Jacquinot
and Dufour (1948) variety. The method used is to force the zero crossings of the
derivatives of the spectral source fringe(s) to occur at a fixed location of the
(mechanical) scan by means of a servo loop which includes piezoelectric material
tubes to provide both the scan and the necessary loop corrections. The spectral
sources are located at the edges of the etalon aperture, and their fringes are detected
away from the main channel by bending the rays associated with them by means of a
wedge. This arrangement is given in Figure 7.34, and it is based on the etalon given
in Figure 7.19. By choosing these subsidiary sources to be in a part of the spectrum
other than the radiation under investigation, and after filtration, there is no measurable leakage from the subsidiary sources into the main channel (Hernandez and
Mills, 1973). Since at least a two-order scan is made during normal operation of this
instrument, two sets of derivatives are included in the servo loop. Typically, the first
servo provides a bias correction for parallelism and spacing at a low voltage drive
point, while the second servo ensures the piezoelectric material change in length with
increasing voltage is the same for the (three) piezoelectric tubes at least one order
away from the first servo location; thus, forcing again both parallelism and equal

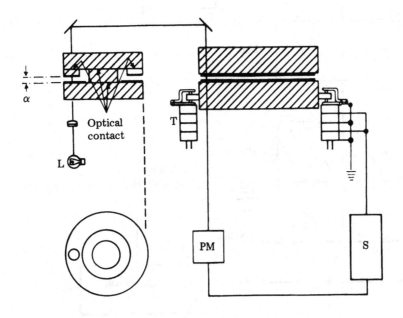

FIGURE 7.33. Ramsay's (1966) white light automatic spacing method.

spacing at that point. The availability of a laboratory reference for each and every measurement, while the measurement is in progress, provides the stability needed for long-term measurements, such as the study over one solar cycle of the small Doppler shifts caused by winds on atmospheric trace emissions (Hernandez and Roble, 1984). This particular method of alignment has proved to be reliable, as the author has operated two of these instruments for a combined 18 year span, with only minor maintenance required.

Gadsden and Williams (1966) reported a method of keeping the etalon spacing constant by making use of the radiation being measured, rather than by the use of an auxiliary source, as done by Ramsay (1962, 1966). The method is based on the harmonic properties of the etalon transmitted radiation.

Sandercock (1970) also has made use of the radiation being measured to achieve both alignment and stability of the etalon gap, by taking advantage of the unshifted scattered radiation of the excitation source in his Brillouin studies. Briefly, his method consists of consecutively introducing a very small wedge in the etalon in

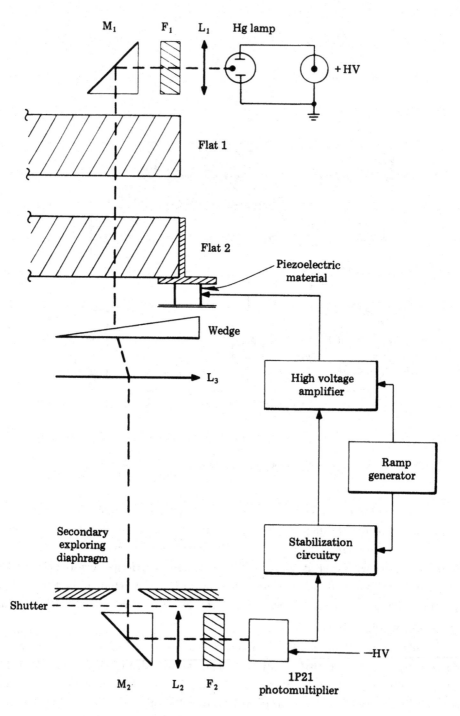

FIGURE 7.34. Spectral line source method to obtain both automatic parallelism and spacing control. Hernandez and Mills (1973).

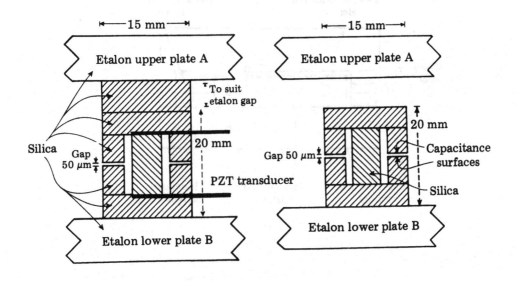

FIGURE 7.35. Capacitive micrometers used to maintain the etalon gap alignment and stability. From Rees *et al.* (1981b).

two orthogonal positions, in step with the scanning of the etalon. This wedge is introduced in both directions about a central position, such that when the change in the signal away from the central position is equal for both orthogonal wedges, the etalon is then in alignment. Many other variants of this basic process have been reported in the literature.

Hicks *et al.* (1974) introduced a servo-controlled etalon, where the active control element determining the etalon gap separation does not depend on the optical properties of the etalon gap. Instead, they used the capacitance micrometers developed by Jones and Richards (1973). With these devices, the change in capacitance with physical spacing is measured, and the servo loop then is used to control this capacitance, hence the etalon gap. These capacitance micrometers, used in conjunction with etalons, are of small physical dimensions (≈ 1 cm^2), because of the limited space available. Thus, to obtain measurable capacitance values, the distance between the capacitance elements must also be made small. When the electrodes are applied directly to the optical flats, the etalon gap must be made small, as in the original Hicks *et al.* (1974) device. For large etalon gaps, it then becomes necessary to change the capacitive micrometer design, such that the capacitance electrode gap remains small. A practical example from Rees *et al.* (1981b) is illustrated in Figure 7.35. As can be seen in the figure, it is necessary to have also a reference capacitance element, physically close to the measurement element, to allow for changes in

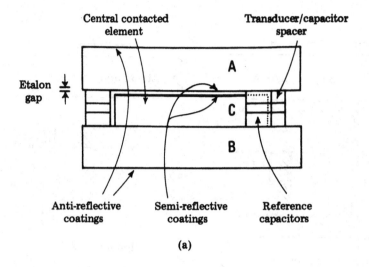

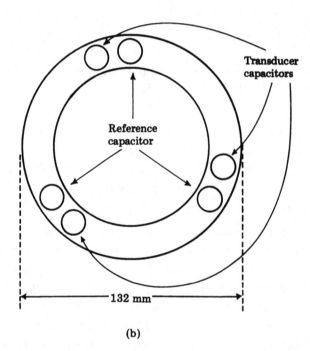

FIGURE 7.36. Layout of a capacitive micrometer stabilized etalon. After Rees *et al.* (1981b).

the dielectric constant of the material between the electrodes. The general layout of an etalon with these capacitance devices is given in Figure 7.36. The capacitive micrometers are an integral part of the etalon, as shown in Figure 7.36, limiting the flexibility of of the device, since changing the gap is best described as inconvenient.

Tests on the linearity and stability of the capacitive micrometers (Gladwin and Wolfe, 1975), show them to be in the $\leqslant 0.2\%$ level; however, there is no published comprehensive study on the behavior of capacitance stabilized etalons for different configurations of the micrometers, in particular, those which use spacers, as in Figure 7.36, as well as their response to changes in composition, temperature and pressure of the material between the electrode gap. Even though capacitive-micrometer stabilized etalons require initial (optical) calibration, the absence of light extraneous to the measurement eliminates the possibility of contaminating the measurements, which is an attractive feature of these etalons.

As has been mentioned earlier in Section 3, the physical distance between the mirrors is defined by custom-made spacers, precision ball bearings and wires. Because of the properties of the coatings used for the mirrors, the optical spacing is slightly different from the physical spacing (Fabry and Perot, 1897; cf. Chapter 3, Section 1). The actual optical spacer can be determined by setting the etalon mirrors as a thin wedge, and then measuring the difference in path between a coated and an uncoated area of the flats (Fabry and Buisson, 1908b). Meissner (1941, 1942) uses the measurement of an emission line by the etalon with various physical spacings in order to determine the contribution of the coating to the optical thickness. Note that this contribution of the coatings to the optical spacing is wavelength-dependent (Perot and Fabry, 1901b; Bauer, 1934; Meissner, 1941, 1942).

6. Flatness of the mirrors

The reasons for testing the surface quality of Fabry - Perot mirrors are varied and sundry, including quality control of newly obtained mirrors, determining the effects of recoating these mirrors, and in some cases, insuring that the mirror mount does not affect the quality of the mirror's surface(s).

In general, the measurement of flatness is a quantitative extension of the alignment process described earlier. The testing techniques and devices developed are the outgrowth of the Fizeau equal-thickness fringe alignment methods. As described by Fabry (1923), the basic measurement consists of placing the two mirrors close together with a small wedge between them. Visual, or photographic (Schwider *et al.*, 1966), measurement of the observed fringes, and determination of the departure of these fringes from straight lines is the basis of this type of measurement. The calibration is built in the measurement, as the separation between fringes is $\lambda/2$. When this test is made with a single line source, the angle of the wedge between the two mirrors must be varied in order to map the etalon aperture. This scheme was simplified by Rasmussen (1946), Tolansky (1948) and Herriott (1961, 1967), who used multiple line sources which would cover a greater area of the mirrors with fringes for a given measurement. Murty (1962) used multiple-pinhole apertures to obtain the

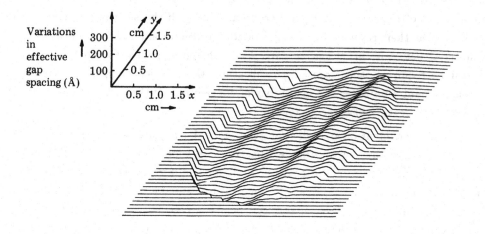

Variations
in
effective
gap
spacing (Å)

FIGURE 7.37. Contour map of the variations in effective gap spacing of an etalon.
(Killeen *et al.*, 1981). Courtesy of T. Killeen, University of Michigan.

same result as Herriott (1961), but with only a single wavelength source. With
these methods of measurement, determination of departures from flatness of about
$\lambda/100$ are attainable with some care.

The second method of measurement was pioneered by Duffieux (1932), and
consists of a visual or photographic measurement of the mirrors through the aperture
used to do the equal-thickness fringe alignment. When using a narrow source (in
practice, a source narrow relative to the etalon free spectral range), the observed
illumination of the mirrors, as the etalon is scanned, does not change evenly, as an
ideally perfect pair of mirrors would be expected to do. This unevenness of illumination can be translated into absolute units, since the relationship between the scan of
the etalon and changes in the gap are well known. Koppelmann and Krebs
(1961a, b) improved on this technique by the inclusion of photometric techniques.
They place a small aperture in back of the etalon exit aperture, and sample the
etalon area by moving the small aperture. The rest of the instrument is a Jacquinot
and Dufour (1948) single aperture spectrometer. This method is essentially a photometric version of Molby's (1936) alignment scheme. A more sophisticated
improvement has been used by Killeen *et al.* (1981), using an image dissector as
the detector. With their arrangement, the total aperture of the etalon is examined at
once and there is no loss of contrast as in the previous method when the etalon is
masked down by the small aperture. A contour map of an etalon measured with the
image dissector arrangement is illustrated in Figure 7.37. The work of Koppelmann

and coworkers (Koppelmann and Schreck, 1969; Koppelmann *et al.*, 1975) and the review of Schultz and Schwider (1976) should be consulted on the details of measurement of surface defects of mirrors used in Fabry-Perot etalons. For practical use, the arrangements of Koppelmann and Krebs (1961a, b) or Killeen *et al.* (1981) are the most convenient for routine tests of mirror departures from flatness, as well as for calibrating the capacitive-micrometer aligned etalons.

7. Reflectivity

In the early days of Fabry-Perot interferometers, an accurate knowledge of the reflectivity of the mirrors was not as important as the need for reasonable transmission through the etalon. Nevertheless, the need eventually did arise and Fabry and Buisson (1908b) reported a method to measure directly the reflectivity by employing a polarizing birefringent element and an analyzing polarizer. Their method takes advantage of the observation that two slightly wedged mirrors, when illuminated by a collimated source, will show a series of images of the source, but of decreasing intensity. The intensity relationship between consecutive images is given by the two reflections of the light involved in making the images. Using a birefringent element, such as a Wollaston prism, to look at these multiple images, will split the images, giving rise to a new set of images slightly displaced from the original ones. When these images are examined with an analyzing polarizer, there exists a null point where the observed intensities are equal for images which have undergone an equal number of reflections. Rotation of the analyzing polarizer will change the relative intensities of the double images and, as such, members of the double images with different number of reflections can be matched in intensity. As Fabry and Buisson (1908b) have shown, the relationship between two images separated by two reflections and the rotation, β, of the polarizer to make these two images of equal intensity is:

$$R^2 = \tan^2(\beta) . \qquad 7.7.1$$

Since the mirrors, in general, have unequal reflectivities, the measured quantity is $R_1 \times R_2$, where the subscripts indicate the individual mirrors.

As discussed in Chapter 2, Section 2, Giacomo (1952) used the response of an etalon to a very wide line to determine the reflectivity of the etalon. The ratio of the radiation observed with and without the etalon in the path is given by Equation 2.2.6), namely:

$$Y = [1 - A(1-R)^{-1}]^2 (1-R) (1+R)^{-1} = \tau^2 (1-R^2)^{-1} , \qquad 7.7.2$$

since $1 = A + \tau + R$. Thus, R is given by:

$$R = (1 - \tau^2 Y^{-1})^{1/2} , \qquad 7.7.3$$

where both Y and τ are measurable quantities. τ is the transmission of each individual mirror and thus $\tau^2 = \tau_1 \times \tau_2$. As Giacomo pointed out, the determination of R by this method has larger uncertainties as the value of the reflectivity approaches unity. This method has the convenience of measuring the complete etalon aperture at once, rather than a small area; thus, the measurement is representative of the etalon as used. Like the Fabry and Buisson (1908b) method, the Giacomo method just described is independent of the quality of the surface of the mirrors.

Another method used in the determination of reflectivity is that sketched in Chapter 3, Section 2, in regard to the determination of the instrumental function. Briefly, the Fourier coefficients, a_k, derived from a measured fringe, are of the form given in Equation 3.2.25), or:

$$a_k = d_k \times s_k , \qquad\qquad 7.7.4$$

where d_k and s_k are the instrumental and source coefficients respectively. If the source is truly monochromatic, the contribution of s_k to a_k is zero, and a measurement of such a source will result directly in d_k. For a practical source, the contribution of s_k can be made as small as desired by increasing the free spectral range, i.e.:

$$\lim_{\Delta\sigma \to \infty} a_k = d_k , \qquad\qquad 7.7.5$$

since the contribution of the source is proportional to the ratio of the source HWHH to the free spectral range, $\Delta\sigma$. Thus, measurements of a source when the free spectral range is very large, are a fair representation of the instrumental function. When such measurements are made with a Jacquinot and Dufour (1948) spectrometer with a very small exploring diaphragm, the effect of this small diaphragm on the determined coefficients will be negligible, and therefore, these coefficients are a good representation of the etalon function alone. For the most common mirror surface defects, namely micropolish inhomogeneities and errors of curvature, the explicit forms of d_k can be obtained from Equations 2.3.1) and 2.3.6), or:

$$(d_k)_G = R^k \exp(-k^2 G^2) , \qquad\qquad 7.7.6$$

$$(d_k)_D = R^k \operatorname{sinc}(k \; 2 \; df_D^*) , \qquad\qquad 7.7.7a$$

where $G = \pi \, df_G^* \, [\ln(2)]^{-1/2}$. For small values of df_D^*, such that $2 \, k \, df_D^* \leqslant 0.5$, the sinc function can be expressed by a standard approximation (Zucker, 1964), and the coefficients for the etalon become:

$$(d_k)_D \simeq R^k \, [\, 1 - 0.1666666664(2\pi \, k \; df_D^*)^2 + ...] . \qquad\qquad 7.7.7b$$

The logarithms of the instrumental Fourier coefficients are then expressed as:

$$\ln(d_k)_G = k \, \ln(R) - k^2 \, G^2 , \qquad\qquad 7.7.8$$

$$\ln(d_k)_D \simeq k \ln(R) - k^2(0.1666666664\ 2\pi\ df_D^*)^2 - ....\qquad 7.7.9$$

Since both Equations 7.7.8) and 7.7.9) have the same second order form, the value of R can be obtained by a least squares solution of equations of the form $a + bx + cx^2$, where $b \equiv \ln(R)$. In practice, the conditions of Equation 7.7.5) are not rigorously met, and the approach taken is to obtain measurements at increasing $\Delta\sigma$ (or decreasing etalon gap). Fourier coefficients are determined from these measurements (Bennett, 1978; Mar *et al.*, 1979), and from these coefficients the values of $\ln(R)$, as outlined above, are obtained. The desired value of R is that value resulting from extrapolation of $\ln(R)$ to zero spacing gap (Mar *et al.*, 1979). Note that the same process can be carried out for the surface defect contribution, as long as it is small for the case of df_D^*. In addition to having the ability to measure the overall etalon aperture, as in the Giacomo (1952) procedure, this method has the advantage of using the etalon and a laboratory source as the only equipment for the determination of reflectivity.

8. Materials

In the previous sections, dealing with substrates, spacers and etalon supports, explicit and implicit assumptions were made in regard to the existence of materials with thermal properties suitable for the construction of etalons. As discussed earlier, the mirror substrates present the most serious limitations in the fabrication of etalons, as these substrates should be transparent, homogeneous and must take and keep a polish. The material that best fits the above requirements is fused silica, at least for the near ultraviolet, visible and near infrared regions of the spectrum.

The thermal expansion coefficient of a given material, α, is defined as (Berthold and Jacobs, 1976):

$$\alpha = \frac{1}{L}\frac{dL}{d\tau} = \frac{1}{\Delta\tau}\left(\frac{\Delta L}{L}\right),\qquad 7.8.1$$

where L is the length and τ is the temperature. Berthold and Jacobs (1976) have reported a comprehensive study of the thermal expansion coefficient of materials normally used in the construction of Fabry-Perot devices. Near room temperature (20 °C), they report a thermal expansion coefficient (in units of 10^{-8} °C^{-1}) of $\sim$ 50 for fused silica, $\sim$ 45 for invar LR35, $\sim$ −20 for Super Invar, $\sim$ 3 for Zerodur, $\sim$ −1.5 for ULE-7971 and $\sim$ −5 for Cer-Vit 101, where the last three are ceramic materials. Some of these results have been confirmed by Killeen *et al.* (1982) with the etalons built for the Dynamics Explorer satellite experiment. As can be seen from the above-mentioned thermal expansion coefficients, the use of ceramic materials for the etalon spacers tends to make the etalon gap rather insensitive to temperature changes (Rees *et al.*, 1982). Careful design of the etalon support will decrease further the effects of temperature changes of the completed instrument (see for instance, Killeen *et al.*, 1982).

9. Selecting a Fabry-Perot

In order to summarize some of the material presented thus far, as well as to provide a rudimentary guide on the selection of a Fabry-Perot spectrometer, an example drawn from the author's investigations will be used.

For this particular exercise it will be assumed that measurements of the night-time upper atmosphere dynamics and physics are desired, specifically near 100 km height. At this altitude chemical reactions give rise to the emission of light at 17924 K due to the ($^1S_0 - {}^1D_2$) transition of atomic oxygen (Chamberlain, 1961). To avoid unnecessary detail, the properties of this radiation will be stated without further justification. The 17924 K emission proceeds from a highly forbidden transition, thus of negligible damping width, while the kinetic temperature in that region of the atmosphere is near 200 K; therefore, the emission can be considered to be an ideal Doppler-broadened line. The expected horizontal motions, i.e., Doppler shifts, are not much larger than 50 m/s, while vertical motions are negligibly small. Also, for the purposes of this exercise, the radiation proceeds from a homogeneous extended source and has no neighboring, or contaminating, emission. The zenith emission rate of the 17924 K line is 100 R (1 Rayleigh = $4\,\pi$ I, where I is expressed in units of 10^6 photons $cm^{-2}\ sec^{-1}\ sterad^{-1}$), but it is necessary to measure this emission at a large zenith angle in order to determine horizontal motions. This method of observation increases the apparent emission rate by a factor of about two, relative to the zenith, since a finite thickness emission layer is being observed at an angle other than perpendicular. This is the so-called van Rhijn (1919) effect. Finally, it is considered necessary to measure simultaneously both the temperature and the horizontal motions with a precision of 10 K and 5 m/s and with a time resolution better or equal to 10 minutes. A Jacquinot and Dufour (1948) single aperture Fabry-Perot spectrometer appears to be suitable for this measurement, since the emission line has no special filtering requirements. For the measurements, the equidistant equal-time sampling method of Chapter 3, Section 2.1 will be employed.

The optimum points of operation for a normalized spectrometer have been derived from Equations 3.2.24) and 3.2.23) and the operational quantities a^*, dg^*, f^*, and N for the optimum, or least uncertainty, operation are given in Figures 3.2 and 3.3 and summarized in Figures 3.4 through Figure 3.8. As discussed in Section 2.1 of Chapter 3, when simultaneous wind and temperature measurements are desired, it is necessary to use the more restrictive parameters needed for temperature determinations. The values for these parameters are found to be as follows: etalon normalized width $a^* = 0.035$ (thus, automatically giving a reflectivity of 0.80);

normalized source width profile $dg^* = 0.11$; scanning aperture normalized width $f^* = 0.065$; and N, or the minimum number of coefficients necessary to unambiguously describe a measurement in its Fourier representation, is equal to 7. Since both the expected temperature and the optimum value of dg^* are given, the value of the free spectral range $\Delta\sigma$ is therefore fixed by the relationships given in Equations 2.1.10a) and 3.2.26), or:

$$\Delta\sigma = dg \; (\; dg^*)^{-1} = \sigma \; (c \; dg^*)^{-1} (\; 2 \; \tau \; k \; A \; M^{-1} \; \ln 2 \;)^{1/2} , \qquad 7.9.1$$

while the order number, n_0, and the etalon spacing, d, are found from Equations 2.1.10a) and 2.1.10b). For the present example, the values of the free spectral range, order number, and etalon spacer are then 0.206 K, 86,960 and 2.424 cm, respectively. The value of T in Equations 3.2.24) and 3.2.33) is equal to $2N = 14$ for critical sampling, while a quantum efficiency of detection $\epsilon = 0.10$ is a reasonable value to assume in this region of the spectrum. Finally, the transmission τ_L, associated with miscellaneous light losses, will be considered first to be equal to unity, which will give the smallest etalon possible to reach the measurement objectives.

The minimum (normalized) uncertainty for temperature determinations is found from either Figure 3.3 or Figure 3.4 to be, in round numbers, equal to 14, while the (normalized) uncertainty for the winds for the same parameters is found in Figure 3.2 or Figure 3.4 to be about 10^9. Note, however, that these numbers are given as standard deviations, not variances as required by Equations 3.2.24) and 3.2.33). Although, from the discussion of Section 2.1 of Chapter 3, it is known that the temperature is the more restrictive of both the desired measurements, the calculations of this exercise will be carried for both temperature and wind determinations. Since the only unknown left is the area of the etalon, A, Equations 3.2.24) and 3.2.33) will be re-arranged to obtain the following:

$$A_\tau = 14^2 \; \tau^2 \; n_o \; (\; \sigma_\tau^2 \; I \; T \; \epsilon \; \tau_L \;)^{-1} , \qquad 7.9.2a$$

$$A_v = (10^9)^2 \; (dg^*)^2 \; (\; \sigma_v^2 \; n_o \; I \; T \; \epsilon \; \tau_L \;)^{-1} , \qquad 7.9.3a$$

where the subscripts for the area, A, indicate temperature and winds respectively. Inserting the previously found values for the quantities in the right-side of the equations, the following areas are obtained for the limiting emission rate determination, i.e., the 100 R zenith measurement for 10 minutes:

$$A_\tau = 14.3 \; cm^2 \quad (\; D = 4.3 \; cm \;), \qquad 7.9.2b$$

$$A_v = 11.7 \; cm^2 \quad (\; D = 3.9 \; cm \;), \qquad 7.9.3b$$

where D is the diameter. The size of the etalon required by the results given above is the smallest possible since unity transmission, i.e., no losses, has been assumed.

In a real experiment, two mirrors will be used to direct the light from arbitrary positions in the sky to the Fabry-Perot spectrometer and a dome will be needed to protect these mirrors from the elements. The etalon normally will be located inside a thermally stabilized vessel which will have at least one window, while the second window would most likely be the lens used to project the fringes onto the scanning aperture. When this vessel must be pressurized, a suitable pressure window should replace the lens. A field-lens is usually placed behind the scanning aperture to collimate the light for selection of the proper wavenumber by an interference filter, with the latter having a typical transmission of near 0.4. Behind this filter there is another lens whose purpose is to project all the light onto a detector, which in turn has been cooled below ambient to reduce its dark emission current. In order to achieve this, the detector must be insulated by a thermal window, which usually consists of a double-pane arrangement. At this point it should be noted that, although it is tempting to remove the two lenses behind the aperture, they do serve two useful purposes. The first lens collimates the light such that the interference filter operates at its most efficient point, as discussed in Chapter 2, Section 5.2, while the second lens not only collects the light for the detector, but also is used to defocus the source and fringe images on the photodetector. This defocusing will remove the effects of both local inhomogeneities in the source and in the quantum efficiency of the detector. Last, for simplicity, it is assumed that the two sky mirrors will reflect 0.96 of the light impinging upon them.

The above enumeration simply shows there exist about 22 surfaces which transmit (or reflect) light, and this count includes the two backside surfaces of the etalon flats. When, for the sake of simplicity, it is assumed that all the transmitting surfaces are uncoated glass, each of these surfaces will have a reflectivity of 0.04, and the light reflected by them is considered to be lost forever. By this assumption, the problem of ghosts, described earlier, is being ignored; however, this is not the place to deal with such detailed information, and the reader is referred to the earlier section on this topic. With the knowledge of the filter transmission and the number of reflecting/transmitting surfaces in the spectrometer, it is now possible to calculate the transmission term τ_L :

$$\tau_L = (\, 0.96 \,)^{22} \times 0.4 = 0.16, \qquad\qquad 7.9.4$$

or the minimum etalon area must be increased about six-fold, i.e.,:

$$\mathbf{A}_\tau = 87.6 \text{ cm}^2 \;\; (\, D = 10.6 \text{ cm} \,) \,, \qquad\qquad 7.9.2c$$

$$\mathbf{A}_v = 72.0 \text{ cm}^2 \;\; (\, D = \;\; 9.6 \text{ cm} \,) \,. \qquad\qquad 7.9.3c$$

Note the large increase in area caused by reflections by the uncoated surfaces, which should serve as an indication of the usefulness of having anti-reflection coatings on as many surfaces as possible.

The area found in Equation 7.9.2c) is the clear aperture of the etalon, which does not include obstructions by such as the etalon adjustment mechanism parts, spacer material, etalon support fingers and holders, self-aligning devices, etc. Since etalon flats are available commercially in 2.5 cm diameter increments, a pair of flats with 12.5 cm diameter would normally satisfy the requirements for the minimum clear aperture set above in Equation 7.9.2c). The required clear aperture found in this equation is that aperture for an ideal defect-free etalon with a 0.80 reflectivity mirror coating; that is, with a finesse N_R, as defined in Equation 2.1.13a), of about 14. Near-ideal behavior for the etalon is obtained when the surface defect finesse is about four times the reflective finesse. Thus, a surface finesse of about 50, corresponding to a flatness of $\lambda/100$ for spherical curvature defects [defined in Equation 2.3.5)], will satisfy the near-ideal etalon requirements.

The previous calculations have implicitly assumed the etalon will behave as an ideal etalon of infinite physical extent. This assumption can be tested with Equation 5.3.14) and Figures 5.23 through 5.25. From the equation it is found that k, or the etalon effective-size, is equal to 1240, which, as seen in the figures, does indeed give near-ideal etalon behavior. In other exercises, when the source irradiance is large, the calculated etalon area may be small enough so that the effective-size considerations just discussed are important.

Similar calculations illustrate that, using the equal-noise measurement method of Chapter 3, Section 2.2, for the same etalon area as found in Equation 7.9.2c), a measurement time of about 20 minutes is required to obtain the same precision of measurement as the equal-time method used earlier. Although the loss of efficiency is nearly a factor of 2, the equal-noise method would deliver the specified precision, even though the emission rate of the source may be fluctuating widely.

The spherical etalon has not been considered here, since the etalon spacer is too small to satisfy the requirements of Equations 5.2.25), 5.2.48) through 5.2.51) and 7.9.2c). For higher-resolution measurements, where etalon spacings greater than about 7 cm are necessary, the spherical etalon becomes competitive, as discussed in Chapter 5, Section 2.

At this point is is useful to compare the single-aperture Jacquinot and Dufour (1948) spectrometer with a Fabry-Perot spectrometer using a multiple-anode device. Since, for optimum operation, the single-aperture covers about 0.13 of an order, i.e., $2 \times f^*$, a multiple anode device can, at best, collect 7.69 times more light per order. As can be seen from Equations 3.2.24) and 3.2.33), the variance of determination is inversely proportional to the irradiance, and, since it is possible to consider the presence of about 8 anodes per order as being equivalent to increasing the irradiance by that value, then the standard deviation of determination will be decreased by the one-half power of the light increase, or 2.77. Thus, using the change in the standard deviation as a measure of the change in efficiency of measurement, a device which will measure a full order will have a gain of 2.77 over the single-aperture spectrometer. Note that the use of multiple-aperture masks, such as that of Sipler and Biondi (1978) given in Figure 7.13, have a gain in efficiency also given by the one-half power of the number of apertures used. Returning to the multiple-anode-device comparison, it was found that about 8 such anodes could be fitted in one order, if they were to be of the optimum size. However, 14 samples are required for the unambiguous description of the measured profile, thus the size of each effective anode must be made smaller than the optimum in order to satisfy this requirement. The new effective aperture size is now less than optimum, thus increasing the error of determination as given in Equations 3.2.24), 3.2.33) and Figures 3.2 through 3.5. By decreasing the aperture size it is found also that the number of coefficients has increased, as well as having changed values for the parameters a^* and dg^*. Inspection of Figure 3.3 shows that the uncertainty therefore will be increased by a factor of about 1.5 in the variance, that is, with a multiple-anode device there is a gain of about 2.26 per order in the standard deviation of determination. This gain can also expressed as a decrease in the diameter of the aperture of an etalon to obtain an arbitrary uncertainty of determination when using a multiple-anode device, rather than a single aperture spectrometer. For the exercise of the night sky measurements used previously, the clear area of the etalon would be reduced from 88.1 cm^2 to 17.2 cm^2, while the diameter is reduced from 10.6 cm to 4.7 cm. Assuming the same degree of obstruction exists for both cases, then an aperture diameter of 6.6 cm would be the minimum aperture diameter that could be used, which represents a sizeable saving in the cost of the etalon flats. This saving must be tempered by the increase in cost and complexity associated with a multiple-anode device and its electronic instrumentation.

Chapter 8

Non-classical Fabry-Perot devices

In the previous chapters, conventional Fabry-Perot interferometers have been discussed in some detail. Other systems, where Fabry-Perot devices are used in combination with other dispersing devices, have been touched upon lightly, but unconventional usage of Fabry-Perot devices has not been discussed to any extent.

1. Fabry-Perot etalons and insect-eye lenses

Insect-eye lenses, i.e., a collection of small-diameter short focal length lenses in a grid (usually monolithic), have been used by Courtès and collaborators (Courtès *et al.*, 1966; Courtès and Georgelin, 1967; Georgelin, 1970a; Monnet, 1970) for imaging applications, mostly astronomical, where conventional fringe imaging on an object gives too sparse a record because of the wide separation between fringes. The use of these insect-eye lenses gives a number of separate fringe images superimposed on the sky object(s), when the foreoptics are properly arranged for imaging purposes (Meaburn, 1976). In effect, the use of these multiple lenses divides the etalon into a number of small discrete etalons, such that each small etalon has its own corresponding object image.

As has been discussed in Chapter 5, Section 3.1, the use of small-effective-size etalons leads to the degradation of the etalon properties, relative to an ideal etalon.

However, for the intended astronomical use of these insect-eye lenses, i.e., Doppler-shift measurements, the degradation in the etalon quality is not considered to be a drawback. As illustrated by Georgelin (1970a), these insect-eye lenses are also very useful in the determination of the degree of parallelism of etalons, as well as the quality of the substrates.

2. Fabry-Perot combined with other dispersive elements

Crossing of a Fabry-Perot etalon with another dispersive element was discussed in Chapter 2, Section 6 in the context of determining absorption spectra, and in Chapter 4, where the dispersing element was used as a filtering device. Note that the multiple-etalon PEPSIOS device (Mack *et al.*, 1963) also accomplishes the same result.

Returning to the combination, or crossing, of a Fabry-Perot etalon and a linear dispersing device (spectrograph), it is found that, when the slit height is small relative to the diameter of the fringes projected on the slit, only those wavenumbers which have integral orders of interference will give rise to bright images. In the limit of a continuum spectral source, Edser-Butler (1898) fringes would be observed; these fringes are fixed in wavenumber for a given optical spacing of the etalon. Chabbal and colleagues (Chabbal and Pelletier, 1965; Chabbal *et al.*, 1967) overcame this fixed location of the fringes by moving the recording element in the spectrograph perpendicular to the direction of dispersion while simultaneously and synchronously changing the optical spacing of the etalon. In effect, this amounts to scanning the Edser-Butler fringes, which are then recorded in separate sections of the recording medium and, thus, all the spectral features present on the source being examined will be measured as long as at least one free spectral range of the longest wavelength of interest is scanned. Proper microdensitometer tracing of the record provides a complete record of the source with the resolving power of the etalon. This combination of spectrograph and Fabry-Perot etalon has been named SIMAC for Spectromètre Interférentielle Multicanal (Chabbal and Pelletier, 1965). When using a spectrometer with a position-insensitive detector, such as a photomultiplier, synchronous scanning of the etalon and grating spectrometer is necessary. Chabbal and Jacquinot (1961) accomplished this simultaneous scanning by changing the common pressure of both instruments, or pressure scanning, by placing them in the same closed container. The line profiles determined from SIMAC or the photoelectric version suffer from distortion caused by the finite width of of the spectrograph (spectrometer) slit. Thompson (1982) has discussed this problem in a quantitative manner.

3. Parallel Fabry - Perot interferometers

Fabry-Perot devices in parallel combinations have not received as much attention as have Fabry-Perot devices in series combinations, as illustrated in Chapter 4. The parallel Fabry-Perot etalons are useful in describing both the basic properties of the Fabry-Perot proper and the properties of the combination. The investigations of Sturkey (1940) and Sturkey and Ramsay (1941) are the primary sources for the study of parallel Fabry-Perot devices.

For convenience, consider a monochromatic wavefront illuminating two etalons, one each in the arms of an ideal Mach-Zehnder interferometer (Zehnder, 1891; Mach, 1892), as shown schematically in Figure 8.1. The etalon spacings are of unequal length and the physical position of the etalons in the arms is such that the beams from one etalon are out of phase, relative to the other, by an arbitrary value ψ when arriving at lens L. For simplicity, these etalons are assumed to be ideal etalons whose amplitudes are described by Equations 2.1.4) and 2.1.5), and their reflectivities are also assumed identical. The phase dispersion, χ of Chapter 2, Section 2.1, is arbitrarily set equal to zero. Then, the transmitted amplitudes of the two etalons, arriving at lens L of the figure, are:

$$A_{t_1}(\phi) = \tau (1 - R e^{i\phi})^{-1} , \qquad 8.3.1a$$

$$A_{t_2}(\delta, \psi) = \tau e^{i\psi} (1 - R e^{i\delta})^{-1} . \qquad 8.3.1b$$

In these expressions the subscripts indicate the two different etalons, and the first etalon (FP_1) is the arbitrary reference etalon from which the phase difference ψ is measured. When lens L, as given in Figure 8.1, collects the beams from both etalons, the projected resultant pattern can be described in classical terms (cf. Chapter 2, Section 3) by the superposition of the intensities, or more appropriately by the superposition of amplitudes. The resultant intensities for both superposition examples are:

$$Y_{t_{1,2}}(\phi,\delta,\psi) = 2^{-1}A_{t_1}(\phi) A_{t_1}^*(\phi) + 2^{-1}A_{t_2}(\delta,\psi) A_{t_2}^*(\delta,\psi) , \qquad 8.3.2a$$

$$\hat{Y}_{t_{1,2}}(\phi,\delta,\psi) = A_{t_{1,2}}(\phi,\delta,\psi) A_{t_{1,2}}^*(\phi,\delta,\psi) . \qquad 8.3.3a$$

In the present discussion the amplitude superposition results will be distinguished by the use of a circumflex, $\hat{Y}$. The superposed amplitudes of $A_{t_1}(\phi)$ and $A_{t_2}(\delta,\psi)$ are given by:

$$A_{t_{1,2}}(\phi,\delta,\psi) = 2^{-1}A_{t_1}(\phi) + 2^{-1}A_{t_2}(\delta,\psi) . \qquad 8.3.4$$

The intensities of Equations 8.3.2a) and 8.3.3a) can be explicitly solved:

$$Y_t(\phi,\delta,\psi) = 2^{-1}[Y(\phi) + Y(\delta)] , \qquad 8.3.2b$$

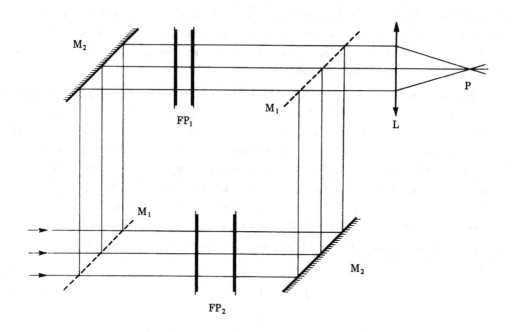

FIGURE 8.1. Configuration used in the description of parallel Fabry-Perot interferometers. The mirrors denoted by M_1 are partially reflecting, while those denoted
by M_2 are totally reflecting. After Sturkey (1940).

$$\hat{Y}_t\,(\phi,\delta,\psi) = 4^{-1}\,[\ Y\,(\phi)\ +\ Y\,(\delta)\ ]\,[\ 1\ +\ \cos\ \psi\ ]$$

$$-\ R\ \ Y\,(\phi)\ \ Y\,(\delta)\ \tau^{-2}\,\sin[\ (\ \phi\ -\ \delta\ )2^{-1}\ ]$$

$$\times\left(\ R\ \ \sin[(\phi-\delta)2^{-1}+\psi]\ -\ \sin\psi\ \cos[(\phi+\delta)2^{-1}]\ \right)\,.\qquad\text{8.3.3b}$$

In the above expressions the number subscripts have been omitted for clarity, and
$Y\,(\theta)$ is defined in the same manner as in Equation 2.1.7a), or:

$$Y\,(\theta)\ =\ \tau^2\,[\ 1\ +\ R^2\ -\ 2R\ \cos\ \theta\ ]^{-1}\,.\qquad\text{8.3.5}$$

Equations 8.3.2b) and 8.3.3b) can be further simplified when it is assumed that:

$$\phi\ =\ \alpha\ \delta\ =\ \delta\ (\ 1\ -\ 2\rho\ )\,.\qquad\text{8.3.6a}$$

After inserting the above into the superposition equations, the following is obtained:

$$Y_t\,[\delta(1-2\rho),\delta,\psi]\ =\ Y\,(\delta)\,[\ 1-2^{-1}\mathbf{F}\ Y\,(\delta)\ \sin\delta\rho\ \ \sin(\delta-\delta\rho)\ ]$$

$$\times\,[1-\mathbf{F}\ Y\,(\delta)\ \sin\delta\rho\ \ \sin(\delta-\delta\rho)\ ]^{-1}\,,\qquad\text{8.3.2c}$$

$$\hat{Y}_t\,[\delta(1-2\rho),\delta,\psi] = 2^{-1}\,Y_t\,[\delta(1-2\rho),\delta,\psi]\,(1+\cos\psi)$$

$$-\,4^{-1}\mathbf{F}\,Y^2(\delta)\,\sin\delta\rho\,[R\,\sin(\delta\rho+\psi)-\sin\psi\,\cos(\delta-\delta\rho)]$$

$$\times\,[1-\mathbf{F}\,Y(\delta)\,\sin\delta\rho\,\sin(\delta-\delta\rho)]^{-1}\,,\qquad\text{8.3.3c}$$

where $\mathbf{F}$, as defined in Equation 2.1.13c), has been used with the assumption of an ideal etalon with zero absorption, i.e., $A=0$. Also, in the above equations, the following identity has been used:

$$Y[\delta(1-2\rho)] = Y(\delta)[1-4\,R\,\,Y(\delta)\,r^{-2}\,\sin\delta\rho\,\,\sin(\delta-\delta\rho)]^{-1}$$

$$= Y(\delta)\,[1-\mathbf{F}\,\,Y(\delta)\,\sin\delta\rho\,\,\sin(\delta-\delta\rho)]^{-1}\,.\qquad\text{8.3.7}$$

The last term of Equation 8.3.3c) shows the effects of the mutual interference of the two etalon beams, which do not appear for the simple classical superposition of intensities. For the use of both Equations 8.3.2) and 8.3.3), the definition of ρ requires further discussion. Assume two etalons with spacers which are different by a small arbitrary amount, such that these etalons are different by Δn orders. In general, the first etalon will have $\delta=2\pi n$, while the second will have $\phi=(2\pi n-2\pi\Delta n)$. Using the definition of Equation 8.3.6a), it is found that:

$$\phi = \alpha\,\delta = 2\,\pi\,n\,(1-\Delta n\,\,n^{-1})\,,\qquad\text{8.3.6b}$$

or $\rho=\Delta n\,(\cdot 2\,n\,)^{-1}$, which can be replaced in Equations 8.3.2c) and 8.3.3c) to give:

$$Y_t\,[n\,(1-\Delta n\,\,n^{-1}),n\,,\psi] = Y(n)$$

$$\times\,[1-2^{-1}\,\mathbf{F}\,\,Y(n)\,\sin(\pi\,\Delta n)\,\sin(2\pi\,n-\pi\,\Delta n)]$$

$$\times\,[1-\mathbf{F}\,\,Y(n)\,\sin(\pi\,\Delta n)\,\sin(2\pi\,n-\pi\,\Delta n)]^{-1}\,,\qquad\text{8.3.2d}$$

$$\hat{Y}_t\,[n\,(1-\Delta n\,\,n^{-1}),n\,,\psi] = 2^{-1}\,Y_t\,[n\,(1-\Delta n\,\,n^{-1}),n\,,\psi]\,[1+\cos\psi]$$

$$-\,4^{-1}\,\mathbf{F}\,Y^2(n)\,\sin(\pi\,\Delta n)\,[1-\mathbf{F}\,\,Y(n)\,\sin(\pi\,\Delta n)\,\sin(2\pi\,n-\pi\,\Delta n)]^{-1}$$

$$\times\,[\,R\,\sin(\pi\,\Delta n+\psi)-\sin\psi\,\cos(2\pi\,n-\pi\,\Delta n)\,]\,.\qquad\text{8.3.3d}$$

The first obvious test for the above equations is found for the configuration consisting of two identical etalons, i.e., both Δn and ψ approach zero:

$$\lim_{\psi,\Delta n\,\to\,0}\,Y_t\,[n\,(1-\Delta n\,\,n^{-1}),n\,,\psi]$$

$$= \lim_{\psi,\Delta n\,\to\,0}\,\hat{Y}_t\,[n\,(1-\Delta n\,\,n^{-1}),n\,,\psi] = Y(n)\,.\qquad\text{8.3.8}$$

The nearly-identical etalons combination, when $\psi=0$, gives the following results:

$$Y_t\,[n\,(1-\Delta n\,\,n^{-1}),n\,,0] = Y_t\,[n\,(1-\Delta n\,\,n^{-1}),n\,,\psi]\qquad\text{8.3.9a}$$

Table 8-1

m	Classical	Y_t	$\hat{Y}_t$
200	0.987	0.987	0.982
100	0.954	0.954	0.935
50	0.851	0.857	0.797
25	0.645	0.693	0.563

$$\hat{Y}_t \left[n \left(1-\Delta n \ n^{-1}\right), n, 0 \right] = Y_t \left[n \left(1-\Delta n \ n^{-1}\right), n, 0 \right]$$

$$- 4^{-1} \ \mathbf{F} \ Y^2(n) \ R \ \sin^2(\pi \Delta n) \left[1 + \mathbf{F} \ Y(n) \ \sin^2(\pi \Delta n) \right]^{-1} . \qquad 8.3.10a$$

These equations can be recognized as the equivalent of an etalon composed of two elemental etalons with Δn orders difference; that is, an imperfect etalon. Therefore, this example can be used to test the classical assumption made to describe surface defects of an etalon, namely, that the superposition of intensities, as described in Chapter 2, Section 3, rather than the more correct superposition of amplitudes, describes an imperfect etalon. An etalon with spherical curvature surface defects, given by a displacement of Δk orders away from an ideal flat surface, can be considered to have a root-mean-square surface variation $\overline{\Delta k} = \Delta k \ (2/3)^{1/2}$. This variation can be identified with Δn of Equations 8.3.9a) and 8.3.10a), since Δk is associated with the lack of surface flatness, given as λ/m in Chapter 2, Section 3, by $\Delta k = m^{-1}$. Therefore:

$$\Delta n = \overline{\Delta k} = m^{-1} \ (2/3)^{1/2} . \qquad 8.3.11$$

Replacing this value of Δn in Equations 8.3.9a) and 8.3.10a) for the two unequal etalon superposition configurations, then a comparison with Equation 2.3.6) for the classical approximation to spherical curvature etalon surface defects, shows no significant differences in the results for high-quality etalons and increasing differences as the etalon quality decreases. A numerical example, using a reflectivity R equal to 0.85, n an integer and varying values of m is shown in Table 8-1. The differences between the classical results and Y_t are small, showing that the present representation of the classical imperfect etalon as a two-elemental etalon combination is reasonable, while the differences between the amplitude superposition, $\hat{Y}_t$, and the classical results are significantly different for the low-quality etalons, or low values of m. Nevertheless, the latter difference is not sufficiently large to abandon the classical representation of etalon surface defects. In fact, it is surprising how well the classical relationship holds even for the extreme example given in the last line of the comparison in Table 8-1.

After this aside on the limitations of the classical approximation to the treatment of surface defects, the parallel combination of Fabry-Perot etalons can be further pursued. A few examples of the behavior of this combination will be highlighted in the following text. The first combination amenable to study consists of two parallel identical etalons, i.e., $\rho = 0$:

$$Y_t\,(\delta,\delta,\psi) = Y\,(\delta)\,, \qquad\qquad\qquad 8.3.12$$

$$\hat{Y}_t\,(\delta,\delta,\psi) = 2^{-1}\,Y\,(\delta)\,(\,1 + \cos\psi\,)\,. \qquad\qquad 8.3.13a$$

The resultant intensity of Equation 8.3.13a) will range from zero to $Y\,(\delta)$, but this result is applicable only to the case of two equally bright beams. When one of the beams is different from the other, by a factor a, the resultant intensity is:

$$\hat{Y}_t\,(\delta,\delta,\psi) = 2^{-1}\,a\,Y\,(\delta)\,(\,1 + \cos\psi\,)$$

$$+\,4^{-1}\,[\,(1-a)\,(R+1-a)\,Y^2(\delta)\,\tau^{-2}\,]\,. \qquad 8.3.13b$$

Another example is obtained when the phase, ψ, is set to zero and $\phi = \delta + \epsilon$, $\epsilon << 1$; that is, an etalon composed of two slightly different spacing elementary etalons:

$$Y_t\,(\delta+\epsilon,\delta,0) = 2^{-1}\,[\,Y\,(\delta+\epsilon) + Y\,(\delta)\,]\,, \qquad 8.3.14a$$

$$\hat{Y}_t\,(\delta+\epsilon,\delta,0) = 2^{-1}\,[\,Y\,(\delta+\epsilon) + Y\,(\delta)\,]$$

$$-\,R^2\,Y\,(\delta+\epsilon)\,Y\,(\delta)\,\tau^{-2}\,\sin^2(2^{-1}\epsilon)$$

$$= Y_t\,(\delta+\epsilon,\delta,0) - R^2\,Y\,(\delta+\epsilon)\,Y\,(\delta)\,\tau^{-2}\,\sin^2(2^{-1}\epsilon)\,. \qquad 8.3.15a$$

$Y\,(\delta+\epsilon)$ can be replaced, by its equivalent expression in Equation 8.3.7), to give:

$$Y_t\,(\delta+\epsilon,\delta,0) = Y\,(\delta)\,[1+Y\,(\delta)\,2\,R\,\tau^{-2}\,\sin(2^{-1}\epsilon)\,\sin(\delta+2^{-1}\epsilon)\,]$$

$$\times\,[1+Y\,(\delta)\,4\,R\,\tau^{-2}\,\sin(2^{-1}\epsilon)\,\sin(\delta+2^{-1}\epsilon)\,]^{-1}\,, \qquad 8.3.14b$$

$$\hat{Y}_t\,(\delta+\epsilon,\delta,0) = \left\{\,Y\,(\delta)\,[1+Y\,(\delta)\,2\,R\,\tau^{-2}\,\sin(2^{-1}\epsilon)\,\sin(\delta+2^{-1}\epsilon)\,]\right.$$

$$\left. -\,[\,R\,Y\,(\delta)\,\tau^{-1}\,\sin(\,2^{-1}\epsilon\,)\,]^2\,\right\}$$

$$\times\,[\,1 + Y\,(\delta)\,4\,R\,\tau^{-2}\,\sin(2^{-1}\epsilon)\,\sin(\delta+2^{-1}\epsilon)\,]^{-1}\,. \qquad 8.3.15b$$

The maxima for $Y\,(\delta+\epsilon,\delta,0)$ and $\hat{Y}(\delta+\epsilon,\delta,0)$ of Equations 8.3.14b) and 8.3.15b) are found at $\delta = 0, -\epsilon$, unless the value of ϵ is small and the two maxima coalesce into a single peak with maxima at $-\epsilon/2$. The value of ϵ at which the two peaks coalesce is a function of the reflectivity (or the width of the peaks). For the ideal etalon, where the absorption A is equal to zero, and for small values of ϵ, the mutual interference term is approximately proportional to $\epsilon^2 (1-R)^{-2} R^2$. Therefore, the mutual interference effect becomes important for increasing etalon defects and increasing reflectivity; that is, the classical approximation of intensity superposition is applicable to high-quality etalons with low reflectivities, as discussed earlier.

Another limiting configuration is obtained from Equations 8.3.2c) and 8.3.3c) when $\phi = \delta + \pi$ and $\psi = 0$, or:

$$Y(\delta+\pi,\delta,0) = Y(\delta) \left[1 + 2^{-1} \mathbf{F} \ Y(\delta) \cos \delta \right]$$

$$\times \left[1 + \mathbf{F} \ Y(\delta) \cos \delta \right]^{-1}$$

$$= 2^{-1} Y(\delta) + 2^{-1} Y(\delta) \left[1 + \mathbf{F} \ Y(\delta) \cos \delta \right]^{-1}$$

$$= 2^{-1} Y(\delta) + 2^{-1} Y(\delta + \pi) . \qquad 8.3.16a$$

The last result could have been obtained directly from Equation 8.3.2b) or from Equation 8.3.7) for $\delta\rho = 2^{-1} \pi$. This result can also be re-stated as follows:

$$Y(\delta+\pi,\delta,0) = (1+R)^2 (1+R^2) (1+R^4-2R \cos2\delta)^{-1}$$

$$= (1+R^2) (1-R)^{-2} (1+\mathbf{F} \sin^2\delta)^{-1} (1+\mathbf{F} \cos^2\delta)^{-1} . \qquad 8.3.16b$$

In other words, two etalons differing in spacing by one-half order appear as a single etalon of twice the original spacing of the original etalons, as well as having an apparent lower reflectivity given by R^2. Note the similarity of Equation 8.3.16b) with Equation 5.2.11b) for the Connes spherical etalon, which also consists of two superposed etalons. The superposed amplitude configuration, under the same conditions, is:

$$\hat{Y}_t (\delta+\pi,\delta,0) = 2^{-1} Y_t (\delta+\pi,\delta,0) - 4^{-1} \mathbf{F} \ Y^2(\delta) R \ [1+\mathbf{F} \ Y(\delta) \cos \delta]^{-1}$$

$$= 2^{-1} Y(\delta) + 2^{-1} Y(\delta+\pi) - 4^{-1}\mathbf{F} \ Y(\delta+\pi) Y(\delta) R . \qquad 8.3.17$$

Other examples of the properties of the superposed amplitudes of two parallel etalons separated by one-half order are:

$$\hat{Y}_t (\delta+\pi,\delta,\pi) = 4^{-1} \mathbf{F} \ Y(\delta) Y(\delta+\pi) R , \qquad 8.3.18$$

$$\hat{Y}_t (\delta+\pi,\delta,\pi/2) = 4^{-1}[Y(\delta)+ Y(\delta+\pi)+\mathbf{F} \ Y(\delta) Y(\delta+\pi) \sin \delta] . \qquad 8.3.19$$

The result of Equation 8.3.18) depends only on the mutual interference of the beams, a result not obtainable with the classical intensity superposition.

With this brief survey of the parallel etalons, it has been possible to test the validity of the classical assumption for the description of the surface defects of an etalon, as well as to describe this seldom-used technique.

4. Simultaneous multiple-line Fabry-Perot operation

The investigations of Barrett and Myers (1971) on the rotational structure of molecular bands with a Fabry-Perot spectrometer make use of the comb properties (cf. overlap of Chapter 2, Section 1) of the device to superimpose the different wavenumber lines and collectively detect them. Barrett and Myers made use of the property of simple linear molecules [nitrous oxide (N_2O) in their example] which show rotational Raman emission lines whose location is approximately given by the following expression:

$$\sigma = \sigma_0 \pm 4 B (J + 3/2) , \qquad 8.4.1a$$

where B is the rotational constant, J is the rotational quantum number and σ_0 is the wavenumber of the exciting radiation. Equation 8.4.1a) is correct for the rigid rotor approximation, but real molecules suffer from centrifugal distortion, which is dependent on the value of the rotational quantum number; the correct expression for the location of the Raman lines is:

$$\sigma = \sigma_0 \pm [(4B - 6D_J) (J+3/2) - 8D_J (J+3/2)^3] . \qquad 8.4.1b$$

The value of the centrifugal distortion constant, D_J, is quite small, $\approx 10^{-6}$, such that the approximation of Equation 8.4.1a) is a good representation of the observed spectrum. From this equation, it can be seen that the Raman lines with the lowest rotational quantum number ($J = 0$) are separated by $6B$ from the exciting, or Rayleigh line, while lines with higher rotational quantum numbers ($J \geq 1$) are separated from each other by $4B$ wavenumbers. Therefore, an etalon with a free spectral range equal to $4B$ is capable of transmitting all the rotational lines, while rejecting the Rayleigh line. When the etalon spacing is varied by one-half order the opposite will be obtained, namely, the Rayleigh line will be transmitted and the Raman lines will be rejected. Note that the amount of rejection is dependent on the etalon contrast, as defined in Equation 2.1.14). The Rayleigh line is used to monitor the exciting source stability, since the intensity of the Raman lines is proportional to the intensity of this source.

Experimentally, the gap of the etalon is set at a free spectral range near $4B$ and then this gap is slowly changed while recording the output of the device. As expected, maximum output will be found when the free spectral range is exactly $4B$ (rigid rotor approximation). Therefore, monitoring the etalon spacer in a subsidiary measurement will provide the value of B. This simple explanation implicitly assumes that phase dispersion of the etalon mirrors [cf. Equation 2.1.3)] is constant, or at worst, very slowly changing. The presence of centrifugal distortion will affect the value of the free spectral range where maximum flux is obtained. Also note that the J value of the maximum intensity Raman lines is dependent on the kinetic temperature of the sample and the value of B of the molecule. Thus the experimental value of B obtained by the Barrett and Myers' method is dependent on the three above-mentioned variables.

Barrett and Myers (1971) have shown there exist other possible values of the free spectral range where overlap of the Raman lines will occur. These etalon free spectral ranges are called primary when all of the Raman lines are transmitted, i.e.:

$$\Delta\sigma = 4B \ (\ 2n - 1 \ )^{-1}, \qquad n = 1, 2, 3...... \qquad\qquad 8.4.2$$

The etalon free spectral range for $n = 1$, or $\Delta\sigma = 4B$, is called the principal free spectral range, and the associated etalon spacing is called the principal spacing. Those free spectral ranges, when not all of the Raman lines are simultaneously transmitted, are called secondary free spectral ranges, and are given by:

$$\Delta\sigma = 2 \ B \ n \ , \qquad n = 3, 4, 5........ \qquad\qquad 8.4.3$$

The similarity between this method and that used in Fourier spectroscopy with a Michelson (1881, 1882) interferometer was noted by Barrett and Myers (1971), who suggested that a Fabry-Perot used in multiplex spectroscopy would not only remove the need for pre-filtering but could, under the right circumstances, make use of the Fellgett (1958) advantage.

5. Multiplex Fabry - Perot

Yoshihara and collaborators (Yoshihara and Kitade, 1978, 1979; Yoshihara *et al.*, 1979, 1980) have shown a general method for the reconstruction of an original spectrum from measurements made with a Fabry-Perot in a multiplex mode. For the sake of completeness, multiplex operation with a Fabry-Perot consists of recording the transmitted flux from a simple etalon when the latter is examining an unfiltered complex spectral source as a function of the etalon spacing. This is just what is done in Fourier spectroscopy with a Michelson (1881, 1882) interferometer, or another two-beam device (Fellgett, 1958).

The development given by Yoshihara and Kitade (1979) will be closely adhered to in the following discussion. The transmission of an ideal etalon can be defined with the aid of Equations 2.1.7a) and 2.1.8) to be:

$$Y_t (\sigma , x) = \tau^2 \left[1 + R^2 - 2R \ \cos(2\pi \sigma x) \right]^{-1} , \qquad\qquad 8.5.1$$

where the symbols have the same meaning given earlier in Chapter 2, except for the new variable x, which is defined as:

$$x = 2 \ \mu \ d \ \cos\Theta \ . \qquad\qquad 8.5.2a$$

When the projected etalon fringes are examined by an infinitesimally small aperture located on the axis, then $\cos\Theta \approx 1$, and x becomes twice the optical separation between the Fabry-Perot mirrors, or:

$$x \simeq 2 \mu d \; . \qquad\qquad 8.5.2b$$

Under these circumstances of a small aperture, and ignoring the constants associated with the use of such an aperture (cf. Chapter 2, Section 4), Equation 8.5.1) can be considered to be the measured flux from the Fabry-Perot spectrometer examining a monochromatic line.

When the Fabry-Perot spectrometer examines an arbitrary source spectrum $\hat{S}(\sigma)$, the measured flux at spacing x is:

$$Y(x) = \int_0^{\sigma_M} \tau^2 \, [1+R^2-2R \, \cos(2\pi \, \sigma \, x) \,]^{-1} \, \hat{S}(\sigma) \, d\sigma \; . \qquad 8.5.3a$$

In this equation σ_M is the upper limit of the spectral components of the source spectrum $\hat{S}(\sigma)$. In practice, this upper limit may be obtained by external filtering or by the sensitivity limits of the detector used. The source spectrum is, for convenience, redefined as:

$$S(\sigma) = \hat{S}(\sigma) \, \tau^2 \, (\, 1 - R^2) \; . \qquad\qquad 8.5.4$$

Then Equation 8.5.3a) can be written as follows:

$$Y(x) = \int_0^{\sigma_M} (1-R^2) \, [1-R^2-2R \cos(2\pi \, \sigma \, x) \,]^{-1} \, S(\sigma) \, d\sigma \; . \qquad 8.5.3b$$

Since, in principle, both $Y(x)$ and the kernel (in brackets) are known, the solution for $S(\sigma)$, and the original spectrum, can be obtained by inversion of Equation 8.5.3b). This can be accomplished (Yoshihara and Kitade, 1979) when the source function is an even function between $-\sigma_M$ and σ_M, which can then be represented by the following Fourier series:

$$S(\sigma) = b_0 + 2 \sum_{n=1}^{n=N} b_n \, \cos[2\pi \, n \, \sigma \, (2\sigma_M)^{-1}] \; . \qquad 8.5.5$$

In this equation N is the number of spectral elements of width $(\, 2X \,)^{-1}$, where X is the maximum value of x in Equation 8.5.2b). Therefore:

$$N = \sigma_M \, / \, (\, 2X \,)^{-1} = 2 \, \sigma_M \, X \; . \qquad\qquad 8.5.6$$

When the distance X is divided into N equally spaced parts, the mth element is then defined as:

$$x_m = m \, (\, 2\sigma_M \,)^{-1} , \qquad 0 \leqslant m \leqslant N \; . \qquad 8.5.7$$

Also the kernel of Equation 8.5.3b) can be replaced by its Fourier identity, as shown in Equation 2.1.7b), or:

$$(1-R^2)\,[1-R^2-2R\,\cos(2\pi\sigma z)\,]^{-1} = 1+2\sum_{k=1}^{k=\infty} R^k \cos(2\pi k\sigma z)\,. \qquad 8.5.8$$

Equation 8.5.3b) is given, after replacement of the kernel of Equation 8.5.3b) and the source function by their Fourier identities, as well as z by z_m of Equation 8.5.7), by:

$$Y_m = Y(z_m) = \int_0^{\sigma_M} \left\{ \left(1 + 2\sum_{k=1}^{k=\infty} R^k \cos[2\pi\,km\,\sigma\,(2\sigma_M)^{-1}]\right) \right.$$
$$\left. \times \left(b_0 + 2\sum_{n=1}^{n=N} b_n \cos[2\pi\,n\,\sigma\,(2\sigma_M)^{-1}]\right) \right\} d\sigma\,. \qquad 8.5.3c$$

In this last equation $Y(z_m)$ has, for convenience, been replaced with Y_m. The properties of Equation 8.5.3c) are better described and understood after separation into individual terms, as follows:

$$Y_m = b_0 \int_0^{\sigma_M} d\sigma + 2\sum_{n=1}^{n=N} b_n \int_0^{\sigma_M} \cos[2\pi\,n\,\sigma\,(2\sigma_M)^{-1}]\,d\sigma$$

$$+ 2b_0 \sum_{k=1}^{k=\infty} R^k \int_0^{\sigma_M} \cos[2\pi\,km\,\sigma\,(2\sigma_M)^{-1}]\,d\sigma$$

$$+ 4 \sum_{k=1}^{k=\infty}\sum_{n=1}^{n=N} R^k b_n \int_0^{\sigma_M} \left\{ \cos[2\pi\,km\,\sigma\,(2\sigma_M)^{-1}] \right.$$
$$\left. \times\ \cos[2\pi\,n\,\sigma\,(\sigma_M)^{-1}] \right\} d\sigma\,. \qquad 8.5.3d$$

In this equation the second term always vanishes, while the third term vanishes except when $m \equiv 0$. The integral of the fourth term can be expressed by its equivalent, namely:

$$2\int_0^{\sigma_M} \cos[2\pi\,km\,\sigma\,(2\sigma_M)^{-1}] \cos[2\pi\,n\,\sigma\,(2\sigma_M)^{-1}]\,d\sigma$$

$$= \int_0^{\sigma_M} \cos[2\pi\,(n+km)(2\sigma_M)^{-1}\sigma]\,d\sigma$$

$$+ \int_0^{\sigma_M} \cos[2\pi\,(n-km)(2\sigma_M)^{-1}\sigma]\,d\sigma\,, \qquad 8.5.9$$

where the first integral always vanishes since $n+km \geq 1$, while the second integral will vanish for $n = km$, $m \neq 0$. Therefore Equation 8.5.3d) has separate solutions depending on whether or not the value of m is zero or greater than zero:

$$Y_0 = b_0\sigma_M\,(1+R)\,(1-R)^{-1}\,, \qquad 8.5.3e$$

$$Y_m = b_0 \sigma_M + 2 \sigma_M \sum_{k=1}^{k \leqslant N/m} R^k b_{km} , \qquad \text{8.5.3f}$$

$$Y_\infty = \lim_{m \to \infty} Y_m = b_0 \sigma_M . \qquad \text{8.5.3g}$$

The solution for the b_{km}, and thus for the source spectrum function $S(\sigma)$ is found by the solution of a set of simultaneous equations of the form:

$$\sum_{m=1}^{m=N} \left\{ (2 \sigma_M)^{-1} Y_m - 2^{-1} b_0 - \sum_{k=1}^{k \leqslant N/m} R^k b_{km} = 0 \right\} , \qquad \text{8.5.10}$$

where b_0 is given by Equation 8.5.3g). Examples of this solution method have been given by Yoshihara and Kitade (1979) for the far infrared.

The practical drawbacks for Fabry-Perot multiplex spectroscopy are the difficulty of obtaining a zero-path measurement, the presence of a phase change in the reflection of the mirror coatings with wavenumber, as well as the variation in the reflectivity and transmissivity of these coatings with wavenumber. The first problem can be conceivably resolved by extrapolation of a separate experiment (Yoshihara and Kitade, 1979), while the other two can be solved by calibration and improved coatings, respectively.

Fourier spectroscopy, as performed with a Michelson (1881, 1882) two-beam interferometer, can be considered to be a special case of the multiple-beam multiplex spectroscopy with a Fabry-Perot interferometer (Yoshihara and Kitade, 1979). A A two-beam Michelson interferogram, $M(x)$; is given by [cf. Equation 8.5.3a)]:

$$M(x) = \int_0^{\sigma_M} \hat{S}(\sigma) \cos(2\pi \sigma x) \, d\sigma . \qquad \text{8.5.11}$$

When $\hat{S}(\sigma)$ is expressed with a similar Fourier series as $S(\sigma)$ of Equation 8.5.5) and x is replaced by x_m of Equation 8.5.7), it is found, with the same arguments used previously, that:

$$M_m = M(x_m) = \int_0^{\sigma_M} b_0 \cos[2\pi m \sigma (2\sigma_M)^{-1}] \, d\sigma$$

$$+ 2 \sum_{n=1}^{n=N} b_n \int_0^{\sigma_M} \cos[2\pi n \sigma (2\sigma_M)^{-1}] \cos[2\pi m \sigma (2\sigma_M)^{-1}] \, d\sigma$$

$$= b_m \sigma_M . \qquad \text{8.5.12}$$

It follows that the inversion of the interferogram is given by:

$$\hat{S}(\sigma) = \sigma_M^{-1} \left\{ M_0 + 2 \sum_{m=1}^{m=N} M_m \cos[2\pi m \sigma (2\sigma_M)^{-1}] \right\} . \qquad \text{8.5.13}$$

This last expression is the same result obtained using the inverse Fourier transform, thus showing the close relationship between the two-beam Michelson interferometer and the multiple-beam Fabry-Perot interferometer.

Bibliography

ABJEAN, R., AND JOHANNIN-GILLES, A. (1970). Interferometre Fabry-Perot enregistreur dans l'ultraviolet a vide ($\lambda > 1800\ A$). *Optics Comm.* **1**, 385 - 387.

ABREU, V. J., KILLEEN, T. L., and HAYS, P. B. (1981). Tristatic high resolution Doppler lidar to study winds and turbulence in the troposphere. *Appl. Opt.* **20**, 2196 - 2202.

ALLARD, N. (1958). Interféromètre de Fabry-Perot enregistreur pour l'étude de la raie verte du ciel nocturne. *J. Phys. Rad.* **19**, 340 - 341.

AMERICAN INSTITUTE OF PHYSICS HANDBOOK. (1963). 6e. Index of refraction for visible light of various solids gases and liquids. (B. H. Billings, ed.), p. 6-96, McGraw-Hill Book Co, New York.

ANDRIESSE, C. D., DE VRIES, J. S., and VAN DER WAL, P. B. (1979). Astrophysics with an infrared Fabry-Perot. *Infrared Phys.* **19**, 375 - 382.

ARMSTRONG, E. B. (1953). A note on the use of the Fabry-Perot etalon for upper atmosphere temperature measurements. *J. Atm. Terr. Phys.* **3**, 274 - 281.

ARMSTRONG, E. B. (1956). The observation of line profiles in the airglow and aurora with a photoelectric Fabry-Perot interferometer. *In* ' The Airglow and the Aurorae ' (E. B. Armstrong and A. Dalgarno, eds.), p. 366 - 373, Pergamon Press, Oxford.

ARMSTRONG, E. B. (1958). Largeur de la raie OI 5 577 du ciel nocturne et des aurores. *J. Phys. Rad.* **19**, 358 - 365.

ASCOLI-BARTOLI, U., BENEDETTI-MICHELANGELI, G. and DE MARCO, F. (1967). An improvement in Fabry-Perot spectrometry. *Appl. Opt.* **6**, 467 - 470.

ATHERTON, P. D., REAY, N. K., RING, J., and HICKS, T. R. (1981). Tunable Fabry-Perot filters. *Opt. Eng.* **20**, 806 - 814.

AUTH, D. C. (1968). Thermal scanning of a Fabry-Perot interferometer. *Phys. Lett.* **27A**, 536 - 537.

BABCOCK, H. D. (1923). A study of the green auroral line by the interference method. *Astrophys. J.* **57**, 209 - 221.

BABCOCK, H. D. (1927). Secondary standards of wavelengths; interferometer measurements of Iron and Neon lines. *Astrophys. J.* **66**, 256 - 282.

BAIRD, K. M. (1958). Techniques nouvelles en métrologie interferérentielle au National Research Council of Canada. *J. Phys. Rad.* **19**, 384 -389.

BAIRD-ATOMIC, INC. (1963). Wide band coherent light modulator. Report ASD-TDR-63-604, Cambridge, MA.

BAKER, E.A.M. and WALKER, B. (1982). Fabry - Perot interferometers for use at submillimetre wavelengths. *J. Phys.(E)* **15**, 25 - 30.

BALLIK, E. A. (1966). The response of scanning Fabry-Perot interferometers to atomic transition probabilities. *Appl. Opt.* **5**, 170 - 172.

BANNING, M. (1947a). Partially reflective mirrors of high efficiency and durability. *J. Opt. Soc. Am.* **37**, 688 - 689.

BANNING, M. (1947b). Practical methods of making and using multilayer filters. *J. Opt. Soc. Am.* **37**, 792 - 797.

BARNES, J. (1904). On the analysis of bright spectrum lines. *Astrophys. J.* **19**, 190 - 211.

BARRELL, H. (1949). New developments in interferometry. *Nature* **164**, 599 - 601.

BARRETT, J. J., and MYERS, S. A. (1971). New interferometric method for studying periodic spectra using a Fabry-Perot interferometer. *J. Opt. Soc. Am.* **61**, 1246 - 1251.

BARRETT, J. J., and STEINBERG, G. N. (1972). Use of air bearings in the construction of a scanning Fabry-Perot interferometer. *Appl. Opt.* **11**, 2100 -2101.

BASEDOW, R. W., and COCKS, T. D. (1980). Piezoelectric ceramic displacement characteristics at low frequencies and their consequences in Fabry-Perot interferometry. *J. Phys. (E)* **13**, 840 - 844.

BATES, B. and BRADLEY, D. J. (1967). Reflectance and transmittance of evaporated aluminium and aluminium : magnesium fluoride films in the ultraviolet (> 1800 A). *J. Opt. Soc. Am.* **57**, 481 - 485.

BATES, B., CONWAY, J. K., COURTS, G. R., MCKEITH, C. D., and MCKEITH, N. E. (1971). A stable high finesse scanning Fabry-Perot interferometer with piezoelectric transducers. *J. Phys. (E)* **4**, 899 - 901.

BATES, B., CONWAY, J. K., MCKEITH, C. D., and YATES, H. W. (1973). Optically contacted, Fabry-Perot interferometer filter for the middle ultraviolet. *Appl. Opt.* **12**, 140 - 142.

BAUER, J. (1934). Die dispersion des Phasensprungs bei der Lichtreflexion an dünnen Metallschichten. *Ann. Physik.* **20**, 481 - 501.

BAUMEISTER, P. W., and STONE, J. D. (1956). Broad-band multilayer film for Fabry-Perot interferometers. *J. Opt. Soc. Am.* **46**, 228 - 229.

BAYER-HELMS, F. (1963a). Analyse von Linienprofilen. I Grundlagen und Messeinrichtungen. *Z. angew. Physik.* **15**, 330 - 338.

BAYER-HELMS, F. (1963b). Analyse von Linienprofilen. II Auswertung registrierter Intensitätsverteilungen eines Fabry-Perot-Interferometers, Tabelle der Verteilungsfunktionen. *Z. angew. Physik.* **15**, 416 - 429.

BAYER-HELMS, F. (1963c). Analyse von Linienprofilen. III. *Z. angew. Physik.* **16**, 44 - 53.

BAYER-HELMS, F. (1963d). Zur Analyse von Linienprofilen: Halbweiten von Voigt-Funktionen. *Z. angew. Physik.* **15**, 532 - 534.

BAYER-HELMS, F. (1964a). Zur Analyse von Linienprofilen: Verteilungsfunktionen eines Fabry-Perot-Interferometers; Auswertung registrierter Intensitätsverteilungen. *PTB Mitt.* **3**, 201 - 207.

BAYER-HELMS, F. (1964b). Zur Analyse von Linienprofilen: Verteilungsfunktionen eines Fabry-Perot-Interferometers; Auswertung registrierter Intensitätsverteilungen. *PTB Mitt.* **4**, 312 - 321.

BENNETT, W. R. (1962). Gaseous optical masers. *Appl. Opt.* **1**, Supp. 1, 24 - 66.

BENNETT, W. R. (1978). Deconvolution of spectral lines by Fourier analysis. *Appl. Opt.* **17**, 3344 - 3345.

BENOIT, J-R, FABRY, Ch.,and PEROT, A. (1913). Nouvelle détermination du rapport des longeurs d'onde fondamentalles avec l'unité métrique. *Travaux et Memoires Bur. Int. Poids Mes.* **15**, 1 - 134.

BENS, A. R., COGGER, L. L., and SHEPHERD, G. G. (1965). Upper atmospheric temperatures from Doppler line widths-III. *Planet. Space Sci.* **13**, 551 - 563.

BERTHOLD, J. W. and JACOBS, S. F. (1976). Ultraprecise thermal expansion measurements of seven low expansion materials. *Appl. Opt.* **15**, 2344 - 2447.

BEST, G. T. (1967). Fabry-Perot interferometers with electronic determination of Doppler widths. *Appl. Opt.* **6**, 287 - 295.

BEVINGTON, P.R. (1969). 'Data Reduction and Error Analysis for the Physical Sciences.' McGraw - Hill Book Co., New York.

BEYSENS, D. (1973). Etude des phénomènes de diffusion de la lumière par spectroscopie Fabry-Pérot a haute résolution et grande sensibilité. *Rev. Phys. Appl.* **8**, 175 - 186.

BIDEAU-MEHU, A., GUERN, Y., ABJEAN, R., and JOHANNIN-GILLES, A. (1976). MgF_2 Fabry-Perot etalon plates for the vacuum ultraviolet. *Appl. Opt.* **15**, 2626 - 2627.

BIDEAU-MEHU, A., GUERN, Y., ABJEAN, R., and JOHANNIN-GILLES, A. (1980). Improved MgF_2 etalon plates for Fabry-Perot interferometry in the V U V from 170 to 138 nm. *J. Phys. (E)*, **13**, 1159 - 1162.

BIONDI, M. A. (1956). High speed, direct recording Fabry-Perot interferometer. *Rev. Sci. Inst.* **27**, 36 - 39.

BLIFFORD, I. H. (1966). Factors affecting the performance of commercial interference filters. *Appl. Opt.* **5**, 105 - 111.

BOGROS, A. (1930). Structure de la raie 6708 du Lithium. *Compt. Rend.* **190**, 1185 - 1187.

BORN, M., and WOLF, E. (1964). 'Principles of Optics'. Pergamon Press, Oxford.

BOULOCH, R. (1893). Dédoublement des franges d'interférence en lumière naturelle. *J. Physique* **2**, 316 - 320.

BOYD, G. D., and GORDON, J. P. (1961). Confocal multimode resonator for millimeter through optical wavelength masers. *Bell System Tech. J.* **40**, 489 - 507.

BOZEC, P., CAGNET, M., DUCHESNE, M., LECONTEL, J-M., and VIGIER, J-P. (1970). Nouvelles expériences d'interférence en lumière faible. *Compt. Rend.* **270**, 324 - 326.

BRACEWELL, R. (1965). 'The Fourier Transform and its Applications'. McGraw - Hill Book Co., New York.

BRADLEY, D. J. (1962a). Electron-optical scanning for a Fabry-Perot spectrometer employing image intensifier detection. *Brit. J. Appl. Phys.* **13**, 83 - 84.

BRADLEY, D. J. (1962b). Parallel movement for high finesse interferometric scanning. *J. Sci. Instrum.* **39**, 41 - 45.

BRADLEY, L. C., and KUHN, H. (1948). Spectrum of Helium-3. *Nature* **162**, 412 - 413.

BRADLEY, D. J. and MITCHELL, C. (1968). Characteristics of the defocused spherical Fabry - Perot interferometer as a quasilinear dispersion instrument for high resolution spectroscopy of pulsed laser sources. *Phil. Trans. Roy. Soc.* **A263**, 209 - 223.

BRADLEY, D. J., BATES, B., JUULMANN, C. O. L. and MAJUMDAR, S. (1964). Time resolved photoelectric spectrography by electron-optical image detection of etalon interferograms. *Appl. Opt.* **3**, 1461 - 1465.

BREWSTER, D. (1814). On the polarisation of light by oblique transmission through all bodies, whether crystallized or uncrystallized. *Phil. Trans.*, 219 - 230.

BREWSTER, D. (1815). On the laws which regulate the polarisation of light by reflexion from transparent bodies. *Phil. Trans.*, 125 - 159.

BRUCE, C.F. (1966). Aperture corrections in Fabry - Perot interferometer measurements. *Rev. Sci. Inst.* **37**, 349 - 353.

BRUCE, C. F., and HILL, R. M. (1961). Wavelengths of Krypton 86, Mercury 198 and Cadmium 114. *Austral. J. Phys.* **14**, 64 - 68.

BUISSON, H., and FABRY, Ch. (1908). Mesures de longueurs d'onde pour l'etablissement d'un système de repères spectroscopiques. *J. Physique* **7**, 169 - 195.

BUISSON, H., and FABRY, Ch. (1910). Mesures de petites variations de longeurs d'onde par la méthode interférentielle. Application a différents problemes de spectroscopie solaire. *J. Physique* **9**, 298 - 316.

BUISSON, H., and FABRY, Ch. (1912). La largeur des raies spectrales et la théorie cinétique des gaz. *J. Physique* **2**, 442 - 464.

BUISSON, H., and FABRY, Ch. (1913). Sur les longueurs d'onde des raies du Krypton. *Compt. Rend.* **156**, 945 - 947.

BUISSON, H., and FABRY, Ch. (1921). Sur le déplacement des raies solaires sous l'action du champ de gravitation. *Compt. Rend.* **172**, 1020 -1022.

BUISSON, H., FABRY, Ch., and BOURGET, H. (1914a). Application des interférences a l'étude de la nébuleuse d'Orion. *J. Physique* **4**, 357 - 378.

BUISSON, H., FABRY, Ch., and BOURGET, H. (1914b). An application of interference to the study of the Orion nebula. *Astrophys. J.* **40**, 241 - 258.

BURGER, H.C., and VAN CITTERT, P.H. (1935). Die Einstellung der Koinzidenz beim Multiplexinterferenzspektroskop. *Physica* **2**, 87 - 96.

BURNETT, C. R. and BURNETT, E. B. (1983). OH PEPSIOS. *Appl. Opt.* **22**, 2887 - 2982.

BURNETT, C. R., and LAMMER, W. E. (1969). Gravity-adjusted Fabry-Perot etalon. *Appl. Opt.* **8**, 2345 - 2346.

BURNS, K., MEGGERS, W. F., and MERRILL, P. W. (1918). Measurements of wave lengths in the spectrum of Neon. *Bull. Bur. Stds.* **14**, 765 - 775.

BURRIDGE, J. C., KUHN, H., PERY, A. (1953). Reflectivity of thin aluminium films and their use in interferometry. *Proc. Phys. Soc. (Lon)* **B66**, 963 - 968.

BUXTON, A. (1937). Note on optical resolution. *Phil. Mag.* **23**, 440 - 442.

CANDLER, C. (1951). 'Modern Interferometers.' Hilger and Watts, Ltd., London.

CANNELL, D. S., and BENEDEK, G. B. (1970). Brillouin spectrum of Xenon near its critical point. *Phys. Rev. Lett.* **17**, 1157 - 1161.

CAPLAN, J. (1975). Temperature and pressure effects on pressure scanned etalons and gratings. *Appl. Opt.* **14**, 1585 - 1591.

CHABBAL, R. (1953). Recherche des meilleures conditions d'utilisation d'un spectromètre photoélectrique Fabry-Perot. *J. Rech. CRNS.* **24**, 138 - 186.

CHABBAL, R. (1957). Calcul du facteur de filtrage intégral d'un spectromètre Fabry-Perot. *J. Rech. CRNS.* **39**, 77 - 106.

CHABBAL, R. (1958a). Recherches expérimentales et théoriques sur la généralisation de l'emploi du spectromètre Fabry-Perot aux divers domaines de la spectroscopie. *Rev. Opt.* **37**, 49-104.

CHABBAL, R. (1958b). Recherches expérimentales et théoriques sur la généralisation de l'emploi du spectromètre Fabry-Perot aux divers domaines de la spectroscopie. *Rev. Opt.* **37**, 336 - 370.

CHABBAL, R. (1958c). Recherches expérimentales et théoriques sur la généralisation de l'emploi du spectromètre Fabry-Perot aux divers domaines de la spectroscopie. *Rev. Opt.* **37**, 501 - 551.

CHABBAL, R. and JACQUINOT, P. (1961). Description d'un spectromètre interférentiel Fabry-Perot. *Rev. Opt.* **40**, 157 - 170.

CHABBAL, R. and PELLETIER, R. (1965). Principe et réalisation d'un spectromètre Fabry-Perot multicanal: Le SIMAC. *Jap. J. Appl. Phys.* **4**, Supp. I, 445 - 447.

CHABBAL, R., and SOULET, M. (1958). Dispositif permettant le déplacement mécanique d'une lame de Fabry - Perot. *J. Phys. Rad.* **19**, 274 - 277.

CHABBAL, R., BIED-CHARRETON, Ph. and PELLETIER, R. (1967). Le SIMAC: Utilisation avec une plaque photographique ou une caméra électronique. *J. Physique* **28**, Supp. C2, 209 - 214.

CHAMBERLAIN, J. W. (1961). 'Physics of the Aurora and Airglow', Academic Press, New York.

CHANTREL, H. (1958). Un double étalon a balayage par pression. *J. Phys. Rad.* **19**, 366 - 370.

CHANTREL, H., COJAN, J. L., GIACOMO, P. (1964). Couches réfléchissantes multidiélectriques pour l'ultraviolet. Mesures interférentielles sur la structure de la raie de resonance du mercure ($\lambda = 2537 \ A$). *J. Physique* **25**, 280 -284.

CHANTREL, H., ANDRE, J. and MERLE, D. (1971). Spectromètre Perot-Fabry a exploration automatique par paliers súccessifs. *Nouv. Rev. Opt.* **2**, 25 - 27.

CHANTRY, G.W. (1982). The use of Fabry - Perot interferometers, etalons and resonators at infrared and longer wavelengths - an overview. *J. Phys. (E)* **15**, 3 - 8.

CHAUX, R., and BOQUILLON, J. P. (1979). Diameter measurements of Fabry-Perot interference rings using CCD linear sensors. *Opt. Comm.* **30**, 239 -244.

CHAUX, R., BOQUILLON, J. P., and MORET-BAILLY, J. (1976). Dispositif de mesure du diametre des anneaux d'un interferometre Perot-Fabry eclaire en impulsions lumineuses. *Opt. Comm.* **17**, 293 - 296.

CHILDS, W. H. J. (1926). The Fabry and Perot parallel plate étalon. *J. Sci. Inst.* **3**, 97 -103 and 129 - 135.

CLARKE, R.N., and ROSENBERG, C.B (1982). Fabry - Perot and open resonators at microwave and millimetre wave frequencies, 2 - 300 GHz. *J. Phys. (E)*, **15**, 9 - 24.

COLOMBEAU, B., and FROEHLY, E. (1976). Speckle technique for accurate adjustments of large spacing Fabry-Perot interferometers. *Opt. Comm.* **17**, 284 - 286.

CONDON, E. U. and SHORTLEY, G. H. (1957). 'The Theory of Atomic Spectra', Cambridge University Press, London.

CONNES, J. (1961). Recherches sur la spectroscopie par transformation de Fourier. *Rev. Opt.* **40**, 45 - 79, 116 - 140, 171 - 190, 231 - 265.

CONNES, P. (1956). Augmentation du produit luminosité × résolution des interféromètres par l'emploi d'une différence de marche indépendante de l'incidence. *Rev. Opt.* **35**, 37 - 43.

CONNES, P. (1958). L'etalon de Fabry - Perot spherique. *J. Phys. Rad.* **19**, 262 - 269.

COOK, A. H. (1960). Developments in interferometry with pressure scanning. *In* National Physical Laboratory Symposium No. 11: 'Interferometry', p.387-406, Her Majesty's Stationery Office, London.

COOK, A.H. (1971). 'Interference of Electromagnetic Waves'. Clarendon Press, Oxford.

COOPER, J. and GREIG, J. R. (1963). Rapid scanning of spectral line profiles using an oscillating Fabry-Perot interferometer. *J. Sci. Instrum.* **40**, 433 - 437.

COOPER, V.G. (1971). Analysis of Fabry - Perot interferograms by means of their Fourier transforms. *Appl. Opt.* **10**, 525 - 530.

COURTES, G., and GEORGELIN, Y. (1967). Montage Pérot-Fabry a multi-lentilles. *J. Physique* **28**, Supp. C2, 218 -220.

COURTES, G., FEHRENBACH, C., HUGHES, E. and ROMANO, J. (1966). Quelques réalisations instrumentales en France. *Appl. Opt.* **5**, 1349 - 1360.

CRNS (1958). Colloque international sur les progrès récents en spectroscopie interférentielle. *J. Phys. Rad.* **19**, 185 - 436.

CRNS (1967). Colloque sur les mèthodes nouvelles de spectroscopie instrumentale. *J. Physique* **28**, Supp. C2, 1 -341.

DAEHLER, M. (1970). A twelve - channel multiple - interferometer Fabry - Perot spectrometer. *Appl. Opt.* **9**, 2529 - 2534.

DAEHLER, M., and ROESLER, F. L. (1968). High contrast in a polyetalon Fabry - Perot spectrometer. *Appl. Opt.* **7**, 1240 - 1241.

DANGOR, A. E., and FIELDING, S. J. (1970). The response of the Fabry-Pérot interferometer to rapid changes in optical length. *J. Phys.(D)* **3**, 413 - 421.

DAVIS, S. P. (1963). Selection of multilayer dielectric coatings for Fabry-Perot interferometry. *Appl. Opt.* **2**, 727 - 733.

DEL PIANO, V. N., and QUESADA, A. F. (1965). Transmission characteristics of Fabry-Perot interferometers and a related electrooptic modulator. *Appl. Opt.* **4**, 1386 - 1390.

DI BIAGIO, B. (1974). Interféromètre de Fabry-Perot a balayage par déformation élastique. *Nouv. Rev. Opt.* **5**, 249 - 256.

DICKE, R. H. (1958). Molecular amplification and generation systems. U. S. Patent 2 851 652.

DITCHBURN, R. W. (1930). VI Notes on resolving power. *Proc. Roy. Irish Acad.* **39**, 58 - 72

DOBROWOLSKI, J. A. (1978). Coatings and Filter. *In* 'Handbook of Optics' (W.G. Driscoll and W. Vaughan, eds), p. 8-1 - 8-124, McGraw-Hill Book Co., New York.

DUFFIEUX, M. (1932). Etude des surfaces très planes pour interféromètre. *Rev. Opt.* **11**, 159 - 167.

DUFFIEUX, M. (1935). Analyse harmonique des franges d'interférence données par les appareils à ondes multiples. *Bull. Soc. Sci. Bret. (Rennes)* **12**, 125 - 132.

DUFFIEUX, P. M. (1939). Analyse harmonique des franges d'interférence données par les appareils a ondes multiples. *Rev. Opt.* **18**, 1 - 19.

DUFFIEUX, P. M. (1969). Origine des appareils interférentiels à ondes multiples. *Appl. Opt..* **8**, 329 - 332.

DUFOUR, Ch. (1945). Utilisation de l'interféromètre Fabry-Perot pour la recherche des satellites faibles. *Rev. Opt.* **24**, 11 - 18.

DUFOUR, Ch. (1948). Technique de l'évaporation des couches minces multiples. *Le Vide* **3**, 480 - 486.

DUFOUR, Ch. (1951). Recherches sur la luminosité, le contraste et la résolution de systèmes interferéntiels a ondes multiples. Utilisation de couches minces complexes. *Ann. Physique* **6**, 5 - 107.

DUFOUR, Ch., and PICCA, R. (1945). Sur l'interféromètre Fabry-Perot. Importance des imperfections des surfaces. *Rev. Opt.* **24**, 19 - 34.

DUPEYRAT, R. (1958). Etude de procédés électriques de "balayage" pour des interféromètres enregistreurs. *J. Phys. Rad.* **19**, 290 - 292.

DUPOISOT, H. and PRAT, R. (1979). High resolvance interference spectrometry with a high factor of merit in the visible spectrum. *Appl. Opt.* **18**, 85 - 90.

DYSON, J. (1970). 'Interferometry as a Measuring Tool'. The Machinery Publishing Co., Ltd., Brighton.

EATHER, R. H., and REASONER, D. L. (1969). Spectrophotometry of faint light sources with a tilting-filter photometer. *Appl. Opt.* **8**, 227 -242.

EDSER, E. and BUTLER, C. P. (1898). A simple method of reducing prismatic spectra. *Phil. Mag.* **46**, 207 - 222.

ENNOS, A. E. (1966). Stresses developed in optical film coatings. *Appl. Opt.* **5**, 51 - 61.

FABELINSKII, I. L., and CHISTYI, I. L. (1976). New methods and advances of high resolution spectroscopy. *Sov. Phys. Usp.* **19**, 597 - 617.

FABRY, Ch. (1904a). On the wave-length of the Cadmium line at λ 5086. *Astrophys. J.* **19**, 116 - 118.

FABRY, Ch. (1904b). Sur les raies satellites dans le spectre du Cadmium. *Compt.. Rend.* **138**, 854 - 856.

FABRY, Ch. (1905a). Sur un noveau dispositif pour l'emploi des méthodes de spectroscopie interférentielle. *Compt. Rend.* **140**, 848 - 851.

FABRY, Ch. (1905b). Sur l'application au spectre solaire des méthodes de spectroscopie interférentielle. *Compt. Rend.* **140**, 1136 - 1139.

FABRY, Ch. (1923). 'Les applications des interférences lumineuses'. Editions de la Revue d'Optique, Paris.

FABRY, Ch., and BUISSON, H. (1908a). Wave-length measurements for the establishment of a system of spectroscopic standards. *Astrophys. J.* **27**, 169 -196.

FABRY, Ch., and BUISSON, H. (1908b). Variation de la surface optique avec la longueur d'onde dans la réflexion sur les couches métalliques minces. *J. Physique.* **7**, 417 - 429.

FABRY, Ch., and BUISSON, H. (1910a). Application de la méthode interférentielle à la mesure de très petits déplacements de raies. Comparaison du spectre solaire avec le spectre d'arc du fer. Comparaison du centre et du bord du soleil. *Astrophys. J.* **31**, 97 - 119.

FABRY, Ch., and BUISSON, H. (1910b). Interférences produites par les raies noires du spectre solaire. *J. Physique.* **9**, 197 - 205.

FABRY, Ch., and BUISSON, H. (1911). Application of the interference method to the study of nebulae. *Astrophys. J.* **33**, 406 - 409.

FABRY, Ch., and BUISSON, H. (1914). Vérification expérimentale du principe de Doppler-Fizeau. *Compt. Rend.* **158**, 1498 - 1499.

FABRY, Ch., and BUISSON, H. (1919). Indications techniques sur les étalons interférentiels a lames argentées. *J. Physique.* **9**, 189 - 210.

FABRY, Ch., and PEROT, A. (1896). Mesure de petites épaisseurs en valeur absolue. *Compt. Rend.* **123**, 802 - 805.

FABRY, Ch., and PEROT, A. (1897). Sur les franges des lames minces argentées et leur application a la mesure de petites épasseurs d'air. *Ann. Chim. Phys.* **12**, 459 - 501.

FABRY, Ch., and PEROT, A. (1898a). Sur un spectroscope interférentiel. *Compt. Rend.* **126**, 331 - 333.

FABRY, Ch., and PEROT, A. (1898b). Sur une méthode de détermination du numéro d'ordre d'une frange d'ordre élevé. *Compt. Rend.* **126**, 1561 - 1564.

FABRY, Ch., and PEROT, A. (1898c). Sur l'étude des radiations du Mercure et la mesure de leurs longeurs d'onde. *Compt. Rend.* **126**, 1706 - 1709.

FABRY, Ch., and PEROT, A. (1898d). Mesure du coefficient de viscosité de l'air. *Ann. Chim. Phys.* **13**, 275 - 288.

FABRY, Ch., and PEROT, A. (1899). Théorie et applications d'une nouvelle méthode de spectroscopie interférentielle. *Ann. Chim. Phys.* **16**, 115 - 144.

FABRY, Ch., and PEROT, A. (1900a). Sur les sources de lumière monochromatiques. *J. Phys. Rad.* **9**, 369 - 382.

FABRY, Ch., and PEROT, A. (1900b). Sur la constitution des raies jaunes du Sodium. *Compt. Rend.* **130**, 653 - 655.

FABRY, Ch., and PEROT, A. (1901a). On a new form of interferometer. *Astrophys. J.* **13**, 265 - 272.

FABRY, Ch. and PEROT, A. (1901b). Measures of absolute wave-lengths in the solar spectrum and in the spectrum of Iron. *Astrophys. J.* **15**, 261 - 273.

FABRY, Ch. and PEROT, A. (1901c). Sur un nouveau modèle d'interféromètre. *Ann. Chim. Phys.* **22**, 564 - 574.

FABRY, Ch. and PEROT, A. (1901d). Longeurs d'onde de quelques raies du fer. *Compt. Rend.* **132**, 1264 - 1266.

FABRY, Ch. and PEROT, A. (1902a). Measures of absolute wave-lengths in the solar spectrum and in the spectrum of Iron. *Astrophys. J.* **15**, 73 - 96.

FABRY, Ch. and PEROT, A. (1902b). Mesures de longueurs d'onde en valeur absolue, spectre solaire et spectre du fer. *Ann. Chim. Phys.* **25**, 98 - 139.

FELLGETT, P. (1958). A propos de la théorie du spectromètre interférentiel multiplex. *J. Phys. Rad.* **19**, 187 - 191.

FIZEAU, H., (1862). Recherches sur les modifications que subit la vitesse de la lumière dans le verre et plusieurs autres corps solides sous l'influence de la chaleur. *Ann. Chim. Phys.* **66**, 429 - 482.

FOURCADE, N. and ROIG, J. (1969). Détermination du profil de la raie verte du Mercure 198 à partir de mesures photométriques sur les anneaux de Perot et Fabry. *Compt. Rend.* **268**, 379 - 382.

FOURCADE, N., ROIG, J., DINGUIRARD, J-P., and GAJAC, I. (1968). Emploi de la transformation de Fourier pour déterminer la distribution spectrale d'une raie atomique, à partir de mesures photométriques sur les anneaux de Pérot et Fabry. Cas d'une raie dissymétrique. *Compt. Rend.* **266B**, 691 - 694.

FOX, A. G. (1970). Optical maser mode selector. U. S. Patent 3 504 299.

FOX, A. G., and LI, T. (1961). Resonant modes in a maser interferometer. *Bell. System Tech. J.* **40**, 453 - 488.

FRESNEL, A. (1826). Mémoire sur la diffraction de la lumière. *In* 'Oeuvres Completes d'Augustin Fresnel' (Publiées par MM. Henri de Senarmont, Emile Verdet et Léonor Fresnel), Vol. I, pp.247 - 382, Imprimerie Impériale, Paris, 1866.

GADSDEN, M., and WILLIAMS, H. M. (1966). Some harmonic properties of an oscillating Fabry-Perot interferometer. *J. Res. NBS* **70C**, 159 - 163.

GAGNE, J.M., HELBERT, J.M., and GESTENKORN, S. (1966). Nouvelle méthode très précise d'étalonnage des spectres enregistrés à l'aide du spectromètre photoélectrique Fabry-Pérot. *Can. J. Phys.* **44**, 681 - 692.

GAGNE, J.M., SAINT-DIZIER, J.P., and PICARD, M. (1974). Méthode d'echantillonnage des fonctions déterministes en spectroscopie: application à un spectromètre multicanal par comptage photonique. *Appl. Opt.* **13**, 581 - 588. Erratum, 1739.

GAULT, W. A. and SHEPHERD, G. G. (1973). Dispersion and refractivity of gases for interferometric pressure scanning. *Appl. Opt.* **12**, 1739 - 1740.

GAUTSCHI, W. (1964). Error Function and Fresnel Integrals. *In* 'Handbook of Mathematical Functions with Formulas, Graphs, and Mathematical Tables' (M. Abramowitz and I. Stegun, eds) p. 295 - 329, United States Government Printing Office, Washington.

GEHRCKE, E. (1905). Uber Interferenzpunkte. *Verh. Deutsch. Physik. Gesselsch.* **7**, 236 - 240.

GEHRCKE, E. (1906). 'Die Anwendung der Interferenzen in der Spektroskopie und Metrologie'. Friedrich Vieweg und Sohn, Braunschweig.

GEHRCKE, E., and LAU, E. (1927). Verschärfte Interferenz. *Zeit. Tech. Physik.* **8**, 157 - 159.

GEHRCKE, E., and LAU, E. (1930). Multiplex-Interferenzspektroskop. *Phys. Zeit.* **31**, 973 - 974.

GEHRCKE, E., and VON BAEYER, O. (1906). Uber die Anwendung der Interferenzpunkte an planparallelen Platten zur Analyse feinster spectrallinien. *Ann. Physik.* **20**, 269 - 292.

GENZEL, L., and SAKAI, K. (1977). Interferometry from 1950 to the present. *J. Opt. Soc. Am.* **67**, 871 - 879.

GEORGELIN, Y. P. (1970a) Application des méthodes interférentielles à la mesure des vitesses radiales. II. Perot-Fabry à lame d'air et à lame de silice. *Astron. Astrophys.* **9**, 436 - 440.

GEORGELIN, Y. P. (1970b) Application des méthodes interférentielles à la mesure des vitesses radiales. III. Images parasites. *Astron. Astrophys.* **9**, 441 - 447.

GERARDO, J. B., VERDEYEN, J. T., and GUSINOW, M. A. (1965). High-frequency laser interferometry in plasma diagnostics. *J. Appl. Phys.* **36**, 2146 - 2151.

GIACOMO, P. (1952). Méthode directe de mesure des caractéristiques d'un systéme interférentiel de Fabry-Perot. *Compt. Rend.* **235**, 1627 -1629.

GIACOMO, P. (1956a). Performances des couches diélectriques multiples en interférometrie. *Rev. Opt.* **35**, 35 - 36.

GIACOMO, P. (1956b). Les couches réfléchissantes multidiélectriques appliquées a l'interféromètre de Fabry-Perot. Etude théorique et expérimentale des couches rélles. *Rev. Opt.* **35**, 317 - 354.

GIACOMO, P. (1958). Propriétés chromatiques des couches réfléchissantes multi-diélectriques. *J. Phys. Rad.* **19**, 307 - 311.

GIRARD, A., and JACQUINOT, P. (1967). Principles of instrumental methods in spectroscopy. *In* 'Advanced Optical Techniques'. (A. C. S. Van Heel, ed.), pp. 73 - 121, North-Holland Pub. Co., Amsterdam.

GLADWIN, M. T. and WOLFE, J. (1975). Linearity of capacitance displacement transducers. *Rev. Sci. Inst.* **46**, 1099 - 1100.

GOBERT, J. (1958). Balayage électromagnétique des anneaux de Perot et Fabry. *J. Phys. Rad.* **19**, 278 - 283.

GOORVITCH, D. (1977). Galatry profile convolved with the Fabry - Perot instrument function. *Appl. Opt.* **16**, 1732 - 1735.

GRADSHTEYN, I. S. and RYZHIK, I. M. (1980). 'Table of Integrals, Series and Products', Academic Press Inc., New York.

GREENLER, R. G. (1957). Interferometric spectrometer for the infrared. *J. Opt. Soc. Am.* **47**, 642 - 646.

GREENLER, R. G. (1958). Un spectromètre interférentiel Fabry - Perot pour l'infra-rouge. *J. Phys. Rad.* **19**, 375 - 378.

GUERN, Y., BIDEAU-MEHU,A., ABJEAN, R., and JOHANNIN-GILLES, A. (1974). Recent developments in multiple beam interferometry in the ultraviolet region 1700-2500 A. *Opt. Comm.* **12**, 66 - 70.

GUNNING, W. (1982). Double-cavity electrooptic Fabry-Perot tunable filter. *Appl. Opt.* **21**, 3129 - 3131.

HAIDINGER, W. (1849). Ueber die schwarzen und gelben Parallel-Linien am Glimmer. *Pogg. Ann. Physik.* **27**, 219 - 228.

HAMY, M. (1906). Sur les franges de réflexion des lames argentées. *J. Physique* **5**, 789 - 809.

HANSEN, G. (1928). Feinstruktur der Spektrallinien. *In* 'Handbuch der Physikalischen Optick' (E. Gehrcke, ed.), pp. 185 - 228. J. A. Barth, Leipzig.

HARIHARAN, P., and SEN, D. (1961). Double-passed Fabry-Perot interferometer. *J. Opt. Soc. Am.* **51**, 398 - 399.

HARLEY, R. T. (1979). Optical alignment of a multipass Fabry-Perot Brillouin scattering spectrometer. *J. Phys. (E).* **12**, 255 - 256.

HARTFIELD, E. and THOMPSON, B. J. (1978). Optical Modulators. *In* ' Handbook of Optics ' (W. G. Driscoll and W. Vaughan, eds), p. 17-1 - 17-24, McGraw-Hill Book Co., New York.

HAYS, P.B., and ROBLE, R.G. (1971). A technique for recovering Doppler line profiles from Fabry-Perot interferometer fringes of very low intensity. *Appl. Opt.* **10**, 193 - 200.

HAYS, P. B., KILLEEN, T. L., and KENNEDY, B. C. (1981). The Fabry-Perot interferometer on Dynamics Explorer. *Space Sci. Inst.* **5**, 395 - 416.

HEAVENS, O. S. (1960). Optical properties of thin films. *Rep. Prog. Phys.* **14**, 1 - 65.

HEAVENS, O. S. (1962). Optical Masers. *Appl. Opt.* **1**, Supp. **1**, 1 - 23.

HECHT, E. and ZAJAC, A. (1974). 'Optics', Addison-Wesley Publishing Co., Reading, Mass.

HEILIG, K., and STEUDEL, A. (1978). New developments of classical optical spectroscopy. *In* 'Progress in Atomic Spectroscopy' (W. Hanle and H. Kleinpoppen, eds.), pp. 263 - 320, Plenum Press, New York.

HERCHER, M. (1968). The spherical mirror Fabry - Perot interferometer. *Appl. Opt.* **7**, 951 - 966.

HERNANDEZ, G. (1966). Analytical description of a Fabry-Perot photoelectric spectrometer. *Appl. Opt.* **5**, 1745 - 1748.

HERNANDEZ, G. (1970). Analytical description of a Fabry-Perot photoelectric spectrometer. II. Numerical results. *Appl. Opt.* **9**, 1591 - 1596.

HERNANDEZ, G. (1974). Analytical description of a Fabry-Perot spectrometer. III. Off-axis behavior and interference filters. *Appl. Opt.* **13**, 2654 - 2661.

HERNANDEZ, G. (1978). Analytical description of a Fabry-Perot spectrometer. IV. Signal noise limitations in data retrieval; winds, temperatures and emission rate. *Appl. Opt.* **17**, 2967 - 2972.

HERNANDEZ, G. (1979). Analytical description of a Fabry-Perot spectrometer. V. Optimization for minimum uncertainties in the determination of Doppler widths and shifts. *Appl. Opt.* **18**, 3826 - 3834.

HERNANDEZ, G. (1980). Measurement of thermospheric temperatures and winds by remote Fabry-Perot spectroscopy. *Opt. Engin.* **19**, 518 - 532.

HERNANDEZ, G. (1982a). Analytical description of a Fabry-Perot spectrometer. VII: TESS, a high luminosity high resolution twin-etalon scanning spectrometer. *Appl. Opt.* **21**, 507 - 513.

HERNANDEZ, G. (1982b). Analytical description of a Fabry-Perot spectrometer. VI: Minimum number of samples required in the determination of Doppler widths and shifts. *Appl. Opt.* **21**, 1695 - 1698.

HERNANDEZ, G. (1982c). Analytical description of a Fabry - Perot spectrometer. V. Optimization for minimum uncertainties in the determination of Doppler widths and shifts; corrigendum. *Appl. Opt.* **21**, 1538.

HERNANDEZ, G. (1985a). Analytical description of a Fabry - Perot spectrometer. VIII. Optimum operation with equidistant equal-noise sampling. *Appl. Opt.* **24**, 2442 - 2449.

HERNANDEZ, G. (1985b). Transient response of optical instruments. *Appl. Opt.* **24**, 928 - 929.

HERNANDEZ, G. (1985c). Fabry-Perot with an absorbing etalon cavity. *Appl. Opt.* **24**, 3062 - 3067.

HERNANDEZ, G. (1985d). Analytical description of a Fabry - Perot spectrometer. IX. Optimum operation with a spherical etalon. *Appl. Opt.* **24**, 3707 - 3712.

HERNANDEZ, G., and MILLS, O.A. (1973). Feedback stabilized Fabry - Perot interferometer. *Appl. Opt.* **12**, 126 - 130.

HERNANDEZ, G., and ROBLE, R.G. (1976). Direct measurements of nighttime thermospheric winds and temperatures 2. Geomagnetic storms. *J. Geophys. Res.* **81**, 5173 - 5181.

HERNANDEZ, G., and ROBLE, R.G. (1979). Thermospheric dynamics investigations with a very high resolution spectrometer. *Appl. Opt.* **18**, 3376 - 3385.

HERNANDEZ, G. and ROBLE, R. G. (1984). The geomagnetic quiet nighttime thermospheric wind pattern over Fritz Peak Observatory during solar cycle minimum and maximum. *J. Geophys. Res.* **89**, 327 - 337.

HERNANDEZ, G., MILLS, O.A., and SMITH, J.L. (1981). TESS: A high-luminosity high-resolution twin-etalon scanning spectrometer. *Appl. Opt.* **20**, 3687 - 3688.

HERNANDEZ, G., SICA, R. J., and ROMICK, G. J. (1984). Equal-noise spectroscopic measurement method. *Appl. Opt.* **23**, 915 - 919.

HERRIOTT, D. R. (1961). Multiple-wavelength multiple-beam interferometric observations of flat surfaces. *J. Opt. Soc. Am.* **51**, 1142 - 1145.

HERRIOTT, D. R. (1967). Some applications of lasers to interferometry. *In* 'Progress in Optics', Vol. VI, (E. Wolf, ed.), pp. 173 - 209, North Holland Pub. Co., Amsterdam.

HICKS, T.R., REAY, N.K., and SCADDAN, R.J. (1974). A servo-controlled Fabry-Perot interferometer using capacitance micrometers for error detection. *J. Phys.(E)*. **7**, 27 - 30.

HILL, R.M. (1963). Some fringe - broadening defects in a Fabry - Perot étalon. *Opt. Acta* **10**, 141 - 152.

HIRSCHBERG, J. G. (1958). Emploi de fentes courbes pour la mise en série d'un réseau de diffraction avec un interféromètre Fabry - Perot balayé par variation de pression. *J. Phys. Rad.* **19**, 256 - 259.

HIRSCHBERG, J.G., and COOKE, F.N. (1970). Fresnel annular zone objective: A new optical element. *Appl. Opt.* **9**, 2807.

HIRSCHBERG, J.G., and FRIED, W.I. (1970). A polarization Fabry-Perot duochromator. *Appl. Opt.* **9**, 1137 - 1139.

HIRSCHBERG, J.G., and PLATZ, P. (1965). A multichannel Fabry-Perot interferometer. *Appl. Opt.* **4**, 1375 - 1381.

HIRSCHBERG, J.G., FRIED, W.I., HAZELTON, L., and WOUTERS, A. (1971). Multiplex Fabry-Perot interferometer. *Appl. Opt.* **10**, 1979 - 1980.

HOCHSTRASSER, U. W. (1964). Orthogonal Polynomials. *In* 'Handbook of Mathematical Functions with Formulas, Graphs and Mathematical Tables', (M. Abramowitz and I. Stegun, eds), p. 771 - 802, United States Government Printing Office, Washington.

HODGKINSON, I. J. (1972). Compensation of Fabry-Perot surface defects. II: Silicon oxide compensating layers. *Appl. Opt.* **11**, 1970 - 1977.

HOEY, M.J., JERMYN, T., and GRIMLEY, H.M. (1970). A scanning Fabry - Perot interferometer for plasma diagnostics. *J. Phys.(E)* **3**, 305 - 310.

HOUSTON, W.V. (1926). The fine structure and the wave lengths of the Balmer lines. *Astrophys. J.* **64**, 81 - 92.

HOUSTON, W.V. (1927). A compound interferometer for fine structure work. *Phys. Rev.* **29**, 478 - 484.

HUMPHREYS, C.J. (1930). Interference measurements in the first spectra of Krypton and Xenon. *J. Res. Bur. Stds.* **5**, 1041 - 1055.

HUMPHREYS, C.J. (1931). Hyperfine structure in the first spectra of Krypton and Xenon. *J. Res. Bur. Stds.* **7**, 453 - 463.

JACKSON, C.V. (1931). Interferometric measurements in the arc spectrum of Iron. *Proc. Roy. Soc.(Lon)* **A130**, 395 - 410.

JACKSON, C V. (1932). Interferometric measurements in the spectrum of Krypton. *Proc. Roy. Soc.(Lon)* **A138**, 147 - 153.

JACKSON, C.V. (1933). Wave-lengths of the red lines of Neon and their use as secondary standards. *Proc. Roy. Soc (Lon)* **A143**, 124 - 135.

JACKSON, C.V. (1936). Wave - length standards in the first spectrum of Krypton. *Phil. Trans. Roy. Soc. (Lon)* **236A**, 1 - 24.

JACKSON, D.A. (1934). The magnetic moment of the nucleus of Caesium. *Proc. Roy. Soc. (Lon)* **A147**, 500 - 513.

JACKSON, D.A. (1961). The spherical Fabry-Perot interferometer as an instrument of high resolving power for use with external or with internal atomic beams. *Proc. Roy. Soc. (Lon)* **A263**, 289 - 308.

JACKSON, D.A., and KUHN, H. (1935). The hyperfine structure of the resonance lines of Potassium. *Proc. Roy. Soc. (Lon)* **A148**, 335 - 352.

JACKSON, D.A., and KUHN, H. (1936). Nuclear mechanical and magnetic moments of K^{39}. *Nature* **137**, 108.

JACKSON, D.A., and KUHN, H. (1937). Intensity ratios of the hyperfine structure components of the resonance lines of Potassium. *Nature* **140**, 276 - 277.

JACKSON, D.A., and KUHN, H. (1938a). The hyperfine structure of the Zeeman components of the resonance lines of Sodium. *Proc. Roy. Soc. (Lon)* **A167**, 205 - 216.

JACKSON, D.A., and KUHN, H. (1938b). Hyperfine structure, Zeeman effect and isotope shift in the resonance lines of Potassium. *Proc. Roy. Soc. (Lon)* **A165**, 303 - 312.

JACQUINOT, P. (1954). The luminosity of spectrometers with prisms, gratings, or Fabry-Perot etalons. *J. Opt. Soc. Am.* **44**, 761 - 765.

JACQUINOT, P. (1960). New developments in interference spectroscopy. *Repts. Prog. Phys.* **23**, 267 - 312.

JACQUINOT, P. (1969). Interferometry and grating spectroscopy: an introductory survey. *Appl. Opt.* **8**, 497 - 499.

JACQUINOT, P., and DUFOUR, Ch. (1948). Conditions optiques d'emploi des cellules photo - électriques dans les spectrographes et les interféromètres. *J. Res. CRNS*, **6**, 91 - 103.

JAHN, H., FELLBERG, G., GLADITZ, B., and SCHEELE, M. (1982). Maximum-likelihood optimization of a Fabry - Perot interferometer for thermospheric temperature and wind measurements. *J. Opt. Soc. Am.* **72**, 386 - 391.

JAVAN, A., BENNETT, W. R. and HERRIOTT, D. R. (1961). Population inversion and continuous optical maser oscillation in a gas discharge containing a $He - Ne$ mixture. *Phys. Rev. Lett.* **6**, 106 - 110.

JENKINS, F. A. and WHITE, H. E. (1976). 'Fundamentals of Optics', 4th Edition, McGraw-Hill Book Co., New York.

JOBIN, A. (1898). Spectroscope interférentiel de MM. A. Perot et Ch. Fabry, construit et présenté par M. Jobin. *Phys. Bull. (Paris Soc.)* **116**, 7 - 9.

JOHNSON, J. R. (1968). A high resolution scanning confocal interferometer. *Appl. Opt.* **7**, 1061 - 1072.

JONES, R. V., and RICHARDS, J. C. S. (1973). The design and some applications of sensitive capacitance micrometers. *J. Phys. (E)* **6**, 589 - 600.

KASTLER, A. (1962). Atomes à l'interieur d'un interféromètre Perot-Fabry. *Appl. Opt.* **1**, 17 - 24.

KASTLER, A. (1974). Transmission d'une impulsion lumineuse par un interféromètre Fabry-Perot. *Nouv. Rev. Opt.* **5**, 133 - 139.

KATZENSTEIN, J. (1965). The axicon - scanned Fabry - Perot spectrometer. *Appl. Opt.* **4**, 263 - 266.

KAUZMANN, W. (1957). 'Quantum Chemistry', Academic Press Inc., New York.

KAYSER, H. (1925). 'Tabelle der Schwingungszahlen'. S. Hirzel, Leipzig.

KHASHAN, M.A. (1979). Analytical determination of linewidths using the Fabry - Perot spectrometer. *Physica* **98C**, 93 - 99.

KHODINSKII, A. N. (1980). Mechanism for adjusting Fabry-Perot interferometers. *Sov. J. Opt. Tech.* **47**, 91 - 92.

KILLEEN, T. L., and HAYS, P. B. (1984). Doppler line profile analysis for a multichannel Fabry-Perot interferometer. *Appl. Opt.* **23**, 612 - 620.

KILLEEN, T. L., DETMANN, D. L., and HAYS, P.B. (1980). Stability of an optically contacted etalon to cosmic radiation. *Appl. Opt.* **19**, 2265 - 2266.

KILLEEN, T. L., HAYS, P. B., and DE VOS, J. (1981). Parallelism maps for optically contacted etalons. *Appl. Opt.* **20**, 2616 - 2619.

KILLEEN, T. L., HAYS, P. B., KENNEDY, B. C., and REES, D. (1982). Stable and rugged etalon for the Dynamics Explorer Fabry-Perot interferometer. 2: Performance. *Appl. Opt.* **21**, 3903 - 3912.

KILLEEN, T. L., KENNEDY, B. C., HAYS, P. B., SYMANOW, D. A. and CECKOWSKI, D. H. (1983). Image plane detector for the Dynamics Explorer Fabry-Perot interferometer. *Appl. Opt.* **22**, 3503 - 3513.

KINDER, W. and TORGE, R. (1971). Zur Technik von Fabry-Perot-Interferometern. *Optik* **34**, 31 - 37.

KOBLER, H. (1963). Servo-techniques in Fabry-Perot interferometry. *Proc. Inst. Rad. Eng. Aust.* **24**, 677 - 684.

KOGELNIK, H. and LI, T. (1966). Laser beams and resonators. *Appl. Opt.* **5**, 1550 - 1567.

KOGELNIK, H. and RIGROD, W. W. (1962). Visual display of isolated optical-resonator modes. *Proc. I. R. E.* **50**, 220.

KOPPELMANN, G. (1969). Multiple - beam interference and natural modes in open resonators. *In* 'Progress in Optics', Vol. VII, (E. Wolf, ed.), pp. 3 - 66, North Holland Pub. Co., Amsterdam.

KOPPELMANN, G. and KREBS, K. (1961a). Eine Registriermethode zur Vermassung des Reliefs höchstebener Oberflächen. *Optik* **18**, 349 - 357.

KOPPELMANN, G. and KREBS, K. (1961b). Sur Technologie des Perot-Fabry-Interferometers. *Optik* **18**, 358 - 372.

KOPPELMANN, G. and SCHRECK, K. (1969). Ein neues Verfahren zur Herstellung von Aufdampfschichten mit vorgegebener Dickenverteilung sur Korrektur der Unebenheiten von Fabry-Perot-interferometer-Spiegeln. *Optik* **29**, 549 - 559.

KOPPELMANN, G., RUDOLPH, H. and SCHRECK, K. (1975). Ein Digitalisiertes interferometrisches Verfahren zur Messung der Abstandsverteilung von Fabry-Perot-Spiegeln. *Optik* **43**, 35 - 52.

KOROLEV, F. A. and GRIDNEV, V. I. (1964). A Fabry-Perot interferometer with diffraction mirrors. *Opt. Spect.* **16**, 181 - 184.

KREBS, K., and SAUER, A. (1953). Uber die Intensitätsverteilung von Spektrallinien im Pérot-Fabry Interferometer. *Ann. Physik.* **13**, 359 - 368.

KUHN, H. (1951). New techniques in optical interferometry. *Rep. Prog. Phys.* **14**, 64 - 94.

LALLEMAND, A. (1962). Photomultipliers. *In* 'Astronomical Techniques', (W.A. Hiltner, ed), pp.126 - 156, University of Chicago Press, Chicago.

LAMB, W. E. (1964). Theory of an optical maser. *Phys. Rev.* **134**, 1429 - 1450.

LANGLEY, K. H., and FORD, N. C. (1969). Attenuation of the Rayleigh component in Brillouin spectroscopy using interferometric filtering. *J. Opt. Soc. Am.* **59**, 281 - 284.

LAU, E. (1930). Multiplex-Interferenzspektroskop. *Zeit. Physik.* **63**, 313 - 317.

LAU, E. (1932). Die Erzielung von Interferenzstreifen hohen Gangunterschiedes durch Zusammenwirken zweier Interferenzsysteme mit verschiedener Dispersion. *Ann. Physik.* **10**, 71 - 80.

LAU, E. and RITTER, E. (1932). Das Multiplex-Interferenzspektroskop im Vergleich mit anderen Interferenzspektroskopen. *Zeitz. Physik* **76**, 190 - 200.

LECULLIER, J. C. and CHANIN, G. (1976). A scanning Fabry-Perot interferometer for the 50-100 μm range. *Infrared Phys.* **16**, 273 - 278.

LENGYEL, B. A. (1971). 'Lasers', John Wiley and Sons, Inc., New York.

LI, T. (1965). Diffraction loss and selection of modes in maser resonators with circular mirrors. *Bell System Tech. J.* **44**, 917 - 932.

LISSBERGER, P. H., and WILCOCK, W. L. (1959). Properties of all-dielectric interference filters. II. Filters in parallel beams of light incident obliquely and in convergent beams. *J. Opt. Soc. Am.* **49**, 126 - 131.

LUMMER, O. (1884). Ueber eine neue Interferenzerscheinung an planparallelen Glasplatten und eine Methode, die Planparallelität solcher Glässer zu prüfen. *Ann. Physik. Chem.* **23**, 49 - 84.

LUMMER, O. (1901a). Eine neue Interferenzmethode zur Auflösung feinster Spectrallinien. *Verh. Deutsch. Physik. Gesselsch.* **3**, 85 - 98.

LUMMER, O. (1901b). Ein neues Interferenz-spectroskop. *Arch. Néerl.* **6**, 773 - 778.

LUMMER, O., and GEHRCKE, E. (1902). Ueber die Interferenz des Lichtes bei mehr als zwei Millionen Wellenlängen Gangunterschied. *Verh. Deutsch. Physik. Gesselsch.* **4**, 337 - 346.

LUMMER, O., and GEHRCKE, E. (1903). Uber die Anwendung der Interferenzen an planparallelen Platten zur Analyse feinster Spektrallinien. *Ann. Physik.* **10**, 457 - 477.

LUMMER, O., and GEHRCKE, E. (1904). Sur la séparation des raies spectrales très voisines. *J. Physique* **3**, 345 - 350.

MACH, L. (1891). Ueber einen Interferenzrefraktor. *Zeitschr. f. Instrkde.* **12**, 89 - 93.

MACK, J.E., MCNUTT, D.P., ROESLER, F.L., and CHABBAL, R. (1963). The PEPSIOS purely interferometric high-resolution scanning spectrometer. 1. The pilot model. *Appl. Opt.* **2**, 873 - 885.

MCNUTT, D.P. (1965). Pepsios purely interferometric high - resolution scanning spectrometer. II. Theory of spacer ratios. *J. Opt. Soc. Am.* **55**, 288 - 292.

MAIMAN, T. H. (1960). Stimulated optical radiation in ruby. *Nature* **187**, 493 - 494.

MAR, S., GIGOSOS, M. A. and MORENO, J. M. (1979). Deconvolution of spectral lines: an improved method. *Appl. Opt.* **18**, 2914-2916.

MARIOGE, J. P. (1971). Flexion symétrique des disques minces et épais. *Nouv. Rev. Opt.* **2**, 143 - 148.

MEABURN, J. (1972). A nebular, two-etalon Fabry-Perot monochromator. *Astron. Astrophys.* **17**, 106 - 112.

MEABURN, J. (1976). 'Detection and Spectrometry of Faint Light'. D. Reidel Publishing Co, Dordrecht.

MEGGERS, W. F. (1921). Interference measurements in the spectra of Argon, Krypton and Xenon. *Sci. Pap. Bur. Stds.* **17**, 193 - 202.

MEGGERS, W.F., and PETERS, C.G. (1918). Measurements on the index of refraction of air for wavelengths from 2218 A to 9000 A. *Bull. Bur. Stds.* **14**, 697 - 740.

MEISSNER, K.W. (1916). Die Gesetzmässigkeiten im Neon-und Argon spektrum. *Phys. Zeit.* **17**, 549 - 552.

MEISSNER, K.W. (1941). Interference Spectroscopy. Part I. *J. Opt. Soc. Am.* **31**, 405 - 427.

MEISSNER, K.W. (1942). Interference Spectroscopy. Part II. *J. Opt. Soc. Am.* **32**, 185 - 211.

MERIWETHER, J. W. (1983). Observations of thermospheric dynamics at high latitudes from ground and space. *Radio Science* **18**, 1035 - 1052.

MERRILL, P.W. (1917). Wave lengths of the stronger lines in the Helium spectrum. *Bull. Bur. Stds.* **14**, 159 - 166.

MERTON, T.R. (1920). On the spectra of isotopes. *Proc. Roy. Soc.(Lon)* **A96**, 388 - 395.

MEWE, R. and DE VRIES, R. F. (1965). Thin spacers for Fabry-Perot etalons. *J.*

Opt. Soc. Am. **55**, 1967.

MICHELSON, A. A. (1881). The relative motion of the earth and the luminiferous ether. *Amer. J. Sci.* **22**, 120 -129.

MICHELSON, A. A. (1882). Interference phenomena in a new form of refractometer. *Phil. Mag.* **13**, 236 - 242.

MIELENZ, K., STEPHENZ, R. B., and NEFFLEN, K. (1964). A Fabry-Perot spectrometer for high resolution spectroscopy and laser work. *J. Res. NBS* **68C**, 1 -6.

MINKOWSKI, R. and BRUCK, H. (1935a). Die Intensitätsverteilung der im Molekularstrahl erzeugten Spektrallinien. *Z. Physik.* **95**, 274 - 283.

MINKOWSKI, R., and BRUCK, H. (1935b). Die Intensitätsverteilung der roten Cd-Linie im Molekularstrahl bei Anregung durch Elektronenstoss. *Z. Physik.* **95**, 284 - 298.

MINKOWSKI, R. and BRUCK, H. (1935c). Wahre und scheinbare Breite von Spektrallinien. *Z. Physik.* **95**, 299 - 301.

MOGHRABI, B. and GAUME, F. (1974). Etude des épaisseurs relatives des deux Fabry-Pérot d'un spectromètre purement interférentiel. *Nouv. Rev. Opt.* **5**, 231 - 236.

MOGHRABI, B. and GAUME, F. (1977). Epaisseurs relatives des Fabry-Pérot d'un spectromètre purement interférentiel II. Cas de deux Fabry-Pérot en série avec un filtre interférentiel. *J. Optics(Paris)* **8**, 271 - 276.

MOLBY, F. A. (1936). Fabry and Perot interferometer adjustment (A precision method). *Rev. Sci. Inst.* **7**, 257.

MONNET, G. (1970). Application des méthodes interférentielles à la mesure des vitesses radiales. I. Montages optiques. *Astron. Astrophys.* **9**, 420 - 435.

MROZOWSKI, S. (1941). Superposition fringes in the internally reflected light from a Fabry-Perot etalon. *J. Opt. Soc. Am.* **31**, 209 - 212.

MULLER, G., and WINKLER, R. (1968). Entwicklung und Einsatz eines zweifach durchstrahlten Druck - Fabry - Perot - Interferometers. *Optik* **28**, 143 - 148.

MURTY, M. V. R. K. (1962). Multiple-pinhole multiple-beam interferometric observation of flat surfaces. *Appl. Opt.* **1**, 364 - 365.

NAGAOKA, H. (1917). On the regularity in the distribution of the satellites of spectrum lines. *Proc. Phys. Soc. (Lon)* **29**, 91 - 119.

NAGAOKA, H. and TAKAMINE, T. (1912). The constitution of Mercury lines examined by an echelon grating and a Lummer-Gehrcke plate. *Proc. Roy. Soc. (Lon)* **A25**, 1 - 30.

NAGAOKA, H., and TAKAMINE, T. (1915). Anomalous Zeeman effect in satellites of the violet line (4359) of Mercury. *Phil. Mag.* **29**, 241 - 252.

NELSON, D. F. and BOYLE, W. S. (1962). A continuously operating ruby optical maser. *Appl. Opt.* **1**, 181 - 183.

NEO, Y.P., and SHEPHERD, G.G. (1972). Airglow observations with a Hadamard photometer. *Planet. Space Sci.* **20**, 1351 - 1355.

NETTERFIELD, R. P., SAINTY, W. G. and SCHAEFFER, R. C. (1980). Coating Fabry-Perot interferometer plates with broadband multilayer dielectric mirrors. *Appl. Opt.* **19**, 3010 -3017.

NEUHAUS, H., and NYLEN, P. (1970). Die Apparatfunktion eines aus Fabry-Pérotschen Etalons zusammengesetzten Systems bei endlicher Apertur. Teil III. *Arkiv för Fysik* **40**, 405 - 412.

NEWTON, SIR ISAAC. (1730). 'Opticks', Fourth edition, W. Innys, London. Republished by Dover Publications, N.Y. (1952).

N. P. L., (1960). National Physical Laboratory, Symposium No. 11: 'Interferometry'. Her Majesty's Stationery Office, London.

NILSON, J. A. and SHEPHERD, G. G. (1961). Upper atmospheric temperatures from Doppler line widths. I. Some preliminary measurements in OI 5577A in aurora. *Planet. Space Sci.* **5**, 299 - 306.

OKANO, S., KIM, J.S., and ICHIKAWA, T. (1980). Design of a multiple - zone aperture and application to a Fabry - Perot interferometer. *Appl. Opt.* **19**, 1622 - 1629.

OLVER, F. W. J. (1964). Bessel Functions of Integer Order. *In* 'Handbook of Mathematical Functions with Formulas, Graphs and Mathematical Tables', (M. Abramowitz and I. Stegun, eds), p. 355 - 433, United States Government Printing Office, Washington.

PARRAT, L. G. (1961). 'Probability and Experimental Errors in Science', Chapter 5, John Wiley and Sons, New York.

PATEK, K. (1967). 'Lasers', CRC Press, Cleveland.

PAULING, L. and WILSON, E. B. (1935). 'Introduction to Quantum Mechanics', McGraw-Hill Book Co. Inc, New York.

PAULS, E. (1932). Uber das Multiplexinterferenzspektroskop. *Physik. Zeitschr.* **33**, 405 - 410.

PELLETIER, E., CHABBAL, R. and GIACOMO, P. (1964). Influence sur l'épaisseur optique de l'interféromètre de Fabry-Perot d'une variation d'épaisseur du revêtement réflecteur. *J. Phys.* **25**, 275 - 279.

PEROT, A., and FABRY, Ch. (1897). Sur les franges des lames minces argentées et leur application a la mesure de petites épasseurs d'air. *Ann. Chim. Phys.* **12**, 459 - 501.

PEROT, A., and FABRY, Ch. (1898a). Sur une nouvelle méthode de spectroscopie interférentielle. *Compt. Rend.* **126**, 34 - 36.

PEROT, A., and FABRY, Ch. (1898b). Etude de quelques radiations par la spectroscopie interférentielle. *Compt. Rend.* **126**, 407 - 410.

PEROT, A., and FABRY, Ch. (1898c). Sur la détermination des numéros d'ordre de franges d'ordre élevé. *Compt. Rend.* **126**, 1624 - 1626.

PEROT, A., and FABRY, Ch. (1898d). Electromètre absolu pour petites différences de potentiel. *Ann. Chim. Phys.* **13**, 404 - 432.

PEROT, A., and FABRY, Ch. (1898e). Sur un voltmètre electrostatique interférentiel pour étalonnage. *J. Phys.* **7**, 650 - 659.

PEROT, A., and FABRY, Ch. (1899a). On the application of interference phenomena to the solution of various problems of spectroscopy and metrology. *Astrophys. J.* **9**, 87 - 115.

PEROT, A., and FABRY, Ch. (1899b). Méthodes interférentielles pour la mesure des grandes épaisseurs et la comparaison des longeurs d'onde. *Ann. Chim. Phys.* **16**, 289 - 338.

PEROT, A., and FABRY, Ch. (1900a). Détermination de nouveaux points de repère dans le spectre. *Compt. Rend.* **130**, 492 - 495.

PEROT, A., and FABRY, Ch. (1900b). Méthode interférentielle pour la mesure des longeurs d'onde dans le spectre solaire. *Compt. Rend.* **131**, 700 - 702.

PEROT, A., and FABRY, Ch. (1901a). Mesures de longueurs d'onde dans le spectre solaire; comparaison avec l'echelle de Rowland. *Compt. Rend.* **133**, 153 - 154.

PEROT, A., and FABRY. Ch. (1901b). Mesure en longueurs d'onde de quelques étalons de longueur a bouts. *Ann. Chim. Phys.* **24**, 119 - 139.

PEROT, A., and FABRY, Ch. (1904a). Sur les longueurs d'onde des raies du spectre solaire et les corrections aux tables de Rowland. *Ann. Chim. Phys.* **27**, 5 - 8.

PEROT, A., and FABRY, Ch. (1904b). Sur la mesure optique de la différence de deux épaisseurs. *Compt. Rend.* **138**, 676 - 678.

PEROT, A., and FABRY, Ch. (1904c). Sur la séparation des raies spectrales très voisines a propos d'un travail récent de MM. Lummer et Gehrcke. *J. Physique* **3**, 28 - 32.

PEROT, A., and FABRY, Ch. (1904d). Rapport sur la nécessité d'établir un nouveau système de longueurs d'onde étalons, présenté au nom de la Société Française de Physique. *J. Physique* **3**, 842 - 850.

PFUND, A. H. (1908). A redetermination of the wave-lengths of standard Iron lines. *Astrophys. J.* **28**, 197 - 211.

PHELPS, F. M. (1965). Making of spacers for Fabry-Perot etalons. *J. Opt. Soc. Am.* **55**, 293 - 295.

PISMIS, P. (1982). La interferometria Fabry-Perot: Aplicaciones astronomicas. *Rev. Mex. Fis.* **29**, 1 - 25.

PLATISA, M. (1974). Azimuthal variations in the brightness of Fabry-Perot axicon fringes. *Opt. Laser. Tech.* **6**, 114 - 116.

POLE, R.V., SPIEKERMAN, A.J., and HANSCH, T.W. (1978). Accurate interferometric wavelength measurements through post-detection signal processing. *Optica Hoy y Mañana* **15**, 463 - 465.

POLE, R.V., SPIEKERMAN, A.J., and HANSCH, T.W. (1980). Interferometric wavelength measurements through post - detection signal processing. *I.B.M. J. Res. Develop.* **24**, 85 - 88.

PROKHOROV, A. M. (1958). Molecular amplifier and generator for submillimeter waves. *Sov. Phys. JETP* **7**, 1140 - 1141.

RAMSAY, B.P., CLEVELAND, E.L., and KOPPIUS, O.T. (1941). Criteria and the intensity - epoch slope. *J. Opt. Soc. Am.* **31**, 26 - 33.

RAMSAY, J.V. (1962). A rapid - scanning Fabry - Perot interferometer with automatic parallelism control. *Appl. Opt.* **1**, 411 - 413.

RAMSAY, J. V. (1966). Control of the spacing of Fabry - Perot interferometers. *Appl. Opt.* **5**, 1297 - 1301.

RANK, D. H. and SHEARER, J. N. (1956). Linear gas mass flow device with applications to interferometry. *J. Opt. Soc. Am.* **46**, 463 - 464.

RASMUSSEN, E. (1946). Testing of interferometer plates. *Physica* **12**, 656 - 660.

RAYLEIGH, LORD. (1879). Investigations in optics, with special reference to the spectroscope. *Phil. Mag.* **8**, 261 - 274

RAYLEIGH, LORD. (1906). Some measurements of wave - lengths with a modified apparatus. *Phil. Mag.* **11**, 685 - 703.

RAYLEIGH, LORD. (1908). Further measurements of wave - lengths, and miscellaneous notes on Fabry and Perot's apparatus. *Phil. Mag.* **15**, 548 - 558.

REES, D., MC WHIRTHER, I., ROUNCE, P. A. and BARLOW, F. E. (1981a). Miniature Imaging photon detectors. II Devices with transparent photocathodes. *J. Phys. (E)* **14**, 229 - 233.

REES, D., MC WHIRTHER, I., HAYS, P. B. and DINES, T. (1981b). A Stable, rugged capacitance-stabilised piezoelectric scanned Fabry-Perot etalon. *J. Phys. (E)* **14**, 1320 - 1325.

REES, D., FULLER-ROWELL, T. J., LYONS, A., KILLEEN, T. L., and HAYS, P. B. (1982). Stable and rugged etalon for the Dynamics Explorer Fabry-Perot interferometer. 1: Design and construction. *Appl. Opt.* **21**, 3896 - 3902.

REYNOLDS, G.T., SPARTALIAN, K., and SCARL, D.B. (1969). Interference effects produced by single photons. *Nuovo Cimento* **61B**, 356 - 364.

RIGROD, W. W., KOGELNIK, H., BRANGACCIO, D. J. and HERRIOTT, D. R. (1962). Gaseous optical maser with concave mirrors. *J. Appl. Phys.* **33**, 743 - 745.

ROESLER, F.L. (1969). Effects of plate defects in a polyetalon Fabry - Perot spectrometer. *Appl. Opt.* **8**, 829 - 830.

ROESLER, F.L. (1974). Fabry - Perot Instruments for Astronomy. *In* 'Methods of Experimental Physics', (N. Carleton, ed.), Vol 12 A, pp. 531 - 569, Academic Press, New York.

ROESLER, F. L., and MACK, J. E. (1967). Pepsios purely interferometric high - resolution scanning spectrometer. IV. Performance of the Pepsios spectrometer. *J. Physique* **28**, Supp. C2, 313 - 320.

ROGERS, G. L. (1969). A device for aligning a large Fabry-Perot etalon. *J. Phys. (E)* **2**, 1006 - 1007.

ROIG, J. (1958). Balayage thermique des anneaux de Perot et Fabry. *J. Phys. Rad.* **19**, 284 - 289.

ROIG, J., FOURCADE, N., DINGUIRARD, J-P. and LETELLIER, S. (1967). Emploi de la transformation de Fourier et de la méthode de P.M.Duffieux pour déterminer la distribution spectrale d'une raie atomique, à partir de mesures photométriques sur les anneaux de Perot - Fabry. Calcul du facteur de cohérence de la raie et du contraste des franges. *Compt. Rend.* **265**, 1303 - 1306.

ROYCHOUDHURI, C. (1975). Response of Fabry-Perot interferometers to light pulses of very short duration. *J. Opt. Soc. Am.* **65**, 1418 - 1426.

ROZUVANOVA, V.A. (1971). Hydropneumatic drive for the scanning system of a Fabry - Perot spectrometer. *Instrum. and Exp. Tech. (USA)* **14**, 584 - 586.

SAHA, M.N. (1917). On the limit of interference in the Fabry - Perot interferometer. *Phys. Rev.* **10**, 782 - 786.

SANDERCOCK, J.R. (1970). Brillouin scattering study of SbSI using a double - passed, stabilised scanning interferometer. *Opt. Comm.* **2**, 73 - 76.

SANDERCOCK, J.R. (1971). The design and use of a stabilised multipassed interferometer of high contrast ratio. *In* 'Light Scattering in Solids'. (M. Balkanski, ed.) pp. 9 - 12, Flammarion Press, Paris.

SANDERCOCK, J. R. (1975). Some recent developments in Brillouin scattering. *RCA Rev.* **36**, 89 - 107.

SCHAWLOW, A. L. and TOWNES, C. H. (1958). Infrared and optical masers. *Phys. Rev.* **112,** 1940 - 1949.

SCHULTZ, G. and SCHWIDER, J. (1976). Interferometric testing of smooth surfaces. *In* 'Progress in Optics', Vol. XIII, (E. Wolf, ed.), pp. 95-167, North Holland Pub. Co., Amsterdam.

SCHUSTER, A. (1904). 'An Introduction to the Theory of Optics', p. 141. Edward Arnold, London.

SCHWIDER, J. (1965). Entkopplungsmöglichkeiten von Fabry - Perot - Interferometern. *Optica Acta* **12**, 65 - 79.

SCHWIDER, J., SCHULZ, G., RIEKHER, R. and MINKWITZ, G. (1966). Ein Interferenzverfahren zur Absolutprüfung von Planflächennormalen I. *Opt. Acta* **13**, 103 - 119.

SEARS, J. E. and BARRELL, H. (1933). IV. A new apparatus for determining the relationship between wavelengths of light and the fundamental standards of length. *Phil. Trans. Roy. Soc. (Lon)* **A231**, 75 - 128.

SERIES, G. W. (1954). The fine structure of the line 4686 A of singly ionized Helium. *Proc. Roy. Soc. (Lon)* **A226**, 377 - 392.

SHEPHERD, G.G. (1960). A Fabry - Perot spectrometer for auroral and airglow observations. *Can. J. Phys.* **38**, 1560 - 1569.

SHEPHERD, G.G. (1967). Applications of the Fabry - Perot spectrometer to upper atmospheric spectroscopy. *J. Physique* **28**, Supp. C2, 301 - 307.

SHEPHERD, G.G. (1969a). Spectroscopic measurements of auroral and airglow temperatures. *Ann. Geophys.* **25**, 841 - 846.

SHEPHERD, G.G. (1969b). Airglow spectroscopic temperatures. *In* 'Atmospheric Emissions', (B.M. McCormac and A. Omholt, eds.), pp. 411 - 422, Van Nostrand Reinhold Co., New York.

SHEPHERD, G.G. (1972). Spectroscopic measurements of upper atmospheric temperature. *In* 'Temperature: Its Measurement and Control in Science and Industry.' (H.H. Plumb, ed.), Volume 4, pp. 2313 - 2327. Instr. Soc. Amer., Pittsburgh.

SHEPHERD, G.G., DEANS, A.J., and NEO, Y.P. (1978). SCIMP - A scanning interferometric multiplex photometer. *Can. J. Phys.* **56**, 681 - 686.

SHEPHERD, G.G., LAKE, C.W., MILLER, J.R., and COGGER, L.L. (1965). A spatial spectral scanning technique for the Fabry - Perot spectrometer. *Appl. Opt.* **4**, 267 - 272.

SHEPHERD, G.G., and PAFFRATH, L. (1967). A moiré fringe method of Fabry - Perot spectrometer scanning. *Appl. Opt.* **6**, 1659 - 1663.

SIPLER, D.P., and BIONDI, M.A. (1978). Equatorial F - region neutral winds from nightglow OI 630.0 nm Doppler shifts. *Geophys. Res. Lett.* **5**, 373 - 376.

SIVJEE, G.G., HALLINAN, T.J., and SWENSON, G.R. (1980). Fabry - Perot - interferometer imaging system for thermospheric temperature and wind measurements. *Appl. Opt.* **19**, 2206 - 2209.

SLATER, P.N., BETZ, H.T., and HENDERSON, G. (1965). A new design of a scanning Fabry - Perot interferometer. *Jap. J. Appl. Phys.* **4**, Supp. I, 440 - 444.

SMITH, P. W. (1965). Stabilized, single frequency output from a long laser cavity. *J. Quantum Electron.* **1**, 343 - 348.

SPARROW, C.M. (1916). On spectroscopic resolving power. *Astrophys. J.* **44**, 76 - 86.

STEEL, W. H. (1967). 'Interferometry'. Cambridge Univ. Press, Cambridge.

STENHOLM, S. (1984). 'Foundations of Laser Spectroscopy'. John Wiley and Sons, New York.

STEYERL, A., and STEINHAUSER, K.-A. (1979). Proposal of a Fabry - Perot - type interferometer for X - Rays. *Zeit. Physik.* **B34**, 221 - 227.

STONER, J.O. (1966). PEPSIOS - Purely interferometric high - resolution spectrometer. III. Calculation of interferometer characteristics by a method of optical transients. *J. Opt. Soc. Am.* **56**, 370 - 376.

STRUTT, R.J. (1919). Bakerian Lecture: A study of the line spectrum of Sodium as excited by fluorescence. *Proc. Roy. Soc. (Lon)* **A96,** 272 - 287.

STUDENT (W. S. GOSSET). (1908). The probable error of a mean. *Biometrika* **6,** 1 - 25.

STURKEY, L. (1940). Fabry - Perot interferometers in a parallel arrangement. *J. Opt. Soc. Am.* **30,** 351 - 354.

STURKEY, L., and RAMSAY, B. P. (1941). A general interferential method. *Phil. Mag.* **31,** 13 - 23.

SVELTO, O. (1982). 'Principles of Lasers'. Plenum Press, New York.

TAKO, T., and OHI, M. (1965). Profile analysis of spectral line by Fabry - Perot interferometer. *Jap. J. Appl. Phys.* **4,** Supp. I, 451 - 454.

THOMPSON, R. C. (1982). An instrumental effect on spectral line profiles. *J. Quant. Spectrosc. Radiat. Transfer* **27,** 417 - 421.

TOLANSKY, S. (1944). New contributions to interferometry. Part IV.- The ghost images and scatter rings of the Fabry-Perot interferometer and their effects in hyperfine structure observations. *Phil. Mag.* **35,** 229 - 241.

TOLANSKY, S. (1946). The intensity efficiency of the Fabry-Perot interferometer both in transmission and reflection. *Phyisica* **12,** 649 -655.

TOLANSKY, S. (1947). 'High Resolution Spectroscopy', Pitman Publishing Corp., New York.

TOLANSKY, S. (1948). 'Multiple-beam Interferometry of Surfaces and Films'. Oxford University Press, Inc., Oxford. (Republished by Dover Publications, New York, 1970)

TOLANSKY, S. (1955). 'An Introduction to Interferometry'. Longmans, Green and Co., London.

TOLANSKY, S. and BRADLEY, D. J. (1960). An oscillating Fabry-Perot interferometer. *In* National Physical Laboratory Symposium No. 11: 'Interferometry', pp. 375-386, Her Majesty's Stationery Office, London.

TURGEON, E.C., and SHEPHERD, G.G. (1962). Upper atmospheric temperatures from Doppler line widths - II. Measurements on the OI 5577 and OI 6300 A lines in aurora. *Planet. Space Sci.* **9**, 295 -304.

TURNER, A. F. (1950). Some current developments in multilayer optical films. *J. Phys.* **11**, 444 - 460.

VANDER SLUIS, K. L., and MCNALLY, J. R. (1956). Fabry - Perot interferometer with finite apertures. *J. Opt. Soc. Am.* **46**, 39 - 46.

VAN RHIJN, P. J. (1919). On the brightness of the sky at night and the total amount of starlight. *Astrophys. J.* **50**, 356 - 375.

VAUGHAN, A.H. (1967). Astronomical Fabry - Perot interference spectroscopy. *Ann. Rev. Astron. Astrophys.* **5**, 139 - 166.

VELICHKO, A.G., KATZ, M.L., and TSOI, V.I. (1971). Optimization of the parameters of a Fabry - Perot interferometer. *Opt. and Spektros.* **30**, 511 - 513.

WILKSCH, P. A. (1985). Instrument function of the Fabry-Perot spectrometer. *Appl. Opt.* **24**, 1502 - 1511.

WILLIAMS, W. E. (1930). 'Applications of Interferometry'. Methuen and Co. Ltd., London.

YOSHIHARA, K. and KITADE, A. (1978). Far infrared spectroscopy by the Fabry-Perot interferometer. *Jap. J. Appl. Phys.* **17**, 1895 - 1896.

YOSHIHARA, K. and KITADE, A. (1979). Far infra-red spectroscopy by the Fabry-Perot interferometer. *Optica Acta* **26**, 1049 - 1056.

YOSHIHARA, K., KAKO, S. and KITADE, A. (1979). Far infrared spectroscopy by the Fabry-Perot interferometer. II *Jap. J. Appl. Phys.* **18**, 2327 - 2328.

YOSHIHARA, K., KITADE, A. and OKADA, K. (1980). Far infrared spectroscopy by the Fabry-Perot interferometer. III *Jap. J. Appl. Phys.* **19**, 2523 - 2524.

YOUNG, A. T. (1974). Photomultipliers: Their cause and cure. *In* 'Methods of Experimental Physics', (N. Carleton, ed), Vol. 12A, pp. 1 - 94, Academic Press, New York.

YOUNG, E.R., and CLARK, K. (1980). Matched tandem etalon camera - MATEC - and its application to auroral observations. *Appl. Opt.* **19**, 2631 - 2637.

ZEEMAN, P. (1908). Beobachtung der magnetischen Auflösung von Spectrallinien mittels der Methode von Fabry und Perot. *Physik. Zeit.* **7**, 209 - 212.

ZEHNDER, L. (1891). Ein neuer Interferenzrefraktor. *Zeitschr. f. Instrkde.* **11**, 275 - 285.

ZIPOY, D. M. (1979). Fabry - Perot inversion algorithm. *Appl. Opt.* **18**, 1988 - 1995.

ZUCKER, R. (1964). Elementary Trascendental Functions. Logarithmic, Exponential, Circular and Hyperbolic Functions. *In* 'Handbook of Mathematical Functions with Formulas, Graphs and Mathematical Tables' (M. Abramowitz and I. Stegun, eds), p. 65 - 225, United States Government Printing Office, Washington.

Glossary

a	HWHH of an ideal Fabry - Perot etalon.
$a(t)$	Response of an etalon to a very short light pulse.
$a(k,0)$	Absorption of measurement.
a^*	HWHH of an ideal Fabry - Perot etalon normalized to $\Delta\sigma$.
a^+	HWHH of a spherical Fabry-Perot etalon normalized to the degenerate free spectral range.
a_n^*	HWHH of n-ideal identical etalons in series.
a	Radius of the entrance aperture of a finite size Fabry - Perot etalon.
$\mathbf{a}$	Ratio of the gaps of two etalons in series.
a_k	Fourier coefficient of an ideal Fabry - Perot.
A	Absorption/scattering coefficient of the mirrors. Amplitude of a beam.
A_x	Amplitude of the wave leaving an etalon.
$A_{j,k}$	Einstein's coefficient of spontaneous emission.
$A(k,0)$	Normalized absorption of measurement.
$\hat{A}(s)$	Frequency response of an etalon.
A	Radius of the back aperture of a finite size Fabry - Perot etalon.
$\mathbf{A}$	Etalon area. Avogadro's number. Area in general.

b	An arbitrary width. Complex transmission of an absorbing species.
b_k	Fourier coefficients of a Fabry - Perot measurement.
b	A reflective parameter used in the description of series etalons.
B	Molecular rotational constant. Also brightness.
$B_{j,k}$	Einstein's coefficient of absorption and induced radiation.
B	Continuous background irradiance.
c	Width of a Lorentzian filter, multi-etalon case ($c = 2\pi s$).
c_k	Cosine Fourier coefficient.
c	Speed of light.
C	Etalon contrast.
$C_{1,2}$	Spherical etalon contrast for either Type 1 or Type 2 rays.
C_k	Effective contrast of a finite size etalon.
C_l	Limiting contrast
C_n	Contrast of a series of n-identical etalons.
$C(z)$	Fresnel integral.
d	Distance, or gap, between Fabry - Perot mirrors.
d_f	HWHH of etalon defect function.
d_{fX}^*	Normalized HWHH of etalon defect function.
dg	HWHH of a Doppler profile.
dg^*	Normalized HWHH of a Doppler profile.
dg^+	HWHH of a Doppler profile normalized to the spherical etalon degenerate free spectral range.
d_k	Fourier coefficient of instrumental function.
D	Fabry - Perot mirror diameter. Also angular dispersion.
D_{sn}	Effective diameter of an etalon.
D_J	Molecular centrifugal distortion constant.
D	Diameter of etalon substrates.
e	Etalon function width.
e^*	Etalon normalized width.
e_k	A term associated with the uncertainties of determination of Doppler shifts.

E	Etalon function.
E_i	Internal energy level of an atom.
f	HWHH of spectrometer aperture function. In general the width of the **f** function. Also the focal length of a lens.
f^*	Normalized HWHH of spectrometer aperture function.
f^+	HWHH of the effective aperture of a spherical etalon normalized to the degenerate free spectral range.
f	A function in the real plane, as opposed to the Fourier plane.
F	A function in the Fourier plane. Interference filter function. Also a flux.
F	A reflective parameter, sometimes called the coefficient of reflective finesse.
FWHH	Full-Width-at-Half-Height.
g	Width of the **g** function. Parameter associated with a spherical laser cavity.
g	A function in the real plane.
G	Doppler width at the $1/e$ point, also a function.
G	Gain.
h	Width of the **h** function. Planck's constant.
h	A function in the real plane.
$H_l(z)$	Hermite polynomial of degree l.
H	Heaviside's step function.
HWHH	Half-Width-at-Half-Height.
I	Line source irradiance.
J	**B I**$^{-1}$. Also rotational quantum number.
$J_n(z)$	Bessel function of the first kind and order n.
k	Etalon effective size, in number of reflections. Transmittance of an absorption species. Also propagating constant of a medium.
k	Boltzmann's constant.
K_X	Value of a residue.
K$_n$	Kernel of an integral equation.
l	Lorentz function HWHH. Stepping distance per reflection in an

etalon gap.

l^* Normalized HWHH of Lorentz function.

L Lorentz function. Length.

$\mathbf{L}$ Luminosity.

M Number of free spectral ranges of an aperture. Also molar mass.

$\mathbf{M}$ Number of beams giving near-ideal etalon behavior.

$\mathbf{M}(x)$ Two-beam (Michelson) interferogram.

n Order of interference.

n_0 Central order of interference.

n_k Etalon back aperture size, in number of reflections.

N Area under an arbitrary source function. Fresnel number or measure of diffraction.

N_i Number of particles per unit volume.

$\mathbf{N}$ Number of useful coefficients in a Fourier series. Also number of spectral elements.

N_D Surface defects finesse.

N_e Etalon finesse.

N_G Global finesse.

N_i Number of free spectral ranges which define N_G.

N_l Limiting finesse.

N_R Reflective finesse.

N_t Measurement or total finesse.

p Pressure, normally refers to a gas. Also a function associated with the P function below.

P A function describing a measurement.

$P(z)$ Complex phase shift associated with the propagation of a beam.

$\mathbf{P}$ Parasitic light.

$q(z)$ Complex intensity parameter.

Q A collection of terms dealing with flux.

$\mathbf{Q}$ A ratio which, in the limit, approaches the contrast in the absence of parasitic light.

r_i	Complex reflection at the boundary of two media.
$\mathbf{r}$	Resolving limit.
R	Reflectivity of Fabry - Perot mirrors.
R_E	Effective reflectivity of an etalon and a Lorentzian function.
$R(z)$	Radius of curvature of a wavefront that intersects the axis in the direction of propagation.
$\dot{R}$	Radius of a displaced aperture.
$\mathbf{R}$	Resolving power.
s	Source width, also a Lorentz HWHH of an interference filter. Generalized frequency in the Fourier plane.
s_k	Sine Fourier coefficient.
$\mathbf{s}_k$	Source function Fourier coefficient.
S	Arbitrary line function.
$S_n(z)$	Solution of the radial function in the presence of diffraction.
$\mathcal{S}$	Preset uncertainty parameter of the equal-noise measurement method.
$\mathbf{S}$	Modified source function.
$\hat{\mathbf{S}}$	Arbitrary source function.
$\mathbf{S}(z)$	Fresnel integral.
t	HWHH of a measured fringe.
t_i	Complex transmission at the boundary of two media. Also the time associated with the ith measurement.
t^*	Normalized HWHH of a measured fringe.
$T(s)$	Transmittance of an etalon.
$\mathbf{t}$	Transit time of light in an etalon gap.
$\mathbf{T}$	Periodicity of an etalon function.
$\mathbf{T}_i$	Relative transmission of an interference filter.
u	Field component of a beam.
U_k	Logarithm of the ratio of a measured coefficient to an instrumental coefficient.
v	Line-of-sight wind velocity. Number of nodes of a standing wave in a

	cavity of arbitrary length.
$v(r,\phi)$	Invariant distribution function of a field component in a cavity after many transits.
$w(z)$	Width of a beam at the 1/e point. The smallest width is denoted by w_0.
w_k	Coefficients for the temperature uncertainty calculations.
W	Area of etalon function P over one order.
W_n	Integral under $E_n \times F$.
$\mathbf{W}_\tau$	A sum associated with the uncertainties of determination of Doppler widths.
x_{0_k}	Displacement from an arbitrary zero.
x	Twice the optical separation of an interferometer in multiplex operation.
$\mathbf{x}_0$	Mean displacement from an arbitrary zero.
$\mathbf{X}$	Maximum value of x above.
y	Position of interference filter maximum relative to axis. Also spectrometer function width.
Y_i	Etalon function. Also transmission of a spectrometer.
$\hat{Y}$	Superposed amplitude transmission of parallel etalons.
Z_X	Pole(s) of complex functions.
α	Angular displacement of an aperture from the central axis. Thermal expansion coefficient. Fractional energy loss per transit, due to diffraction, of a wave in a cavity.
α_i	Slit angular width.
β	Phase shift per transit, due to diffraction, of a wave in a cavity.
β_i	Slit angular height.
γ	Proportionality constant relating temperature to Doppler width. Retardation of a wave at an arbitrary point inside an etalon cavity.
δ	General phase lag in transcendental equations.
δt	Light pulse width.
$\delta(x)$	Dirac delta function.

$\delta^{1/2}(x)$	Null function.
Δ	Difference between two quantities.
$\Delta\sigma$	Free spectral range.
$\Delta\sigma_n$	Wavenumber difference between two neighboring longitudinal modes.
Δt	Time between coherent pulses necessary to attain near-ideal fringes.
ϵ	Quantum efficiency of a detector. Absorption of a sample. Also a small value.
η_x	Dimensionless number used as the transverse coordinate of a mode.
Θ	Angle measured away from the axis.
λ	Wavelength.
μ	Index of refraction.
ν	Frequency of radiation.
$\Pi(x)$	Rectangular function.
ρ	Radius of the useful confocal etalon. Radius of a fringe in angular units. Also a very small quantity.
σ	Wavenumber. Inverse wavelength.
σ_B	Cross-section for absorption of an atomic species.
$\sigma_{n,m,l}$	Wavenumber associated with a cavity mode.
σ_M	Upper limit of the spectral components of a source.
σ^2	Variance of determination.
τ	Transmission. Also temperature.
τ_A	Loss of transmission associated with the absorption coefficient A.
τ_0	Peak transmission of an interference filter at normal incidence.
τ_E	Etalon function shorthand.
$\tau_e(\sigma)$	Etalon function.
$\tau_{e,h}$	Etalon function.
Φ_i	Flux.
Φ	Phase lag between beams; phase in general.
Φ_k	Phase lag in confocal etalon.
χ	Phase change upon reflection.
ψ	Blaze angle of grating.

ψ_k Phase lag of the kth beam.

$\Psi(x,y,z)$ Complex function representing the differences between a coherent beam and a plane wave.

ω Lorentzian interference filter FWHH.

ω_0 FWHH of an interference filter at normal incidence.

Ω Solid angle.

$\text{III}(x)$ Dirac comb function.

$*$ Convolution operator.

$\oplus$ Convolution operator for specific properties of functions.

$\blacksquare$ Cross correlation operator.

$\bullet$ Scalar multiplication operator.

$\supset$ Operator to indicate a Fourier transformation.

Author index

Subject index

A

A, *see* etalon, -aperture area
A , *see* absorption
Aberration, spherical, 129ff
Absorber
  -strong, 182ff
  -weak, 181ff
Absorption, 5, 11, 18, 93, 157, 179ff
  /scattering, 11, 18, 157, 179
  -apparent, 182
  -cell, 182
  -cross-section, 186ff
  -gain, 181ff
  -in the cavity, 179ff
  -measurement, 180, 182
  -sample, 180, 182
  -spectra, 53ff
Air
  -bearings, *see* scanning -mechanical

  -index of refraction, 230ff
  -*see*, refractive index
Alignment, 245
  -aids, 247ff
  -automatic, 6, 238, 248ff
    -capacitive micrometer, 253ff
    -superposition, 248ff
    -spectral sources, 249ff
  -dynamic, *see* -automatic
  -precision, 246
  -reference, 249ff
  -static, 222, 245ff
    -equal-inclination, 246
    -equal-thickness, 245
    -methods, 245ff
    -multiple etalons, 246
    -superposition, 246
    -white light, 245, 248
    -*see* -automatic

Printed in the United States
By Bookmasters